Problems for
General Chemistry
and Qualitative Analysis

Problems for General Chemistry and Qualitative Analysis

SECOND EDITION

C. J. NYMAN & G. B. KING

Professors of Chemistry, Washington State University

John Wiley & Sons, Inc.,

New York   London   Sydney   Toronto

Preface to First Edition

In writing this problem book, the authors had the following primary objectives in mind: (1) to furnish the student simple and direct methods for solving numerical problems that illustrate chemical principles and to provide him with many examples of the types of problems encountered in his general chemistry course; (2) to provide the instructor with supplementary problems and exercises for use in tutorial and recitation sections and for homework assignments.

Essentially, this book is a compilation of supplementary problems that we have used in general chemistry classes over the past several years. An attempt has been made to include all the types of problems normally encountered in elementary chemistry courses. Problems of a wide range of difficulty are included to provide considerable flexibility in assignments to students of varying capabilities. We believe there is no more effective self-teaching device than for a student to study the detailed solutions to numerous problems and then work out many similar problems on his own.

The subject matter has been divided into *chapters*, wherein the principles needed for solving a given type of problem are presented briefly at the start of each major section. This introduction is followed by *examples* of problems embodying the principles discussed, and for which detailed solutions are provided. The mole method is emphasized and used in the solution of these sample problems whenever appropriate. In other instances the factor-unit method has been employed. Care has been taken to carry dimensional units through all the mathematical operations and to express answers in the proper number of significant figures. At the end of the section of examples is a set of *problems* involving the principles illustrated. Later, at the conclusion of each chapter, a set of *general problems* covering all the types of problems in the chapter is listed. Usually problems are arranged in order of increasing difficulty. Appendix XI contains answers to all problems. Students should not expect to duplicate these answers exactly; they may find that their answers differ somewhat in the last significant figure reported.

In the Appendix are found tables of constants for various equilibria, a table giving the vapor pressure of water as a function of temperature, and a four-place table of logarithms. A detailed discussion of significant figures is included here, and it is highly recommended that instructors introduce students to this concept at the earliest possible time. For those students who may need a review of fundamental mathematical operations, sections of the Appendix are devoted to simple algebraic manipulations, the use of exponents, and the use of logarithms. Answers to problems presented in these sections are also given in Appendix XI.

In no sense is the material presented in this problem book intended as a substitute for the textbook, where more detailed information and explanations of principles are given. The student should use this problem book in close conjunction with his textbook and should refer to it for the extensive elucidation of principles sometimes necessary for a clear understanding of certain topics. Our introductory discussions have, of necessity, been made concise.

We gratefully acknowledge the assistance and suggestions of members of the Washington State University chemistry staff, including the many graduate students who have used these problems in tutorial and recitation classes over the past several years.

C. J. NYMAN

G. B. KING

Pullman, Washington
December 1965

Preface to Second Edition

A number of changes will be found in this edition. Certain sections, notably molecular and ionic equilibria, have been expanded. A new chapter has been included on thermodynamics.

Most of the problems have been changed, and some new examples added.

C. J. Nyman
G. B. King

Pullman, Washington

Contents

APPENDIX

Problems for
General Chemistry
and Qualitative Analysis

CHAPTER 1

Introduction

Every student in beginning chemistry will be expected to solve many numerical problems since most of the principles of chemistry are best illustrated by the use of mathematical concepts. The student is apt to find that the most effective way to learn these principles is to work many, many problems. To do this, he must be familiar with the elementary operations of mathematics; therefore he should carefully and thoroughly review them before attempting chemical problems. These operations are briefly treated in the Appendix, page 243.

To solve any problem, certain logical steps should be followed:

1. Read the problem carefully to get a clear understanding of just what is desired.

2. Tabulate the data that are given, including the units in which the data are expressed. At this point you will know what is given and what is unknown.

3. Develop a plan for solving the problem. To do this you must determine the chemical principle which applies.

4. If a chemical reaction is involved, write the balanced equation.

5. If there is a mathematical relationship between the items of data given, set up the proper mathematical formulation.

6. Work out the problem by direct reasoning from the principles that apply, or by substituting items of data into the mathematical formulation and solving for the unknown. After getting an answer, consider it carefully to determine if it is *possible* and *reasonable*.

7. Some problems involve several steps, and in reality are two or three consecutive problems. Always keep in mind the objective of the problem and the units in which the answer is to be expressed.

8. When arriving at an answer, make certain you obtain the proper number of significant figures. The necessary rules to be followed are outlined in Appendix IV.

It is suggested that the student make the utmost use of pencil and paper in solving problems. Following the procedure outlined above, write down in sequence the details of the steps involved. The solution should be recorded neatly on paper in such a manner that the procedure followed is obvious to anyone examining it. Recording unconnected bits of information "helter-skelter" on a sheet of paper and then expecting someone else to follow your line of reasoning is unrealistic and often unacceptable. Remember that a clearly and concisely presented solution to a problem makes a good impression on the instructor, and similarly, a poorly presented one makes a bad impression. Great effort should be made to cultivate proper work habits; they will reap benefits not only in presenting the solutions to homework and examination questions in chemistry, but also in presenting solutions to all types of problems encountered by the student throughout his entire career.

CHAPTER 2

Units of Measurement

A. METRIC SYSTEM

The metric system of measurement is now used universally in scientific work. This is a decimal system in which changes are made from one unit to another merely by moving the decimal point an appropriate number of places. The prefixes to the names of the various units, their abbreviations, and the factors relating each to a standard quantity are shown in the following table.

Name	Abbreviation	Factor	Name	Abbreviation	Factor
pico	p	10^{-12}	deci	d	10^{-1}
nano	n	10^{-9}	kilo	k	10^{3}
micro	μ	10^{-6}	mega	M	10^{6}
milli	m	10^{-3}	giga	G	10^{9}
centi	c	10^{-2}	tera	T	10^{12}

Not all these prefixes will be encountered in introductory chemistry, but the abbreviations are in general use in more advanced areas of science and in engineering. The more commonly encountered units in chemistry and other sciences are as follows.

Linear Measurement

The basic unit of length is the *meter* (about 39.37 inches).

1 kilometer (km)	=	1000 meters (m)
1 decimeter (dm)	=	0.1 meter
1 centimeter (cm)	=	0.01 meter
1 millimeter (mm)	=	0.001 meter
1 micron (μ)*	=	0.000001 meter (10^{-6} meter)
1 millimicron (mμ)*	=	0.000000001 meter (10^{-9} meter)
1 angstrom (Å)*	=	0.0000000001 meter (10^{-10} meter)

* Common unit not following metric system rules of naming.

3

Weight Measurement

The basic unit of weight is the *gram* (g) (about 1/454 pound).

1 kilogram (kg)	=	1000 g
1 milligram (mg)	=	0.001 g
1 microgram (μg)	=	0.000001 g
1 nanogram (ng)	=	1×10^{-9} g
1 picogram (pg)	=	1×10^{-12} g

Volume Measurement

The basic unit of volume is the *liter* (about 1.06 quarts).

1 milliliter (ml) = 0.001 liter = 1 cubic centimeter (cm^3)
1 liter* = 1000.000 ml = 1000.000 cubic centimeters (cm^3)

* In 1964 the *liter* was reduced slightly in size (27 parts per million) so that 1 liter = $1000 \ cm^3$ and 1 ml = $1 \ cm^3$. Before this date the liter had been defined as the volume of one kilogram of water at 4°C (temperature of maximum density) which turned out to be $1000.027 \ cm^3$.

B. NUMBERS AND UNITS

Most numbers that appear in problems are measurements of something and as such must include the units in which the quantities are expressed. Only occasionally do we encounter numbers that are unitless. Students should acquire the habit of always writing down the units following the numerical values when appropriate. These units should be carried along with the numbers through all the mathematical operations in multiplication, division, etc. In a formulation involving several terms, cancellations should be made wherever possible; this applies to units as well as numbers. Throughout this book, efforts have been made to follow this practice in the solutions of problems worked as Examples. If the student considers both the units and the numbers in his own study of each Example, the procedure will quickly become obvious to him.

It is often necessary to change from one set of units to another, and to do this the factors relating the units must be known; for example, if we are to convert 100 g to kilograms, we must know that 1 kg = 1000 g. We must then multiply the number of grams by the number of kilograms per gram as is done in the following illustration:

$$100 \text{ g} = 100 \text{ g} \times \frac{1 \text{ kg}}{1000 \text{ g}} = 0.1 \text{ kg}$$

Suppose we want to convert 50 mg to kilograms. Since we know that 1 mg = 0.001 g and 1 g = 0.001 kg, we can write

$$50 \text{ mg} = 50 \text{ mg} \times \frac{0.001 \text{ g}}{\text{mg}} \times \frac{0.001 \text{ kg}}{\text{g}} = 0.00005 \text{ kg}$$

These two conversions are good illustrations of changes from one unit to another by the *factor-unit* method.

Example 2-a. Add the following masses and express the answer decimally in grams: 0.00200 kg, 450 mg, 30.0 cg, 0.11 dg.

Solution. Each unit must be converted to grams or some other convenient common unit prior to addition.

$$
\begin{aligned}
0.00200 \text{ kg} &= 2.00 \text{ g} \\
450 \text{ mg} &= 0.45 \text{ g} \\
30.0 \text{ cg} &= 0.300 \text{ g} \\
0.11 \text{ dg} &= \underline{0.011 \text{ g}} \\
& 2.761 \text{ g}
\end{aligned}
$$

When it is rounded to the proper number of significant figures, the number becomes 2.76 g.

Example 2-b. Determine the number of liters in a box with the dimensions: 3.02 m long, 25 cm deep, and 41 mm wide.

Solution. Since 1 liter = 1000 cm^3, convert each dimension to centimeters. The volume is then obtained in cubic centimeters by multiplying length × width × depth. Finally, multiply by 1 liter/ 1000 cm^3 to obtain the number of liters.

$$\text{Volume} = 302 \text{ cm} \times 25 \text{ cm} \times 4.1 \text{ cm} = 30,955 \text{ cm}^3$$

This number must be rounded to 31,000 cm^3 to obtain the proper number of significant figures.

$$\text{Volume} = 31,000 \text{ cm}^3 \times \frac{1 \text{ liter}}{1000 \text{ cm}^3} = 31 \text{ liters}$$

Example 2-c. A drop of oleic acid having a volume of 0.054 cm^3 is spread out on a surface to a uniform thickness of exactly 10 Å. What surface area in square meters is covered?

Solution.

$$\text{Area} = \frac{\text{Volume}}{\text{Thickness}}$$

Since the area is wanted in square meters, the volume should be expressed in m^3 and thickness in m.

$$\text{Area} = \frac{\text{Volume } (m^3)}{\text{Thickness } (m)} = \frac{0.054 \text{ cm}^3 \times (0.010 \text{ m/cm})^3}{10 \text{ Å} \times 10^{-10} \text{ m/Å}}$$

$$= \frac{5.4 \times 10^{-8} \text{ m}^3}{1.0 \times 10^{-9} \text{ m}} = 54 \text{ m}^2$$

PROBLEMS (Metric System)

2-1. Add the following weights and express the answer decimally in grams: 1.000000 kg, 327 mg, 15.0 cg, 434.000 g, 26.00 dg.

2-2. Add the following weights and express the sum decimally as grams: 0.3 cg, 40 mg, 0.50 dg, 0.001000 kg, 0.150 g.

2-3. Add the following units of linear measurement and express the sum decimally as meters: 150 cm, 3.00000 km, 340 mm, 0.900 dm, 33.00 m.

2-4. Assuming 1 liter of paint is spread uniformly on a wood surface to a thickness of 100 mμ, what surface area in square meters would be covered?

2-5. One cubic millimeter of oil spread out on water forms a film which covers an area of 1 m^2. What is the thickness of the oil film in Å units?

2-6. Calculate the number of liters in 1 m^3.

2-7. Make the calculations indicated:

(a) 10 μ = _____ m	(f) 4.0 lb = _____ g
(b) 0.001 mg = _____ kg	(g) 8.9 cm = _____ in.
(c) 0.10 ml = _____ liters	(h) 5.3 liters = _____ qt
(d) 1 mm^3 = _____ m^3	(i) 227 mg = _____ lb
(e) 50 dm^3 = _____ liters	(j) 1 mile = _____ m

2-8. Violet rays of light near the lower end of the visible spectrum have a wavelength of about 0.000045 cm. Express this wavelength in microns, millimicrons and angstrom units.

C. TEMPERATURE SCALES

The relationships between the Fahrenheit, centigrade, and Kelvin (absolute) scales of temperature are shown in Fig. 2-1. Note that between the fixed points, the freezing and boiling points of water, there are 180 divisions or degrees on the Fahrenheit scale and 100 divisions or degrees on the centigrade and Kelvin (absolute) scales. Hence 1 centigrade degree = $\frac{180}{100}$ = $\frac{9}{5}$ = 1.8 Fahrenheit degrees or 1 Fahrenheit degree = $\frac{5}{9}$ centigrade degree. To convert centigrade temperature to Fahrenheit

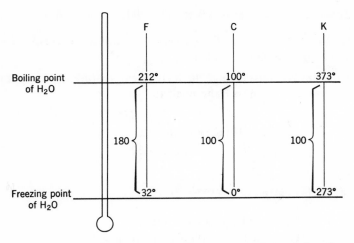

Fig. 2-1 Comparison of temperature scales: F = Fahrenheit, C = Centigrade or Celsius, K = Kelvin or absolute.

temperature, multiply the centigrade reading (°C) by $\frac{9}{5}$, then add 32, that is,

$$°F = \frac{9}{5}°C + 32 \tag{1}$$

or to convert from Fahrenheit to centigrade, first subtract 32 from the Fahrenheit reading, then multiply by $\frac{5}{9}$, that is,

$$°C = (°F - 32)\frac{5}{9} \tag{2}$$

On the centigrade and Kelvin scales, the degrees are the same size, but the Kelvin reading (°K) will be 273 higher than the centigrade reading, that is,

$$°K = °C + 273 \tag{3}$$

Example 2-d. What is the Fahrenheit equivalent of 20°C?

Solution. Each centigrade degree is equivalent to $\frac{9}{5}$ Fahrenheit degrees. Therefore, 20 degrees or divisions on the centigrade scale correspond to $20 \times \frac{9}{5} = 36$ divisions or degrees on the Fahrenheit scale. However, since these degrees or divisions are measured above the freezing point of water taken as 32, the latter must be added; therefore

$$°F = \frac{9}{5}(20) + 32 = 36 + 32 = 68$$

Example 2-e. Iron begins to glow red at about 900°F. What is this temperature on the centigrade scale?

Solution. $°C = (°F - 32) \times \frac{5}{9} = (900 - 32)\frac{5}{9} = 482$

Example 2-f. What is the Fahrenheit temperature corresponding to absolute zero on the Kelvin scale?

> *Solution.* We must change the Kelvin temperature first to centigrade temperature, then to Fahrenheit temperature.
>
> $$°K = °C + 273$$
> $$\text{If } °K = 0, \text{ then } °C = -273$$
> $$°F = \tfrac{9}{5}(-273) + 32 = -459$$

PROBLEMS (Temperature Conversions)

2-9. Make the following temperature conversions:

(a) 45°C to Fahrenheit   (f) 0°F to centigrade
(b) 98.6°F to centigrade   (g) −40°C to Fahrenheit
(c) −273°C to Kelvin   (h) −459°F to Kelvin
(d) 173°K to Fahrenheit   (i) 212°F to Kelvin
(e) −112°F to centigrade   (j) 0°K to Fahrenheit

2-10. Make the temperature conversions as indicated by the blanks in the following table.

	Fahrenheit	Centigrade	Absolute (K)
(a)	32°	—	—
(b)	212°	—	—
(c)	—	80°	—
(d)	—	—	673°
(e)	0°	—	—
(f)	1328°	—	—
(g)	—	—	233°
(h)	—	200°	—
(i)	−184°	—	—
(j)	−40°	—	—
(k)	—	—	0°

2-11. At what temperature will a centigrade and a Fahrenheit thermometer have the same reading?

D. DENSITY AND SPECIFIC GRAVITY

Density is a measure of the mass of a substance in relation to its volume, and is defined

$$\text{Density} = \frac{\text{Mass}}{\text{Volume}}$$

Any units of mass and volume may be used, but they must always be specified. For example, if 10 ml of a liquid weighs 8 g, its density is 8 g/10 ml = 0.8 g/ml. In chemical work, densities are usually expressed as grams per milliliter or grams per cubic centimeter.

Specific gravity is a *number* which is the ratio of the masses of equal volumes of the substance in question and some standard substance. In other words, it is a ratio of the density of the substance in question and the density of a standard substance, both densities being expressed in the same units. For liquids and solids, the standard chosen is usually water with a density of 1 g/ml, and when the comparison is to water in this unit, the specific gravity of a substance is *numerically* equal to its density. For example, if 10 ml of a certain metal weighs 190 g, its density is 190 g/10 ml = 19 g/ml. Its specific gravity is 19 since specific gravity equals

$$\frac{\text{Density of metal}}{\text{Density of water}} = \frac{19 \text{ g/ml}}{1 \text{ g/ml}} = 19$$

Example 2-g. A cube of platinum (Pt) metal 10.0 cm along each edge has a mass of 21.5 kg. What is the density of Pt in g/cm³?

Solution.

$$\text{Volume of cube} = (10.0 \text{ cm})^3 = 1000 \text{ cm}^3$$

$$\text{Mass in grams} = 21.5 \cancel{\text{kg}} \times \frac{1000 \text{ g}}{\cancel{\text{kg}}} = 21{,}500 \text{ g}$$

$$\text{Density} = \frac{\text{Mass}}{\text{Volume}} = \frac{21{,}500 \text{ g}}{1000 \text{ cm}^3} = 21.5 \text{ g/cm}^3$$

Example 2-h. The specific gravity of concentrated sulfuric acid, H_2SO_4, is 1.85. What volume of H_2SO_4 would have a mass of 925 g?

Solution.

$$\text{Specific gravity} = \frac{\text{Density of } H_2SO_4}{\text{Density of } H_2O}$$

The density of H_2O is 1.00 g/ml, hence

$$\text{Density of } H_2SO_4 = (1.85)(1.00 \text{ g/ml}) = 1.85 \text{ g/ml}$$

$$\text{Volume} = \frac{\text{Mass}}{\text{Density}} = \frac{925 \text{ g}}{1.85 \text{ g/ml}} = 5.00 \times 10^2 \text{ ml}$$

Example 2-i. A rectangular piece of metal measuring 20.0 cm × 70.0 mm × 3.0 dm weighs 46.2 kg. What is the density of the metal in grams per cubic centimeter?

Solution.

$$\text{Volume of metal} = 20.0 \text{ cm} \times 7.00 \text{ cm} \times 30 \text{ cm} = 4200 \text{ cm}^3$$

$$\text{Density} = \frac{\text{Mass}}{\text{Volume}} = \frac{46.2 \text{ kg} \times 1000 \text{ g/kg}}{4200 \text{ cm}^3} = 11.0 \text{ g/cm}^3$$

Example 2-j. Assuming that atoms of mercury, Hg, are spherical in the liquid state, calculate the percentage of unoccupied space in liquid mercury if the radius of a mercury atom is 1.56 Å. The density of liquid mercury is 13.6 g/ml. Volume of a sphere $= \frac{4}{3}\pi r^3$ where r is the radius.

Solution.

$$\text{Volume of 1 atom} = \frac{4}{3}\pi(1.56 \times 10^{-8} \text{ cm})^3 = 1.59 \times 10^{-23} \text{ cm}^3$$

Volume of 1 g atom (atomic volume)

$$= (1.59 \times 10^{-23} \text{ cm}^3/\text{atom}) \times (6.02 \times 10^{23} \text{ atoms/g atom})$$
$$= 9.57 \text{ cm}^3/\text{g atom}$$

$$\text{Theoretical density} = \frac{\text{Atomic weight}}{\text{Atomic volume}} = \frac{200.6 \text{ g/g atom}}{9.57 \text{ cm}^3/\text{g atom}} = 21.0 \text{ g/cm}^3$$

By comparing the theoretical density with the actual density the % of unoccupied space may be calculated thus:

$$\% \text{ Unoccupied space} = \frac{\text{Theoretical density} - \text{Actual density}}{\text{Theoretical density}} \times 100$$

$$\% \text{ Unoccupied space} = \frac{21.0 - 13.6}{21.0} \times 100 = 35.2\%$$

PROBLEMS (Density)

2-12. A piece of galena (PbS) weighing 231.0 g displaces 30.8 ml of water. What is the density of galena (PbS)?

2-13. The specific gravity of mercury, Hg, is 13.60 relative to water. What volume of Hg would have the same weight as 816 ml of H_2O?

2-14. Lead has a density of 11.4 g/cm^3. What is the weight of a lead brick measuring 20.0 cm long $\times$ 8.0 cm wide $\times$ 6.00 cm high?

2-15. Assuming that bromine, Br_2, molecules are spherical, determine the percentage of vacant space in liquid bromine which has a density of 2.93 g/ml. Assume the radius of a Br_2 molecule to be 2.28 Å.

GENERAL PROBLEMS

2-16. A metal tank has the dimensions 80 cm × 150 mm × 0.20 m. What is the volume in liters?

2-17. What weight of alcohol can be contained in the tank whose volume is 640 liters if the density of alcohol is 0.75 g/ml?

2-18. What weight of mercury, density 13.6 g/ml, will occupy the same volume as 43.2 g of sodium chloride, density 2.16 g/ml?

2-19. The volume of a cylinder is given by the formula $\pi r^2 h$ where r is the radius and h is the height. Calculate the capacity of the cylinder in liters if the radius is 2 dm and the height is 0.5 m.

2-20. How many grams of alcohol would a spherical container with a diameter of 40 cm inside measurement hold if the density of the alcohol is 0.8 g/ml? Volume of sphere $= \frac{4}{3}\pi r^3$ where $r =$ radius.

2-21. Copper metal has a density of 8.94 g/cm³. What is the weight of a cube of copper which measures 0.15 m along each edge?

2-22. What is the temperature difference in °F between −150°C and 53°K?

2-23. One cubic decimeter of titanium weighs 4.50 kg. Calculate the specific gravity of titanium.

2-24. Concentrated (100%) sulfuric acid has a specific gravity of 1.85 or a density of 1.85 g/ml. What volume of 100% H_2SO_4 should be measured out for an experiment requiring 0.50 kg of the acid?

2-25. An iceberg weighing 4.6×10^{10} kg and with a density of 0.92 g/ml will have what volume above the water line? *Hint:* A floating object displaces its own weight of water.

2-26. Determine the temperature at which
(a) a centigrade and a Fahrenheit thermometer will read numerically the same but opposite in sign.
(b) a Fahrenheit thermometer will read 200 degrees higher than a centigrade thermometer.
(c) the Kelvin thermometer has the same reading as a Fahrenheit thermometer.

2-27. Derive a formula which defines Fahrenheit temperature in terms of absolute temperature.

2-28. Water has a density of 1.0 g/cm³. What is the density of water in pounds per cubic foot?

2-29. The spherical atomic nucleus of calcium has a diameter of 1.0×10^{-12} cm, and 6.02×10^{23} of these nuclei weigh 40.1 g. Calculate the approximate density of this nuclear matter in tons per cubic centimeter. The volume of a sphere is $\frac{4}{3}\pi r^3$ where r is the radius.

2-30. From the beginning of time to 1970 the world production of gold, density 19.3 g/ml, has been estimated to be approximately two billion ounces.

Assuming that all this gold is packed together in a cube, what would be the approximate length of a side of the cube in meters?

2-31. 0.001 milliliter of human blood is diluted to 1 liter. On microscopic examination, the diluted sample is found to contain 8 red corpuscles per cubic millimeter. What is the count per cubic centimeter of the original blood?

2-32. Assuming that sea water contains 0.040 mg of gold per cubic meter, what weight of sea water in metric tons (1000 kg) would contain 1 ounce of gold (about 28 g)? Assume the density of sea water to be 1.03 g/ml.

2-33. Exactly 1 liter of paint is used to cover an area of 25 m². Assuming the coat of paint has uniform thickness, what is the thickness in microns?

2-34. Gold may be hammered to a thickness of 0.00001 in. Starting with 1 ounce of gold (28 g), what area in square feet of gold foil 0.000010 in. thick could be obtained? The density of gold is 19 g/cm³.

2-35. The density of liquid cesium is 1.87 g/ml and its atomic radius is 2.62 Å. Calculate the percentage of vacant space in liquid Cs, assuming spherical atoms.

2-36. Bromine is recovered commercially from sea water which contains 65 parts of this element per million. Assuming 100% recovery and that sea water has a density of 1.03 g/cm³, calculate the volume in liters of sea water required to produce one pound (0.454 kg) of bromine.

2-37. A certain hypothesis proposes that the interior of the earth is composed of three principal layers. The outer layer is primarily a crust of rock that has a density of about 3.0 g/ml and is 10 miles thick (assume two significant figures). The next layer is composed of molten silicates of density about 4.4 g/ml and is approximately 1800 miles thick. The inner core, thought to be composed of molten iron with some cobalt and nickel, has a radius of 2150 miles and a density which averages about 11.0 g/ml. From this information (a) estimate the mass of the earth in tons, and (b) the average density of the earth in grams per milliliter.

2-38. One gram atom of any element contains 6.02×10^{23} atoms. Assuming that every individual on the earth's surface (population 3 billion) is engaged in counting the atoms in one gram atom of an element at the rate of 2 per second and working 24 hours per day and 365 days a year, how long would it take to complete the job?

CHAPTER 3

Atomic and Molecular Weights

A. ATOMIC, MOLECULAR, FORMULA, AND EQUIVALENT WEIGHTS

Atomic Weight (At Wt)

The atomic weight of an element is a number which expresses the relative mass of one atom of the element as compared to an atom of carbon, C^{12}, which has been chosen as the standard and assigned the mass of 12. Fluorine, for example, has an atomic weight of 19. This means that one atom of fluorine is $\frac{19}{12}$ as heavy as one atom of C^{12}; aluminum has an atomic weight of 27, which means that one atom of aluminum is $\frac{27}{12}$ as heavy as one atom of C^{12}, etc.

Molecular Weight (MW)

The molecular weight refers to the relative mass of a molecule of a substance as compared to the mass of one atom of C^{12}. A molecule is represented by a chemical formula that shows the number and types of atoms of which the substance is composed. The weight of the molecule is obtained by adding the atomic weights of all atoms in the formula for the substance as shown in Example 3-a.

Example 3-a. Sulfuric acid, H_2SO_4, is composed of two H atoms, one S atom, and four O atoms with atomic weights 1, 32, and 16, respectively. What is the molecular weight of H_2SO_4?

Solution.

Two atoms of H weigh 2 × the atomic weight of H = 2 × 1 = 2
One atom of S weighs 1 × the atomic weight of S = 1 × 32 = 32
Four atoms of O weigh 4 × the atomic weight of O = 4 × 16 = 64
$$\text{MW } H_2SO_4 = \overline{98}$$

Therefore, one molecule of H_2SO_4 is $\frac{98}{12}$ as heavy as one atom of C^{12}.

Formula Weight (FW)

In dealing with ionic compounds and others where discrete molecules do not exist, it is customary to use the term formula weight rather than molecular weight. It is obtained in the same way as a molecular weight, that is, the sum of the weights of all atoms in the formula for the substance. Formula weight may be used for all substances whether ionic or molecular. Thus the formula weight of H_2SO_4 is 98; the formula weight of NaCl is 58.5, etc. The distinction between formula weight and molecular weight lies entirely in the desirability of using precise terminology. The term molecular weight should be reserved for those substances which do in fact exist in the form of molecules as represented by the chemical formula. Insofar as chemical calculations are concerned, formula weights and molecular weights are handled by identical procedures.

Equivalent Weight (EW)

Equivalent weights of elements are experimental quantities determined in the laboratory that can lead to the calculation of atomic weights. The equivalent weight of an element is that weight which combines with or replaces 1* part of hydrogen by weight; for example, in the formation of the compound hydrogen chloride, 35.5 parts of chlorine combine with 1 part of hydrogen by weight and hence the equivalent weight of chlorine is 35.5. In the formation of sodium chloride, 23 parts of sodium by weight combine with 35.5 parts of chlorine, and since 1 part of hydrogen also combines with 35.5 parts of chlorine, 23 parts of sodium are equivalent to 1 part of hydrogen by weight.

In the formation of the compound water, 16 parts of oxygen combine with 2 parts of hydrogen by weight, and so oxygen exhibits an equivalent weight of 8. With the exception of peroxides, oxygen always exhibits an equivalent weight of 8 in its binary compounds. Often it is convenient to determine the equivalent weight of other elements by determining the

---

* Actually 1.008, with four significant figures.

weight that combines with an equivalent weight of oxygen. For example, 23 grams of sodium are combined with 8 grams of oxygen in the compound sodium oxide; therefore sodium has an equivalent weight of 23.

Many elements exhibit more than one equivalent weight; for example, sulfur has an equivalent weight of $5\frac{1}{3}$ in sulfur trioxide, 8 in sulfur dioxide, and 16 in hydrogen sulfide. It should be obvious from this that the equivalent weight of an element must depend on the reaction which the element undergoes, or, in other words, on the formation of a particular compound.

Atomic weights and equivalent weights are related by a factor called the valence—that small whole number which, when multiplied by the equivalent weight, produces the atomic weight.

$$\text{Equivalent weight} \times \text{Valence} = \text{Atomic weight}$$

The valence may be considered the number of H atoms which one atom of an element combines with or replaces. The atomic weight of sulfur (32) can be calculated from the equivalent weights cited above, for example, $5\frac{1}{3} \times 6 = 32$; $8 \times 4 = 32$; $16 \times 2 = 32$. In this case, the small whole numbers relating the equivalent and atomic weights are 6, 4, and 2. Equivalent weights of compounds are discussed in Chapters 9 and 12.

Example 3-b. 10.90 parts of zinc displace 0.3360 parts of hydrogen by weight from acids. What is the equivalent weight of zinc?

> *Solution.* The equivalent weight of zinc is that number of parts by weight which displace 1.008 parts of hydrogen. Therefore,
>
> $$\text{Equivalent weight of Zn} = \frac{10.90 \text{ parts Zn}}{0.3360 \text{ parts H}} \times 1.008 \text{ parts H}$$
>
> $$= 32.69 \text{ parts Zn}$$

Example 3-c. Zinc oxide is analyzed and found to contain 80.34 parts of zinc for every 19.66 parts of oxygen by weight. What is the equivalent weight of zinc? What is the valence of zinc if its atomic weight is 65.38?

> *Solution.* Calculate the weight of zinc which combines with 8.000 parts of oxygen.
>
> $$\text{Equivalent weight of Zn} = \frac{80.34 \text{ parts Zn}}{19.66 \text{ parts O}} \times 8.000 \text{ parts O}$$
>
> $$= 32.69 \text{ parts Zn}$$
>
> $$\text{Valence} = \frac{\text{Atomic weight}}{\text{Equivalent weight}} = \frac{65.38}{32.69} = 2$$

PROBLEMS (Atomic, Molecular, and Equivalent Weights)

3-1. Determine the molecular or formula weight of each of the following substances. (See the inside front cover for a table of atomic weights.)

    (a) Li_2O (d) GeH_4 (g) POF_3
    (b) H_3PO_4 (e) CuO (h) $NaVO_3$
    (c) $COCl_2$ (f) KNH_2 (i) $Ag_2S_2O_3$

3-2. A certain atom is $\frac{189}{27}$ as heavy as a C^{12} atom. What is its atomic weight?

3-3. Determine the molecular or formula weight of each of the following:

    (a) $MgSO_4$ (d) B_2O_3 (g) Li_2CO_3
    (b) $H_4P_2O_7$ (e) CaI_2 (h) H_2SiO_3
    (c) Ag_2SeO_3 (f) KH_2PO_4 (i) $CaCN_2$

3-4. The formula weight of the compound $K_3M(NO_2)_6$ is 452. What is the atomic weight of M?

3-5. Six parts of aluminum are found to dissolve in hydrochloric acid to displace $\frac{2}{3}$ part of hydrogen by weight. (a) What is the equivalent weight of aluminum? (b) If its atomic weight is 27, what is its valence?

3-6. Carbon dioxide is found to be composed of 6 parts of carbon for every 16 parts of oxygen by weight. (a) What is the equivalent weight of carbon? (b) If the atomic weight of carbon is 12, what is the valence of carbon?

3-7. The atomic weight of bismuth is known to be 209.0. If it dissolves in acid to form a compound in which its valence is 3, what is the equivalent weight of bismuth?

3-8. An oxide of scandium is found to be composed of 71.9 parts of scandium for every 38.4 parts of oxygen. (a) What is the equivalent weight of scandium? (b) If the atomic weight is 44.96, what is the valence of scandium?

3-9. Vanadium metal dissolves in perchloric acid to form a perchlorate salt and displace hydrogen gas. If 0.590 parts of vanadium displace 0.035 parts of hydrogen by weight, what is the equivalent weight of vanadium?

B. GRAM ATOMIC, GRAM MOLECULAR, GRAM FORMULA, AND GRAM EQUIVALENT WEIGHTS

Atomic, molecular, formula, and equivalent weights are simply numbers which express *relative masses* and as such do not carry units of mass. In practice, units may be assigned and although any unit may be used, ordinarily the unit employed is the gram.

A *gram atomic weight* or *gram atom* (abbreviated *g atom*) is that number of grams numerically equal to the atomic weight; for example, there are 32 g of sulfur per gram atom, 238 g of uranium per gram atom, and 23 g of sodium per gram atom, etc. It has been determined that a gram atom of any element contains 6.023×10^{23} atoms. This number is termed Avogadro's number. Consequently, a gram atom is the mass in grams of one Avogadro's number of atoms of that element.

A *gram molecular weight* or *gram formula weight* is that number of grams numerically equal to the molecular or formula weight; for example, there are 98 g of sulfuric acid (MW 98) per gram molecular weight, and 342 g of sugar ($C_{12}H_{22}O_{11}$, MW 342) per gram molecular weight. One gram molecular weight of any substance contains one Avogadro's number of molecules and one gram formula weight contains one Avogadro's number of formula units.

The term *mole* is widely applied to mean one gram molecular weight, one gram formula weight, and occasionally one gram atomic weight. In this book, this abbreviation will be adopted for the terms gram molecular weight and gram formula weight. If a unit other than grams is intended, this will be specified. The term mole can appropriately be applied to mean one gram atom of an element; for example, in the case of helium gas (At Wt 4), there are 4 g of He per mole and in the case of argon gas (At Wt 40), there are 40 g of argon per mole.

The term *gram ion* is sometimes used instead of gram mole or mole in dealing with ionic quantities. One gram ion is simply the weight in grams of one Avogadro's number of the ions in question. For example, one gram ion of chloride ions would be 35.5 g; of sulfate ions, 96 g, etc. In this book, we will ordinarily use the term moles when referring to quantities of either ions or molecules. That is, 35.5 g of Cl^- will be referred to as one mole of chloride ions; 96 g of SO_4^{--} will be termed one mole of sulfate ions, etc.

A *gram equivalent weight* or a *gram equivalent* is that number of grams numerically equal to the equivalent weight; for example, a gram equivalent of hydrogen is 1.008 grams. The term "gram equivalent" is often contracted to *equivalent*.

The *atomic mass unit* (abbreviated *amu*) is a unit employed when dealing with masses of individual atoms. An atomic mass unit is defined as $\frac{1}{12}$ the mass of the C^{12} atom, or in other words, one C^{12} atom has a mass exactly equal to 12 amu. Since there are 12.00 g for 6.023×10^{23} C^{12} atoms and since there is one C^{12} atom per 12 amu, the mass in grams of one atomic mass unit is given by

$$\text{Mass in grams of 1 amu} = \frac{12.00 \text{ g}}{6.023 \times 10^{23} \text{ C}^{12} \text{ atoms}} \times \frac{1 \text{ C}^{12} \text{ atom}}{12 \text{ amu}}$$

$$= 1.660 \times 10^{-24} \text{ g/amu}$$

The preceding terms are perhaps best understood from illustrative examples.

Example 3-d. 196 g of H_3PO_4 would represent how many (a) moles of H_3PO_4, (b) gram atoms of each element, (c) atoms of each element?

> **Solution.** (a) First determine the molecular weight of H_3PO_4, and then divide the number of grams of H_3PO_4 by the number of grams of H_3PO_4 per mole. The atomic weights of H, P, and O are 1, 31, and 16, respectively.

$$\begin{array}{llr} \text{Three atoms of hydrogen weigh} & 3 \times 1 = & 3 \\ \text{One atom of phosphorus weighs} & 1 \times 31 = & 31 \\ \text{Four atoms of oxygen weigh} & 4 \times 16 = & 64 \\ & \text{MW} = & \overline{98} \end{array}$$

$$\text{Number of moles of } H_3PO_4 = \frac{196 \text{ g } H_3PO_4}{98 \text{ g } H_3PO_4/\text{mole}} = 2.00 \text{ moles}$$

(b) The number of gram atoms is obtained by multiplying the number of moles by the number of gram atoms per mole. Therefore 196 g H_3PO_4 represent

$$2.00 \text{ moles } H_3PO_4 \times \frac{3 \text{ g atoms H}}{\text{mole } H_3PO_4} = 6.00 \text{ g atoms H}$$

$$2.00 \text{ moles } H_3PO_4 \times \frac{1 \text{ g atom P}}{\text{mole } H_3PO_4} = 2.00 \text{ g atoms P}$$

$$2.00 \text{ moles } H_3PO_4 \times \frac{4 \text{ g atoms O}}{\text{mole } H_3PO_4} = 8.00 \text{ g atoms O}$$

(c) The number of atoms of each element is obtained by multiplying the number of gram atoms by Avogadro's number. Therefore 196 g H_3PO_4 represent

$$6.00 \text{ g atoms H} \times \frac{6.02 \times 10^{23} \text{ H atoms}}{1 \text{ g atom H}} = 3.61 \times 10^{24} \text{ H atoms}$$

$$2.00 \text{ g atoms P} \times \frac{6.02 \times 10^{23} \text{ P atoms}}{1 \text{ g atom P}} = 1.20 \times 10^{24} \text{ P atoms}$$

$$8.00 \text{ g atoms O} \times \frac{6.02 \times 10^{23} \text{ O atoms}}{1 \text{ g atom O}} = 4.82 \times 10^{24} \text{ O atoms}$$

Example 3-e. 28.5 g of $Al_2(SO_4)_3$ would represent how many (a) moles of $Al_2(SO_4)_3$, (b) gram atoms of each element?

Solution. (a) Obtain the formula weight of $Al_2(SO_4)_3$ and divide the number of grams by the number of grams per mole. The atomic weights of Al, S, and O are 27, 32, and 16, respectively.

$$
\begin{aligned}
\text{Two atoms of Al weigh} && 2 \times 27 &= 54 \\
\text{Three atoms of S weigh} && 3 \times 32 &= 96 \\
\text{Twelve atoms of O weigh} && 12 \times 16 &= 192 \\
&& FW &= \overline{342}
\end{aligned}
$$

$$\text{Number of moles of } Al_2(SO_4)_3 = \frac{28.5 \text{ g}}{342 \text{ g/mole}} = \tfrac{1}{12} \text{ mole}$$

(b) Since 28.5 g is $\tfrac{1}{12}$ mole of $Al_2(SO_4)_3$, then the number of gram atoms of each element is obtained by multiplying by the number of g atoms per mole.

$$\tfrac{1}{12} \text{ mole } Al_2(SO_4)_3 \times \frac{2 \text{ g atoms Al}}{1 \text{ mole } Al_2(SO_4)_3} = \tfrac{1}{6} \text{ g atom Al}$$

$$\tfrac{1}{12} \text{ mole } Al_2(SO_4)_3 \times \frac{3 \text{ g atoms S}}{1 \text{ mole } Al_2(SO_4)_3} = \tfrac{1}{4} \text{ g atom S}$$

$$\tfrac{1}{12} \text{ mole } Al_2(SO_4)_3 \times \frac{12 \text{ g atoms O}}{1 \text{ mole } Al_2(SO_4)_3} = 1 \text{ g atom O}$$

PROBLEMS (Gram Atoms, Gram Moles, Etc.)

3-10. What fraction of a mole is 0.405 g of HBr?

3-11. How many moles of K_3PO_4 are represented by 848 g of the compound?

3-12. How many moles are represented by exactly 100 g of each of the following substances?
(a) $BaSO_4$ (c) HN_3 (e) CuS
(b) Ca_3N_2 (d) LiOH   (f) $Mg(ClO_4)_2$

3-13. How many gram atoms of nitrogen are present in exactly 100 g of each of the following substances?
(a) NH_3 (c) N_2O_5 (e) $CaCN_2$
(b) HNO_2 (d) NH_4NO_3 (f) $Sr(NO_2)_2$

3-14. What weight in grams is represented by each of the following?
(a) 0.40 mole of SO_2 (c) 5.14 moles of H_2SeO_4
(b) 0.20 mole of Na_3PO_4 (d) 3.00 moles of PH_3

3-15. What is the mass in grams of a single
(a) H atom of isotopic weight 1.008?
(b) Ga atom of isotopic weight 70.93?
(c) Th atom of isotopic weight 232.0?

3-16. How many atoms of each element are present in 2.0 moles of pyrophosphoric acid, $H_4P_2O_7$?

3-17. How many atoms of each element are there in 12.6 g of nitric acid, HNO_3?

3-18. What is the gram equivalent weight of phosphorus in a compound containing only oxygen and phosphorus if there are 49.6 g of phosphorus for every 64 g of oxygen?

3-19. What is the gram equivalent weight of chromium in a compound prepared by displacing 0.074 g of hydrogen from hydrochloric acid with 1.91 g of chromium?

3-20. The isotopic mass of a single atom of element M is 3.16×10^{-22} g. What is its atomic weight?

C. ATOMIC WEIGHT DETERMINATION FROM MASS SPECTROMETRIC DATA

Most elements in the natural state are mixtures of isotopes or nuclides, and the chemical atomic weight is the weighted average of all nuclides present in the sample of the element. A nuclide is an atom having a particular number of protons and neutrons; two or more nuclides having the same number of protons but different numbers of neutrons are isotopes. The composition of the mixture is usually expressed in terms of atom percentage, which means the percentage of an isotope of a given mass number in the mixture.

Atom percentage of a given isotope

$$= \frac{\text{number of atoms of a given isotope}}{\text{total number of isotopic atoms}} \times 100$$

A mass spectrometer is an instrument that measures the exact mass of each nuclide and its relative abundance. From such data the chemical atomic weight may be calculated.

Example 3-f. Strontium, as it exists naturally, is composed of four isotopes. These are listed below with their relative abundances expressed in atom per cent. From this information, calculate the atomic weight of naturally occurring strontium.

Sr Isotope	Exact Mass	% Abundance
84	83.913	0.560
86	85.909	9.86
87	86.909	7.02
88	87.906	82.56
		100.00

Solution. The contribution of each isotope to the atomic weight will be in proportion to its relative abundance. Therefore,

$$Sr^{84} \text{ contributes } 83.913 \times 0.00560 = 0.47$$
$$Sr^{86} \text{ contributes } 85.909 \times 0.0986 \ = 8.47$$
$$Sr^{87} \text{ contributes } 86.909 \times 0.0702 \ = 6.10$$
$$Sr^{88} \text{ contributes } 87.906 \times 0.8256 \ = 72.58$$
$$\text{Chemical atomic weight} = \overline{87.62}$$

The calculated atomic weight based on $C^{12} = 12.0000$ is 87.62, the number listed inside the front cover of this book.

Example 3-g. Chlorine which exhibits an atomic weight of 35.453 is composed of two isotopes, Cl^{35} and Cl^{37}. The exact mass of Cl^{35} is 34.969 and the exact mass of Cl^{37} is 36.966. With this information, calculate the percentage of each isotope in the naturally occurring mixture.

Solution. Let $X = $ the fraction of Cl^{35} atoms. Since only two isotopes of Cl are found naturally, the quantity $(1 - X)$ equals the fraction of Cl^{37} atoms. The sum of the contributions of each species to the atomic weight must equal the chemical atomic weight. Consequently,

$$(X)(34.969) + (1 - X)(36.966) = 35.453$$
$$34.969X + 36.966 - 36.966X = 35.453$$
$$-1.997X = -1.513$$
$$X = 0.7576$$
$$\text{and} \quad (1 - X) = 0.2424$$

Therefore the percentage of Cl^{35} is 75.76 and the percentage of Cl^{37} is 24.24.

PROBLEMS (Atomic Weights from Isotopic Abundance)

3-21. The naturally occurring form of the element fluorine consists of a single nuclide of mass 18.9984. What is the atomic weight of fluorine?

3-22. The element boron is composed of two isotopes. B^{10} constitutes 19.60% of all boron atoms and has a mass of 10.01294. B^{11} constitutes 80.40% and has a mass of 11.00931. Calculate the atomic weight to the proper number of significant figures.

3-23. Potassium has a chemical atomic weight of 39.10. It is composed of three isotopes of mass 38.96, 39.96, and 40.96. The isotope of mass 39.96 is present in a negligibly small amount insofar as making a significant contribution to the atomic weight. What are the percentages of the other two isotopes?

3-24. Ordinary oxygen gas consists of three isotopes with the abundances listed below. What is the atomic weight of oxygen gas?

O Isotope	Exact Mass	% Abundance
16	15.99491	99.759
17	16.99914	0.037
18	17.99916	0.204

GENERAL PROBLEMS

3-25. Determine the molecular or formula weight of each of the following substances (see inside front cover for atomic weights):
(a) HPO_3 (c) $(NH_4)_2CO_3$ (e) $Cr_2(CO_3)_3$
(b) $Ba(OH)_2$ (d) $RbOH$ (f) $MgNH_4AsO_4$

3-26. 58.0 g of H_2SeO_4 represents how many moles of H_2SeO_4?

3-27. 447 g of KCl represents how many moles of KCl?

3-28. How many moles of KCl are there in 44.7 g of KCl?

3-29. (a) How many moles are there in 50.0 g of each of the following?
(i) H_3PO_4 (iii) NH_4NO_3
(ii) $CuSO_4$ (iv) $Fe_2(CO_3)_3$
(b) Determine the number of gram atoms of each element in 50.0 g of the compounds in (a).

3-30. How many grams of strontium are there in 0.025 moles of $Sr(OH)_2$?

3-31. Determine the number of grams in each of the following:
(a) 4.00 moles H_2SO_4 (c) 1.25 moles $K_2Cr_2O_7$
(b) 0.32 mole $KClO_3$ (d) 0.0025 mole $C_{12}H_{22}O_{11}$

3-32. Determine the number of moles in 0.100 kg of each of the following:
(a) $LiOH$ (c) H_2TeO_4
(b) $BaCO_3$ (d) H_3AsO_4

3-33. Determine the number of grams in each of the following:
(a) 0.10 mole HI (c) 3.0 moles $Be(OH)_2$
(b) 0.55 mole $CdSO_4$ (d) 10.0 moles $HClO_4$

3-34. How many gram atoms of oxygen are there in 9.8 g of H_2SO_4?

3-35. 27.0 g of Al represents how many (a) gram atoms, (b) atoms?

3-36. How many gram atoms of each element are present in 11.7 g of NaCl?

3-37. 6.02×10^{24} atoms of sodium would be how many gram atoms?

3-38. 3.01×10^{22} molecules of HCl would be how many moles?

3-39. Calculate the weight in grams of 3.011×10^{24} molecules of H_2SO_4.

3-40. How many moles of CCl_4 will contain 1.6 g atoms of chlorine?

3-41. What weight in grams of H_2SO_4 will contain 3.2 g atoms of oxygen?

3-42. 4.9 g of H_2SO_4 contains how many (a) moles of sulfuric acid, (b) gram atoms of hydrogen, (c) gram atoms of oxygen, (d) molecules of sulfuric acid, (e) grams of sulfur?

3-43. 37.8 g of HNO_3 contains how many (a) moles of HNO_3, (b) gram atoms of nitrogen, (c) grams of nitrogen, (d) molecules of nitric acid, (e) atoms?

3-44. The average mass of zinc atoms in naturally occurring zinc metal is 1.085×10^{-22} g. What is the atomic weight of zinc?

3-45. Element X has been found to consist of a number of isotopes whose atoms have an average mass of 1.972×10^{-22} g. What is the atomic weight of X?

3-46. 5.44 g of Mn dissolve in hydrochloric acid to produce manganous chloride and 0.200 g of hydrogen gas. What is the equivalent weight of Mn?

3-47. If the atomic weight of an element is 88.9 and its valence is 3, what weight of the metal would combine with 16.0 g of oxygen?

3-48. It is found that 2.19 g of a certain metal combines with 0.400 g of oxygen. What is the equivalent weight of the metal?

3-49. 2.30 g of gallium (a metal with an atomic weight of 69) when dissolved in excess HCl releases 0.10 g of hydrogen gas. What is the valence of gallium?

3-50. An oxide of iron is found on analysis to contain 72.1 g of iron for every 27.9 g of oxygen. What is the equivalent weight of iron in this oxide?

3-51. An oxide of antimony is found to contain 75.27% antimony. What is the equivalent weight of antimony in this compound?

3-52. Determine the equivalent weight of gallium, Ga, from the following data: 69.7 g of gallium combines with 240 g of bromine; 40.0 g of bromine combines with 0.500 g of hydrogen; 0.500 g of hydrogen combines with 4.00 g of oxygen.

3-53. Natural gallium, whose atomic weight is 69.72, is a mixture of the two isotopes, Ga^{69} and Ga^{71}. The exact masses of these isotopes are 68.93 amu and 70.93 amu, respectively. Determine the percentage of each isotope in the naturally occurring mixture.

3-54. Determine the atomic weight of lithium which is a mixture of 7.5% of Li^6 and 92.5% Li^7. The exact masses of these isotopes are 6.01 amu and 7.02 amu, respectively.

3-55. Calculate the atomic weight of the element germanium which consists of five isotopes whose exact masses and per cent abundances are as follows:

Mass	% Abundance
69.925	20.53
71.922	27.43
72.923	7.76
73.922	36.54
75.921	7.74

3-56. Antimony consists of two isotopes whose exact masses are 120.904 and 122.904. The atomic weight of antimony is 121.754. What is the percentage of each isotope in the natural element?

3-57. A sample of uranium is prepared in the laboratory by mixing U^{235} (mass 235.04) and U^{238} (mass 238.05). What would the atomic weight of this synthetic sample appear to be if 60.00% of the atoms were U^{238} and the balance were U^{235}?

3-58. Thallium metal, whose atomic weight is 204.384, consists of two isotopes, Tl^{203} and Tl^{205}. The exact mass of Tl^{203} is 202.972 and the exact mass of Tl^{205} is 204.975. What is the isotopic ratio of Tl^{205} to Tl^{203}?

3-59. Natural occurring copper, atomic weight 63.54, is made up of two isotopes weighing 62.93 and 64.93 respectively. Determine the percentage of each isotope in the natural occurring mixture.

3-60. Determine the atomic weight of molybdenum from the following data:

Mass of isotope	% Abundance
91.91	15.84
93.90	9.04
94.90	15.72
95.90	16.53
96.91	9.46
97.91	23.78
99.91	9.63

CHAPTER 4

Formulas and Percentage Composition

A. PERCENTAGE COMPOSITION FROM FORMULAS

A chemical formula is a good source of information in that it gives the number and types of atoms present in the compound. With the aid of atomic weights, the percentage of each element present may be determined. The percentage of any element is the number of parts by weight of that element in 100 parts of the compound.

Example 4-a. Calculate the percentage of each element in sucrose, $C_{12}H_{22}O_{11}$. The atomic weights of C, H, and O are 12, 1, and 16, respectively.

> *Solution.* First determine the weight of each element in one molecular weight of the compound. Then divide the weight of each element by the molecular weight of the compound to obtain the fraction of each element and multiply the fraction by 100 to obtain the percentage.

$$\text{Twelve atoms of carbon weigh} \quad 12 \times 12 = 144$$
$$\text{Twenty-two atoms of hydrogen weigh} \quad 22 \times 1 = 22$$
$$\text{Eleven atoms of oxygen weigh} \quad 11 \times 16 = \overline{176}$$
$$\text{MW} = \overline{342}$$

$$\text{Per cent carbon} = \frac{144}{342} \times 100 = 42.1\%$$

$$\text{Per cent hydrogen} = \frac{22}{342} \times 100 = 6.4\%$$

$$\text{Per cent oxygen} = \frac{176}{342} \times 100 = \underline{51.5\%}$$

$$\textit{Total} \quad 100.0\%$$

Example 4-b. What weight of Fe is contained in 2000 pounds of pure Fe_2O_3 ore? The atomic weights of Fe and O are 56 and 16, respectively.

Solution. First determine the percentage of Fe in the compound, and then multiply the weight of the compound by the fraction of Fe to obtain the weight of Fe.

$$\text{Per cent Fe in } Fe_2O_3 = \frac{2 \times 56}{160} \times 100 = 70\%$$

Weight of Fe in 2000 lb Fe_2O_3 = 0.70 × 2000 lb = 1400 lb

Example 4-c. A silver coin weighing 2.50 g is analyzed by dissolving in dilute nitric acid and then precipitating and weighing AgCl. If the precipitate of AgCl weighs 2.99 g, what is the percentage of silver in the coin?

Solution.

$$\% \text{ Ag in AgCl} = \frac{Ag(107.9)}{AgCl(143.4)} \times 100 = 75.2$$

Weight of Ag in coin = (0.752)(2.99 g) = 2.25 g

$$\% \text{ Ag in coin} = \frac{2.25 \text{ g}}{2.50 \text{ g}} \times 100 = 90.0\%$$

PROBLEMS (Percentage Composition)

4-1. Calculate the percentage composition of each of the following compounds:
(a) NaBr   (b) As_2O_3 (c) C_6H_6

4-2. Calculate the percentage composition of each of the following compounds:
(a) $KMnO_4$ (b) $La_2(SO_4)_3$ (c) $Zr_3(PO_4)_4$

4-3. What mass of nitrogen is contained in 50 g of each of the following?
(a) KNO_3 (c) NH_4NO_3
(b) NH_4Cl (d) $(NH_4)_2HPO_4$

4-4. If 3.2 g of the compound ammonium nitrite, NH_4NO_2, decomposes to give only H_2O and N_2 gas, what mass of N_2 is produced?

4-5. 4.65 g of a magnesium alloy is analyzed for Mg by precipitation and weighing $MgNH_4PO_4$. If 3.94 g of the latter is obtained, what is the percentage of Mg in the alloy?

4-6. 1.320 g of an ore containing barium carbonate ($BaCO_3$) yielded 1.248 g of $BaSO_4$ in an analysis for barium. What is the percentage of barium carbonate in the sample?

B. FORMULAS FROM PERCENTAGE COMPOSITION DATA

An empirical formula, which shows the relative number of atoms of each element in a formula unit, may be determined from the percentage composition data of a pure compound. To convert the empirical formula to a molecular formula, an experimental measurement must be made of the molecular weight. Such measurements are discussed in Chapters 6 and 10.

The following steps may be followed in obtaining the empirical and molecular formulas from percentage composition data.

1. Assume that you have 100 g of the compound. This gives a number of grams of each element numerically equal to its percentage in the compound.

2. Determine the number of gram atoms of each element.

3. Determine the highest common divisor for the number of gram atoms.

4. Divide each of the numbers of gram atoms by the highest common divisor. The quotients obtained should be very close to whole numbers. Round to the nearest whole number.

5. Write these whole numbers as subscripts of the elements in an empirical formula.

6. If the molecular weight is known, determine how many empirical formula weights are required to obtain the molecular weight. Use this factor to multiply the number of atoms in the empirical formula to obtain the number of atoms of each element in the true molecular formula.

Examples 4-d and 4-e illustrate these steps in detail. Study them carefully.

Example 4-d. Analysis of a compound of boron and hydrogen showed 78.2% B and 21.8% H. An experimentally determined value of the molecular weight was 27.6. What are the empirical and molecular formulas of the compound?

> *Solution.* Assume that you have 100.0 g of compound. Then 100.0 g of compound will contain 78.2 g B and 21.8 g H. Converting these weights to gram atoms:
>
> $$78.2 \text{ g B} \times \frac{1 \text{ g atom B}}{10.8 \text{ g B}} = 7.24 \text{ g atoms B}$$
>
> $$21.8 \text{ g H} \times \frac{1 \text{ g atom H}}{1.01 \text{ g H}} = 21.6 \text{ g atoms H}$$

The ratio of gram atoms of H to gram atoms of B in 100 g of the compound must be the same as the ratio of H to B atoms in one molecule of the compound. That is, the empirical formula could be written

$$B_{7.24}H_{21.6}$$

However, atoms in formulas usually have whole number subscripts, and to obtain them, we must divide through by the highest common divisor which in this case is 7.24. The formula then is

$$B_{\underset{(7.24)}{(7.24)}}H_{\underset{(7.24)}{(21.6)}} = BH_{2.98} \quad \text{or} \quad BH_3$$

BH_3 is therefore the *empirical* formula for this compound, and the formula weight is $10.8 + 3.0 = 13.8$. The molecular formula will either be the same as the empirical formula or some multiple of it such as B_2H_6 or B_3H_9, etc. Because the molecular weight of the compound was determined experimentally to be 27.6 (or 2 × the empirical formula weight), the *molecular* formula must be B_2H_6.

Example 4-e. A compound is found to contain 11.2% N, 3.2% H, 41.2% Cr, and 44.4% O. Determine the empirical formula.

Solution. Calculate the number of gram atoms of each element in 100 g of the compound.

$$11.2 \text{ g N} \times \frac{1 \text{ g atom N}}{14.0 \text{ g N}} = 0.8 \text{ g atom N}$$

$$3.2 \text{ g H} \times \frac{1 \text{ g atom H}}{1.0 \text{ g H}} = 3.2 \text{ g atoms H}$$

$$41.2 \text{ g Cr} \times \frac{1 \text{ g atom Cr}}{52.0 \text{ g Cr}} = 0.8 \text{ g atom Cr}$$

$$44.4 \text{ g O} \times \frac{1 \text{ g atom O}}{16.0 \text{ g O}} = 2.8 \text{ g atoms O}$$

The first three numbers are divisible by 0.8 an integral number of times, but the last number, 2.8, when divided by 0.8 gives $3\frac{1}{2}$. Hence the highest common divisor of all these numbers would be $\frac{1}{2}$ of 0.8 or 0.4. The empirical formula can be written as

$$N_{\underset{(0.4)}{(0.8)}}H_{\underset{(0.4)}{(3.2)}}Cr_{\underset{(0.4)}{(0.8)}}O_{\underset{(0.4)}{(2.8)}} \quad \text{or} \quad N_2H_8Cr_2O_7$$

Example 4-f. A hydrate of ferric thiocyanate, $Fe(SCN)_3$, was found to contain 19.0% H_2O. What is the formula for the salt hydrate?

Solution. This question is analogous to asking how many moles of H_2O there are per mole of $Fe(SCN)_3$. In principle this problem is worked in a manner similar to Examples 4-d and 4-e. First calculate the number of moles of H_2O and $Fe(SCN)_3$ in 100 g of the compound. Then determine the number of moles of H_2O per mole of $Fe(SCN)_3$.

$$19.0 \text{ g } H_2O \times \frac{1 \text{ mole } H_2O}{18.0 \text{ g } H_2O} = 1.056 \text{ moles } H_2O$$

$$81.0 \text{ g } Fe(SCN)_3 \times \frac{1 \text{ mole } Fe(SCN)_3}{280.1 \text{ g } Fe(SCN)_3} = 0.352 \text{ mole } Fe(SCN)_3$$

Number of moles of H_2O/mole $Fe(SCN)_3 = 1.056/0.352 = 3.00$

The formula for the compound is therefore $Fe(SCN)_3 \cdot 3 \ H_2O$.

PROBLEMS (Empirical and Molecular Formulas)

4-7. A compound is analyzed and found to contain 47.42% Mo and 52.57% Cl. What is the empirical formula?

4-8. A pure compound of mercury and iodine results on reaction of 32.50 g of mercury with 41.12 g of iodine. What is the empirical formula of the compound?

4-9. A compound of chlorine and bromine is formed by direct reaction of the two elements. By a careful experiment, it is found that 7.10 g of chlorine react with 16.0 g of bromine. (a) What is the empirical formula for the compound? (b) If the molecular weight is 115, what is the molecular formula of the compound?

4-10. A compound of aluminum and chlorine is composed of 9.0 g of aluminum for every 35.5 g of chlorine. (a) What is the empirical formula? (b) If the molecular weight is 267, what is the molecular formula?

4-11. Determine the empirical formula of each of the following compounds from the percentage composition data.

Weight Percentage

Compound	C	H	O	N
(a)	52.2	13.0	34.8	—
(b)	—	2.13	68.0	29.8
(c)	48.6	8.1	43.3	—
(d)	58.5	4.07	26.0	11.4
(e)	49.3	9.6	21.9	19.2

4-12. Gallium perchlorate, $Ga(ClO_4)_3$, forms a hydrate containing six molecules of water per formula unit of $Ga(ClO_4)_3$. What is the percentage of water in the hydrate?

4-13. The formula for a hydrate of hydrazine, N_2H_4, is $N_2H_4 \cdot H_2O$. What is the percentage of water in the compound?

4-14. Gadolinium chloride, $GdCl_3$, forms a hydrate containing 70.7% $GdCl_3$. What is the formula for the hydrate?

4-15. Perchloric acid, $HClO_4$, forms a crystalline hydrate melting at $50°C$. It contains 15.2% H_2O and 84.8% $HClO_4$. What is the formula of the hydrate?

4-16. In 1933 Professor Linus Pauling of the California Institute of Technology predicted the existence of a certain compound of the rare gas xenon and fluorine. Recently the compound has been prepared almost simultaneously by research workers in several different laboratories. The compound was found to consist of 53.5% Xe and 46.5% F. (a) What is its empirical formula? (b) If its molecular weight is 245, what is its molecular formula?

GENERAL PROBLEMS

4-17. Calculate the percentage of oxygen in each of the following compounds:
(a) ZnO (c) $Zn(NO_3)_2$ (e) $Ba_3(AlO_3)_2$
(b) $Na_2Cr_2O_7$ (d) $Fe_2(SO_4)_3$ (f) TiO_2

4-18. What weight of oxygen may be obtained by electrolysis of H_2O when 0.540 kg of H_2O is completely converted to hydrogen and oxygen?

4-19. Assuming 100% conversion, what weight of S would be needed to produce 1.96 kg of H_2SO_4?

4-20. How many grams of As are there in 85 g of As_2S_3?

4-21. A compound is found to contain 15.6% Mg, 48.3% As, and 36.1% O. What is the empirical formula of the compound?

4-22. A pure compound is analyzed and found to contain 15.9% B and 84.1% F. What is the empirical formula of the compound?

4-23. A substance which has an empirical formula of CH_2 and a known molecular weight of 84 must have what molecular formula?

4-24. From the following analytical data, determine the empirical formula for each of the following compounds:
(a) 80% C; 20% H   (d) 31.0% B; 69.0% O
(b) 88.8% Cu; 11.2% O   (e) 17.2% Cr; 35.2% Cl; 47.6% O
(c) 30.8% K; 44.1% O; 25.2% S

4-25. A compound of aluminum was found experimentally to have a molecular weight of 144. By chemical analysis, it was shown to consist of 37.4% Al, 50.0% C, and 12.6% H. What is its molecular formula?

4-26. Determine the percent water of hydration of each of the following hydrates:

(a) $Na_2SO_4 \cdot 10\ H_2O$ (d) $In(NO_3)_3 \cdot 3\ H_2O$

(b) $CuSO_4 \cdot 5\ H_2O$ (e) $NH_4Cr(SO_4)_2 \cdot 12\ H_2O$

(c) $Pb(ClO_4)_2 \cdot 3\ H_2O$ (f) $La(BrO_3)_3 \cdot 9\ H_2O$

4-27. If 10.00 g of a hydrate of $NiSO_4$ loses 4.49 g H_2O when heated, what is the formula for the hydrate?

4-28. Determine the formula of a hydrated salt which has the composition 27.8% Mn, 35.9% Cl, and 36.4% H_2O.

4-29. If 5.0 g of a hydrate of Na_2CO_3 on being heated leaves a residue which weighs 1.85 g, determine the formula of the hydrated salt.

4-30. $ZnCl_2$ combines with NH_3 to form a complex salt. If 1.000 g of $ZnCl_2$ combines with 0.497 g of NH_3, what is the formula for the complex?

4-31. If, on being heated, 10.00 g of a hydrate of $Li_2B_4O_7$ leaves a residue of $Li_2B_4O_7$ which weighs 6.53 g, determine the formula of the hydrated salt.

4-32. On heating 4.64 g of a compound, 4.32 g of silver and 0.32 g of oxygen were obtained. What is the simplest formula of the compound?

4-33. Potassium chlorate, when heated, yields oxygen and potassium chloride. If 4.90 g of potassium chlorate yields 1.92 g of oxygen and 2.98 g of potassium chloride, determine the simplest formula for the potassium chlorate.

4-34. A compound formed between xenon, oxygen, and fluorine analyses $Xe = 70.8\%$, $O = 8.7\%$. What is the empirical formula?

4-35. If 1.00 g of Al metal is burned in oxygen and the product weighs 1.88 g, determine the empirical formula of the oxide from these data.

4-36. On heating solid calcium carbonate, $CaCO_3$, gaseous carbon dioxide, CO_2, and solid calcium oxide, CaO, are formed. A 5.0 g sample of $CaCO_3$ is heated until 4.5 g of solid remains. What per cent of the $CaCO_3$ has decomposed?

4-37. O_2 gas may be obtained by heating certain oxygen-containing compounds. On heating, potassium chlorate, $KClO_3$, releases all of its oxygen; potassium nitrate, KNO_3, releases one-third of its oxygen; and lead dioxide, PbO_2, releases one-half of its oxygen. If costs per pound of the three compounds are respectively $1.50, $0.75, and $0.30, which is the cheapest source of oxygen?

4-38. The following nitrogen compounds are frequently added to the soil as fertilizers: (a) $NaNO_3$, (b) $CaCN_2$, (c) $(NH_4)_2SO_4$, (d) NH_4NO_3. The approximate costs per pound of these in bulk quantities are $0.0210, $0.0280, $0.0180, and $0.0350, respectively. For each of these compounds, determine the cost to furnish 1.000×10^3 kg of nitrogen to the soil.

4-39. In a laboratory experiment to determine the formula for the compound formed between Cu and S, a student heated a weighed amount of copper with an excess of sulfur. After the reaction was completed, the excess sulfur was volatilized off by heating until the crucible and its contents had achieved a constant weight. The following data were obtained:

Mass of crucible . . . . 21.383 g
Mass of crucible plus Cu . . . . 29.324 g
Mass of crucible plus the Cu–S compound . . . . . 31.328 g

What is the simplest formula for the compound?

4-40. The first true chemical compound of a rare gas element was prepared in 1962 by Professor Neil Bartlett of the University of British Columbia. He succeeded in preparing a yellow compound by direct reaction of xenon gas with platinum hexafluoride, which, on chemical analysis, was found to be composed of 29.8% Xe, 44.3% Pt, and 25.9% F. What is the empirical formula for the yellow compound?

4-41. $MgCO_3$ when heated yields MgO and CO_2. If 25.0 g of $MgCO_3$ is heated until the residue weighs 20.0 g, what percentage of the $MgCO_3$ remains undecomposed?

CHAPTER 5

Methods of
Balancing Equations

A balanced equation is one in which the same number of atoms of each element and the same total of charges appear on the left and right sides of the equation. For this to be true, the proper coefficients or numbers appearing immediately before the formulas in the equation must be used. The ease of obtaining the proper coefficients to give a balanced equation depends to a large extent on the number of reactants and products. Equations involving only a few reactants and products may be balanced by simple inspection; this includes some simple oxidation-reduction reactions which, by definition, involve changes in oxidation numbers of some of the substances. Often, however, oxidation-reduction equations are difficult to balance by inspection; fortunately systematic methods are available for balancing them. These will be considered following a brief discussion of the inspection method of balancing equations.

A. BALANCING BY INSPECTION

1. Molecular Equations

A good general rule to follow is: Starting with the most complicated formula, balance elements by adjusting coefficients of the formulas as necessary, leaving hydrogen and oxygen until last.

Example 5-a. Balance the equation

$$Al(OH)_3 + H_2SO_4 \longrightarrow Al_2(SO_4)_3 + H_2O$$

Solution. Pick out the most complicated formula which in this case is obviously $Al_2(SO_4)_3$. Since two Al atoms are present in this formula, place the coefficient 2 in front of $Al(OH)_3$ on the left side. Likewise three S atoms (or three SO_4 radicals) appear in the formula, hence a coefficient 3 is needed for H_2SO_4 on the left side. Next balance H atoms; since twelve H atoms now appear on the left side, a coefficient of 6 is required for H_2O on the right. The balanced equation becomes:

$$2\ Al(OH)_3 + 3\ H_2SO_4 \longrightarrow Al_2(SO_4)_3 + 6\ H_2O$$

Example 5-b. Balance the equation

$$BaBr_2 + H_3PO_4 \longrightarrow Ba_3(PO_4)_2 + HBr$$

Solution. Start with $Ba_3(PO_4)_2$ and observe that three Ba appear in the formula; therefore use a coefficient of 3 for $BaBr_2$. Likewise two P atoms (or two PO_4 radicals) call for a coefficient of 2 for H_3PO_4. Now six Br atoms on the left call for a coefficient of 6 for HBr on the right. The balanced equation becomes:

$$3\ BaBr_2 + 2\ H_3PO_4 \longrightarrow Ba_3(PO_4)_2 + 6\ HBr$$

2. Ionic Equations

The same general rules employed for balancing molecular equations apply to ionic equations, but, *in addition to balancing atoms*, the *charges must also be balanced.*

In writing ionic equations, write as ions those substances which exist as ions in solution; write complete formulas for nonionized substances, solids, and gases. Generally only those substances undergoing a net chemical change are included in ionic equations.

Example 5-c. Balance the equation $Al + H^+ \longrightarrow Al^{+++} + H_2$.

Solution. If a coefficient of 2 is used for H^+, the equation is balanced atomically but not electrically, since there would be two positive charges on the left and three positive charges on the right. Inspection reveals that an even number of H^+, and thus positive charges, must be used on the left because every molecule of H_2 on the right has two atoms of hydrogen. Since the Al^{+++} ion has three positive charges, the smallest coefficient for Al must be 2, so that an even number of charges are obtained on the right. The coefficient of H^+ must be 6 for the charges to balance. Hence the balanced equation becomes:

$$2\ Al + 6\ H^+ \longrightarrow 2\ Al^{+++} + 3\ H_2$$

EQUATIONS (Balance by Inspection)

5-1. $KNO_3 \rightarrow KNO_2 + O_2$

5-2. $Al + HCl \rightarrow AlCl_3 + H_2$

5-3. $Fe_2O_3 + CO \rightarrow Fe_3O_4 + CO_2$

5-4. $MgO + HBr \rightarrow MgBr_2 + H_2O$

5-5. $K + H_2O \rightarrow KOH + H_2$

5-6. $Li_2O_2 + H_2O \rightarrow LiOH + O_2$

5-7. $CaH_2 + H_2O \rightarrow Ca(OH)_2 + H_2$

5-8. $ZnCO_3 + HNO_3 \rightarrow Zn(NO_3)_2 + CO_2 + H_2O$

5-9. $BaCl_2 + Na_3AsO_4 \rightarrow Ba_3(AsO_4)_2 + NaCl$

5-10. $Al + NaOH + H_2O \rightarrow NaAl(OH)_4 + H_2$

5-11. $NH_4NO_3 + Ca(OH)_2 \rightarrow Ca(NO_3)_2 + NH_3 + H_2O$

5-12. $C_2H_2 + O_2 \rightarrow CO_2 + H_2O$

5-13. $Cu(NO_3)_2 \rightarrow CuO + NO_2 + O_2$

5-14. $PbS + O_2 \rightarrow PbO + SO_2$

5-15. $NaMnO_4 + H_2SO_4 \rightarrow Na_2SO_4 + Mn_2O_7 + H_2O$

5-16. $P_4 + O_2 + H_2O \rightarrow H_3PO_4$

5-17. $C_4H_{10} + O_2 \rightarrow CO_2 + H_2O$

5-18. $Th(NO_3)_4 + K_3PO_4 \rightarrow Th_3(PO_4)_4 + KNO_3$

5-19. $CaNCN + H_2O \rightarrow CaCO_3 + NH_3$

5-20. $PBr_3 + H_2O \rightarrow H_3PO_3 + HBr$

5-21. $PbCO_3 + H^+ \rightarrow Pb^{++} + CO_2 + H_2O$

5-22. $NH_4^+ + OH^- \rightarrow NH_3 + H_2O$

5-23. $Cr + H^+ \rightarrow Cr^{++} + H_2$

5-24. $As_2S_5 + S^{--} \rightarrow AsS_4^{---}$

5-25. $Zn + OH^- + H_2O \rightarrow Zn(OH)_4^{--} + H_2$

5-26. $Ba_3(PO_4)_2 + H^+ \rightarrow H_3PO_4 + Ba^{++}$

5-27. $CdS + H^+ \rightarrow H_2S + Cd^{++}$

5-28. $Fe_2O_3 + H^+ \rightarrow Fe^{+++} + H_2O$

5-29. $CdS + As^{+++} \rightarrow As_2S_3 + Cd^{++}$

5-30. $Th^{++++} + H_2PO_4^- \rightarrow Th_3(PO_4)_4 + H^+$

B. BALANCING OXIDATION-REDUCTION EQUATIONS

Oxidation and reduction are mutually dependent processes. Oxidation involves an increase in oxidation number and an apparent loss of electrons by an atom, group of atoms, or an ion. Similarly, reduction involves a decrease in oxidation number and an apparent gain of electrons by a substance. The substance whose oxidation number is increased is said to be oxidized; the substance whose oxidation number is decreased is said to be reduced. The substance which appears to gain electrons is called the oxidizing agent, and the substance which appears to lose electrons is called the reducing agent.

Oxidation numbers are the apparent charges that atoms would have if the electrons in the compound were distributed among atoms in a very arbitrary fashion. These numbers are a useful aid in following electron shifts in oxidation-reduction reactions and are assigned as follows.

(a) The oxidation numbers of atoms of elements are zero.

(b) In ionic substances, the oxidation numbers of monatomic ions are equal to their charges.

(c) In a covalent molecule or a complex ion, the oxidation number is the apparent charge left on an atom when the bonding electrons are assigned completely to the more electronegative of the two bonded atoms; electrons shared between atoms of the same element are divided equally between them. In all such cases, the assignment must be such that there is a conservation of charge on the molecule or ion.

From the preceding rules, Li ion and Cl ion in LiCl are assigned the values $+1$ and -1, respectively. Oxygen in H_2O is assigned the value -2 and hydrogen the value $+1$ because oxygen is the more electronegative of the two atoms. These are the usual values for H and O except in a few instances; O is -1 in peroxides and H is -1 in hydrides.

Using customary oxidation numbers of the simple ions, the less common oxidation numbers of other atoms may be assigned. In $HClO_4$, the contribution of H to the apparent charge is $+1$ and the contribution of four O atoms is -8. For the sum of the oxidation numbers to be zero, as required for a neutral molecule, the total apparent charge for Cl must be $+7$. In the complex ion $[PtCl_6]^{--}$, Cl is given its usual value of -1, and then for the sum of the oxidation numbers to equal the -2 charge of the ion, the Pt atom must be assigned the oxidation number of $+4$.

Oxidation-reduction equations are balanced on the principle of *conservation of charge*, that is, the number of electrons lost by the reducing agent must equal the number of electrons gained by the oxidizing agent. One method of balancing these equations is known as the *Change in Oxidation Number Method* and the other is known as the *Ion-Electron Half-Reaction Method*.

1. Balancing from Oxidation Number Changes

The following steps are useful in balancing equations from oxidation number changes.

(a) Determine which elements change in oxidation number.

(b) Determine the amount of change per formula unit of each substance.

(c) Make the gain of electrons equal the loss of electrons by adjusting the coefficients of the formulas for the reducing agent and oxidizing agent in the equation.

(d) Complete the balancing of the equations by inspection, keeping in mind that the ratio of oxidizing to reducing agent as determined above must be maintained.

These steps are illustrated in the examples below.

Example 5-d. Balance the equation

$$ZnS + HNO_3 \rightarrow ZnSO_4 + NO + H_2O$$

Solution. In this reaction, sulfur changes in oxidation number from -2 in ZnS to $+6$ in $ZnSO_4$; a loss of eight electrons per formula unit of ZnS. Meanwhile, nitrogen changes in oxidation number from $+5$ in HNO_3 to $+2$ in NO; a gain of three electrons per molecule of HNO_3. The total gain of electrons must equal the total loss and to balance them, the number 3 is made the coefficient of ZnS and the number 8 becomes the coefficient of HNO_3 in the final equation. Thus 3 ZnS lose twenty-four electrons and 8 HNO_3 gain twenty-four electrons.

$$
\begin{array}{c}
\overbrace{\qquad\qquad}^{\text{(lose } 8\,e^-) \times 3 = 24} \\
\underset{\text{(oxidation)}}{} \\
\underset{-}{|}2 \quad\;\; +5 \qquad +|6 \qquad +2 \\
3\ ZnS + 8\ HNO_3 \rightarrow 3\ ZnSO_4 + 8\ NO + 4\ H_2O \\
\underbrace{\qquad\qquad}_{\substack{\text{(gain } 3\,e^-) \times 8 = 24 \\ \text{(reduction)}}}
\end{array}
$$

Example 5-e. Balance the equation

$$H_2S + K_2Cr_2O_7 + HCl \rightarrow S + CrCl_3 + KCl + H_2O$$

Solution. Sulfur apparently loses two electrons per molecule of H_2S; each Cr atom in $K_2Cr_2O_7$ apparently gains three electrons, so the electron gain per formula unit of $K_2Cr_2O_7$ is $2 \times 3 = 6$. Hence, to balance electrons, we need a ratio of 6 H_2S to 2 $K_2Cr_2O_7$, or simply 3 to 1. Having fixed the ratio of H_2S and $K_2Cr_2O_7$, the remainder of the balancing follows by inspection.

$$
\begin{array}{c}
\overbrace{\qquad\qquad}^{\text{(lose } 2\,e^-) \times 3 = 6} \\
\underset{\text{(oxidation)}}{} \\
-|2 \quad\;\; +6 \qquad\qquad\qquad |0 \quad\; +3 \\
3\ H_2S + K_2Cr_2O_7 + 8\ HCl \rightarrow 3\ S + 2\ CrCl_3 + 2\ KCl + 7\ H_2O \\
\underbrace{\qquad\qquad}_{\substack{\text{gain } 3\,e^-/\text{Cr} \\ \textit{or } \text{gain } 6\,e^-/K_2Cr_2O_7 \\ \text{(reduction)}}}
\end{array}
$$

Example 5-f. Balance the equation

$$AsO_3^{---} + Cr_2O_7^{--} + H^+ \rightarrow AsO_4^{---} + Cr^{+++} + H_2O$$

Solution.

$$\overbrace{\hspace{4cm}}^{\text{(lose } 2\,e^-) \times 3 = 6}$$

(oxidation)

$+|3 \qquad\qquad +6 \qquad\qquad +|5 \qquad\qquad +3$

$3\,AsO_3^{---} + Cr_2O_7^{--} + 8\,H^+ \rightarrow 3\,AsO_4^{---} + 2\,Cr^{+++} + 4\,H_2O$

$$\underbrace{\hspace{5cm}}_{\text{(gain } 6\,e^-/Cr_2O_7^{--})}$$

(reduction)

Example 5-g. Balance the equation

$$Sb_2S_5 + H^+ + NO_3^- \rightarrow HSbO_3 + S + NO + H_2O$$

Solution. Lose $2\,e^-/S$ *or* $10\,e^-/Sb_2S_5$; gain $3\,e^-/HNO_3$.

$$\overbrace{\hspace{5cm}}^{\text{(lose } 10\,e^-) \times 3 = 30}$$

(oxidation)

$-|2 \qquad\qquad +5 \qquad\qquad\qquad |0 \qquad +2$

$3\,Sb_2S_5 + 10\,H^+ + 10\,NO_3^- \rightarrow 6\,HSbO_3 + 15\,S + 10\,NO + 2\,H_2O$

$$\underbrace{\hspace{5cm}}_{\text{(gain } 3\,e^-) \times 10 = 30}$$

(reduction)

Example 5-h. Occasionally a reaction in which more than two elements change in oxidation number is encountered. We need only obtain the *net change in each substance* to arrive at the proper coefficients. Balance the equation

$$As_2S_3 + Mn(NO_3)_2 + K_2CO_3 \rightarrow K_3AsO_4 + K_2SO_4$$
$$+ K_2MnO_4 + NO + CO_2$$

Solution.

Apparent electron loss by As per formula weight $As_2S_3 = 2 \times 2 = 4$ electrons lost.

Apparent electron loss by S per formula weight $As_2S_3 = 3 \times 8 = 24$ electrons lost.

Net change per formula weight $As_2S_3 = 28$ electrons lost.

Apparent electron loss by Mn per formula weight $Mn(NO_3)_2 = 1 \times 4 = 4$ electrons lost.

Apparent electron gain by N per formula weight $Mn(NO_3)_2 = 2 \times 3 = 6$ electrons gained.

Net change per formula weight $Mn(NO_3)_2 = 2$ electrons gained.

To balance the electron gain and loss, we arrive at the ratio 28 to 2 or 14 $Mn(NO_3)_2$ to 1 As_2S_3. The rest of the equation is balanced by inspection.

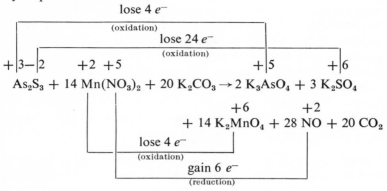

EQUATIONS (Oxidation Number Change)

Balance the following equations by considering oxidation number changes.

5-31. $Bi(OH)_3 + K_2Sn(OH)_4 \rightarrow Bi + K_2Sn(OH)_6$

5-32. $FeSO_4 + H_2SO_4 + KMnO_4$
$$\rightarrow Fe_2(SO_4)_3 + K_2SO_4 + MnSO_4 + H_2O$$

5-33. $H_2S + KMnO_4 + HCl \rightarrow S + KCl + MnCl_2 + H_2O$

5-34. $Sb + HNO_3 \rightarrow Sb_2O_5 + NO + H_2O$

5-35. $MnO_2 + PbO_2 + HNO_3 \rightarrow Pb(NO_3)_2 + HMnO_4 + H_2O$

5-36. $HMnO_4 + AsH_3 + H_2SO_4 \rightarrow H_3AsO_4 + MnSO_4 + H_2O$

5-37. $Bi + HNO_3 \rightarrow Bi(NO_3)_3 + NO + H_2O$

5-38. $Ag + H^+ + NO_3^- \rightarrow Ag^+ + NO + H_2O$

5-39. $Sn^{++} + H^+ + IO_3^- \rightarrow Sn^{++++} + I^- + H_2O$

5-40. $Cr_2O_7^{--} + C_2H_5OH + H^+ \rightarrow Cr^{+++} + HC_2H_3O_2 + H_2O$

5-41. $As_4 + H_2O + NO_3^- \rightarrow AsO_4^{---} + NO + H^+$

5-42. $Fe_2(SO_4)_3 + NaI \rightarrow FeSO_4 + Na_2SO_4 + I_2$

5-43. $Ag_3AsO_4 + Zn + H^+ \rightarrow AsH_3 + Ag + Zn^{++} + H_2O$

5-44. $Sb_2S_3 + H^+ + NO_3^- \rightarrow Sb_2O_5 + SO_4^{--} + NO + H_2O$

5-45. $Cr_2S_3 + Mn(NO_3)_2 + Na_2CO_3$
$$\rightarrow Na_2CrO_4 + Na_2MnO_4 + Na_2SO_4 + CO_2 + NO$$

2. Ion-Electron Half-Reaction Method

In this method the reaction is separated into two half-reactions, one of which shows the oxidation reaction and the other the reduction reaction.

Each step is balanced atomically, then electrically by adding electrons. Formulas for the reacting species are written to represent as closely as possible those present in large amounts; formulas for the products are written similarly. It is often necessary to add H_2O, H^+, or OH^- to the half-reactions to balance them, and it should only be remembered that OH^- is not present in large amount in acid solution and that H^+ is not present in large amount in basic solution. Convenient rules to follow are (a) To balance oxygen atoms, add water if in acidic solution, and OH^- if in basic solution, to the side of the half-reaction deficient in O atoms; (b) Balance H atoms after O balancing has been completed. To obtain the net overall reaction, the two half-reactions are added after balancing the electron gain and loss. This method is illustrated in detail in Example 5-i.

Example 5-i. Balance the equation for the oxidation of Cu by NO_3^- in acid solution. Copper is oxidized to Cu^{++} and NO_3^- is reduced to NO.

Solution. Write two skeleton half-reactions, one for the oxidation half-reaction and one for the reduction half-reaction.

$$Cu \longrightarrow Cu^{++} \quad \text{(oxidation)}$$
$$NO_3^- \longrightarrow NO \quad \text{(reduction)}$$

Balance the atoms next. It is observed that in the oxidation half-reaction there is one Cu atom on each side of the equation. In the reduction reaction there are two more O atoms on the left than on the right, and in acid solution, O with a -2 oxidation state must become H_2O. Therefore $4 H^+$ are required on the left and the equation is now written as

$$NO_3^- + 4 H^+ \longrightarrow NO + 2 H_2O$$

Both half-reactions are now balanced atomically; next the charge must be balanced by adding negative electrons. The two half-reactions thus become

$$Cu \longrightarrow Cu^{++} + 2 e^- \quad \text{(oxidation)}$$
$$3 e^- + 4 H^+ + NO_3^- \longrightarrow 2 H_2O + NO \quad \text{(reduction)}$$

Multiply the first step by 3 and the second step by 2 to obtain 6 electrons lost and gained.

$$3 Cu \rightarrow 3 Cu^{++} + 6 e^-$$
$$6 e^- + 8 H^+ + 2 NO_3^- \rightarrow 4 H_2O + 2 NO$$

On adding these two half-reactions, the electrons cancel and the net ionic equation for the reaction becomes

$$3 Cu + 8 H^+ + 2 NO_3^- \rightarrow 3 Cu^{++} + 4 H_2O + 2 NO$$

Example 5-j. Balance the equation

$$Fe^{++} + H^+ + MnO_4^- \rightarrow Fe^{+++} + Mn^{++} + H_2O$$

Solution.

$$5\,Fe^{++} \rightarrow 5\,Fe^{+++} + 5e^- \quad \text{(oxidation)}$$
$$5e^- + 8\,H^+ + MnO_4^- \rightarrow 4\,H_2O + Mn^{++} \quad \text{(reduction)}$$

$$\overline{5\,Fe^{++} + 8\,H^+ + MnO_4^- \rightarrow 5\,Fe^{+++} + Mn^{++} + 4\,H_2O}$$

Example 5-k. Balance the equation for the reaction between $Bi(OH)_3$ and SnO_2^{--} which occurs in basic solution.

$$Bi(OH)_3 + SnO_2^{--} \rightarrow Bi + SnO_3^{--}$$

Solution. Separate the equation into an oxidation half-reaction and a reduction half-reaction. Remember that, in basic solution, no H^+ should appear in either half-reaction.

$$Bi(OH)_3 \rightarrow Bi \quad \text{(reduction)}$$
$$SnO_2^{--} \rightarrow SnO_3^{--} \quad \text{(oxidation)}$$

To balance atoms for the reduction half-reaction, add 3 OH^- to the right side of the equation. There is a deficiency of O atoms on the left side of the oxidation half-reaction, and so hydroxide ions are added to the left to provide them. Twice as many OH^- ions must be added as oxygen atoms needed because the H^+ separated from one OH^- must form water by reaction with a second OH^-. When balanced atomically, these equations become

$$Bi(OH)_3 \rightarrow Bi + 3\,OH^-$$
$$2\,OH^- + SnO_2^{--} \rightarrow SnO_3^{--} + H_2O \quad \text{(correct)}$$
$$OH^- + SnO_2^{--} \rightarrow SnO_3^{--} + H^+ \quad \text{(wrong)}$$

The latter equation is incorrect because H^+ is not present in large amount in basic solution. On balancing charge, these half reactions become

$$Bi(OH)_3 + 3\,e^- \rightarrow Bi + 3\,OH^- \quad \text{(reduction)}$$

and

$$2\,OH^- + SnO_2^{--} \rightarrow SnO_3^{--} + H_2O + 2\,e^- \quad \text{(oxidation)}$$

After multiplying the first equation by 2 and the second by 3 to make electrons gained equal to electrons lost, these equations are added to obtain the final balanced equation.

$$\cancel{6e^-} + 2\,Bi(OH)_3 \rightarrow 2\,Bi + 6\,OH^-$$
$$6\,OH^- + 3\,SnO_2^{--} \rightarrow 3\,SnO_3^{--} + 3\,H_2O + \cancel{6e^-}$$
$$\overline{2\,Bi(OH)_3 + 3\,SnO_2^{--} \rightarrow 3\,SnO_3^{--} + 2\,Bi + 3\,H_2O}$$

Note that the OH^- ions cancel; such cancellations should be made whenever possible.

Example 5-1. Balance the equation for the oxidation of AsO_3^{---} by MnO_4^- in basic solution:

$$MnO_4^- + AsO_3^{---} + H_2O \rightarrow MnO_2 + AsO_4^{---} + OH^-$$

Solution. Separate the completed equation into half-reactions and balance each step separately. Since the reaction is in basic solution, hydrogen ion should not appear in either half-reaction. It is seen from the equation that AsO_3^{---} is oxidized to AsO_4^{---} and MnO_4^- is reduced to MnO_2. With this information the two half-reactions are

$$3[AsO_3^{---} + 2\,OH^- \rightarrow AsO_4^{---} + H_2O + 2\,e^-]$$

$$\text{(oxidation)}$$

$$2[3\,e^- + MnO_4^- + 2\,H_2O \rightarrow MnO_2 + 4\,OH^-] \quad \text{(reduction)}$$

$$\overline{2\,MnO_4^- + 3\,AsO_3^{---} + H_2O \rightarrow 2\,MnO_2 + 3\,AsO_4^{---} + 2\,OH^-}$$

EQUATIONS (Ion-Electron Half-Reactions)

5-46. Balance by the ion-electron method. Balance each half-reaction atomically and electrically; then add the two steps to obtain the net ionic equation.

(a) $C_2O_4^{--} \rightarrow CO_2$
    $H^+ + MnO_4^- \rightarrow H_2O + Mn^{++}$

(b) $Bi_2S_3 \rightarrow Bi^{+++} + S$
    $H^+ + NO_3^- \rightarrow H_2O + NO$

(c) $Fe^{++} \rightarrow Fe^{+++}$
    $H^+ + Cr_2O_7^{--} \rightarrow H_2O + Cr^{+++}$

(d) $Cl^- + HgS \rightarrow HgCl_4^{--} + S$
    $H^+ + NO_3^- \rightarrow H_2O + NO$

(e) $Br_2 \rightarrow Br^-$

$S_2O_3^{--} + H_2O \rightarrow SO_4^{--} + H^+$

(f) $Zn \rightarrow Zn^{++}$

$HAsO_2 + H^+ \rightarrow AsH_3 + H_2O$

(g) $Pd + Cl^- \rightarrow PdCl_6^{--}$

$H^+ + Cl^- + NO_3^- \rightarrow NOCl + H_2O$

5-47. In the following only the final unbalanced and incomplete equation is given. Separate each reaction into two half-reactions, one oxidation and the other reduction. Balance each step, then add the two steps. Use OH^-, H^+, and H_2O where necessary to balance each half-reaction. Collect all terms and make cancellations where possible.

(a) $Al + H^+ \rightarrow Al^{+++} + H_2$

(b) $As + H^+ + NO_3^- \rightarrow AsO_4^{---} + NO_2$

(c) $MnO_4^- + SnO_2^{--} \rightarrow SnO_3^{--} + MnO_2$ (basic solution)

(d) $Zn + NO_3^- \rightarrow NH_4^+ + Zn^{++}$ (acid solution)

(e) $Ag_2S + NO_3^- \rightarrow Ag^+ + S + NO$ (acid solution)

(f) $I_2 + S_2O_3^{--} \rightarrow I^- + S_4O_6^{--}$

(g) $C_2H_5OH + Cr_2O_7^{--} \rightarrow CH_3CHO + Cr^{+++}$ (acid solution)

(h) $Fe^{++} + H^+ + NO_3^- \rightarrow Fe^{+++} + NO$

(i) $CoS + H^+ + NO_3^- + Cl^- \rightarrow Co^{++} + S + NOCl$

(j) $MnO_4^- + NO_2 + H_2O \rightarrow Mn^{++} + NO_3^- + H^+$

(k) $Fe^{++} + H_2O_2 + H^+ \rightarrow Fe^{+++} + H_2O$

(l) $H_2O_2 + MnO_4^- + H^+ \rightarrow O_2 + Mn^{++} + H_2O$

(m) $Zn + CNS^- \rightarrow Zn^{++} + H_2S + HCN$

(n) $Co(OH)_2 + O_2^{--} \rightarrow Co(OH)_3 + OH^-$

GENERAL OXIDATION-REDUCTION EQUATIONS

Balance by any method.

5-48. $Zn + SbO_4^{---} + H^+ \rightarrow Zn^{++} + SbH_3 + H_2O$

5-49. $AsH_3 + Ag^+ + OH^- \rightarrow AsO_4^{---} + H_2O + Ag$

5-50. $Cr_2O_7^{--} + Fe^{++} + H^+ \rightarrow Cr^{+++} + Fe^{+++} + H_2O$

5-51. $Cr_2O_7^{--} + I^- + H^+ \rightarrow Cr^{+++} + I_2 + H_2O$

5-52. $Cr_2O_7^{--} + H_2C_2O_4 + H^+ \rightarrow Cr^{+++} + CO_2 + H_2O$

5-53. $Cr(OH)_3 + Cl_2 + OH^- \rightarrow CrO_4^{--} + Cl^- + H_2O$

5-54. $KMnO_4 + H_2SO_4 + HCl \rightarrow K_2SO_4 + MnSO_4 + H_2O + Cl_2$

5-55. $KMnO_4 + H_2SO_4 + PH_3 \rightarrow K_2SO_4 + MnSO_4 + H_2O + H_3PO_4$

5-56. $Br_2 + OH^- \rightarrow Br^- + BrO_3^- + H_2O$

5-57. $Br_2 + NH_3 \rightarrow NH_4Br + N_2$

5-58. $I^- + Br_2 + H_2O \rightarrow IO_3^- + H^+ + Br^-$

5-59. $AuCl_4^- + AsH_3 + H_2O \rightarrow H_3AsO_3 + H^+ + Au + Cl^-$

5-60. $Sb^{+++} + Zn + H^+ \rightarrow Zn^{++} + SbH_3$

5-61. $Zn + H^+ + NO_3^- \rightarrow Zn^{++} + NH_4^+ + H_2O$

5-62. $Fe + HNO_3 \rightarrow Fe(NO_3)_3 + NO + H_2O$

5-63. $KMnO_4 + H_2C_2O_4 + H_2SO_4 \rightarrow K_2SO_4 + MnSO_4 + CO_2 + H_2O$

5-64. $KMnO_4 + MnSO_4 + KOH \rightarrow K_2SO_4 + MnO_2 + H_2O$

5-65. $[PtCl_6]^{--} + Sn^{++} \rightarrow Sn^{++++} + Cl^- + [PtCl_4]^{--}$

5-66. $TiCl_4 + Zn \rightarrow TiCl_3 + ZnCl_2$

5-67. $Mn^{++} + BiO_3^- + H^+ \rightarrow Bi^{+++} + H_2O + MnO_4^-$

5-68. $Re + NO_3^- + H^+ \rightarrow HReO_4 + NO + H_2O$

5-69. $Na_2S + Ag + H_2O + O_2 \rightarrow NaOH + Ag_2S$

5-70. $SO_3^{--} + H_2O + I_2 \rightarrow SO_4^{--} + H^+ + I^-$

5-71. $VO^{++} + MnO_4^- + H_2O \rightarrow VO_4^{---} + Mn^{++} + H^+$

5-72. $Sb + HNO_3 \rightarrow Sb_2O_5 + NO_2 + H_2O$

5-73. $MnO_2 + BiO_3^- + H^+ \rightarrow HMnO_4 + Bi^{+++} + H_2O$

5-74. $NiS + H^+ + Cl^- + NO_3^- \rightarrow Ni^{++} + S + NOCl + H_2O$

5-75. $F_2 + H_2O \rightarrow HF + O_3$

5-76. $Ce^{+++} + Cr_2O_7^{--} + H^+ \rightarrow Ce^{++++} + Cr^{+++} + H_2O$

5-77. $AsO_3^{---} + MnO_4^- + H^+ \rightarrow AsO_4^{---} + Mn^{++} + H_2O$

5-78. $Sb_2S_3 + HNO_3 \rightarrow Sb_2O_5 + NO_2 + S + H_2O$

5-79. $As_2S_3 + K_2CO_3 + KNO_3 \rightarrow K_3AsO_4 + K_2SO_4 + NO + CO_2$

5-80. $HBr + H^+ + MnO_4^- \rightarrow Br_2 + H_2O + Mn^{++}$

5-81. $FeS + H^+ + NO_3^- \rightarrow Fe^{+++} + S + NO + H_2O$

5-82. $Am^{+++} + S_2O_8^{--} + H_2O \rightarrow AmO_2^+ + SO_4^{--} + H^+$

5-83. $VO^+ + H^+ + Cr_2O_7^{--} \rightarrow VO_3^- + Cr^{+++} + H_2O$

5-84. $PbSO_4 + Na_2CO_3 + C \rightarrow Pb + Na_2SO_4 + CO_2$

5-85. $As_2S_5 + Cl_2 + H_2O \rightarrow H_3AsO_4 + H_2SO_4 + HCl$

5-86. $PuO_2^+ + Fe^{++} + H^+ \rightarrow Pu^{++++} + Fe^{+++} + H_2O$

CHAPTER 6

Gas Laws

When dealing with gases it is usually more convenient to measure volumes rather than masses for determining the number of moles of gas involved; consequently, it is important to know the relationship between the volume of a gas and moles. We also need to know how the volume of a gas responds to changes in temperature and pressure. The gas laws describe the relationships between all these variables, and an understanding of the laws enables us to make interconversions of volumes to moles, etc. In this section, we shall work problems based on the principal laws to which real gases conform quite closely.

Usually gas volumes are referred to *standard conditions* of temperature and pressure (STP) which are: a pressure of 760 mm of Hg, or 1 atmosphere, and a temperature of 0°C or 273°K.

A. BOYLE'S LAW

For a given mass of gas at constant temperature the volume is inversely proportional to the pressure.

Expressed mathematically, $V =$ constant $\times 1/P$ or $PV =$ constant. It may also be stated in the form $P_1 V_1 = P_2 V_2$, where V_1 is the volume of the gas at a pressure P_1, and V_2 is the volume of the gas at a pressure P_2.

Example 6-a. 500 ml of gas at a pressure of 600 mm of Hg will occupy what volume if the pressure is increased to 750 mm of Hg at constant temperature?

Solution.

Original conditions	Final conditions
$P_1 = 600$ mm	$P_2 = 750$ mm
$V_1 = 500$ ml	$V_2 =$ new volume

45

(a) We may reason this way: the new volume will be the volume we start with (in this case 500 ml) multiplied by a correction factor to account for the change in pressure, that is,

$$V_2 = 500 \text{ ml} \times \text{correction factor}$$

The correction factor will be a ratio of the two pressures, that is, 600 mm/750 mm or 750 mm/600 mm. Since the pressure is increasing, the volume must decrease (Boyle's law); therefore, the correction factor must be the ratio less than one, that is, 600/750. Consequently,

$$V_2 = 500 \text{ ml} \times \frac{600 \text{ mm}}{750 \text{ mm}} = 400 \text{ ml}$$

(b) Or, we may substitute in the equation $P_1V_1 = P_2V_2$

$$(600 \text{ mm})(500 \text{ ml}) = (750 \text{ mm})(V_2)$$

$$V_2 = \frac{(600 \text{ mm})(500 \text{ ml})}{(750 \text{ mm})} = 400 \text{ ml}$$

Example 6-b. What pressure must be applied to 2.0 liters of a gas at 1.0 atmosphere pressure to compress it to 0.80 liter?

Solution.

Original conditions	Final conditions
$V_1 = 2.0$ liters	$V_2 = 0.80$ liter
$P_1 = 1.0$ atm	$P_2 = $ new pressure

We reason that the new pressure will be the original pressure multiplied by a correction for the volume change.

$$P_2 = 1.0 \text{ atm} \times \text{correction for volume}$$

The volume correction must be the ratio of volumes, that is, 2.0 liters/0.80 liter or 0.80 liter/2.0 liters. Since the volume in this problem is decreasing (from 2.0 liters to 0.80 liter) the pressure must increase (Boyle's law); hence we must use the ratio of volumes greater than one, that is, 2.0 liters/0.80 liter.

$$P_2 = 1.0 \text{ atm} \times \frac{2.0 \text{ liters}}{0.80 \text{ liter}} = 2.5 \text{ atm}$$

PROBLEMS (Boyle's Law)

6-1. By what factor must the pressure on a gas be decreased to quadruple the volume?

6-2. 440 ml of H_2 gas, originally at a pressure of 570 mm of Hg, will occupy what volume at standard pressure?

6-3. Oxygen gas in a cylinder is under a pressure of 3600 lb/inch². The volume of the cylinder is 5.00 ft³. If all of the gas in the cylinder is allowed to expand to 15.0 lb/inch² pressure, what will be the new volume of gas?

6-4. 1.00 liter of O_2 gas at 760 mm of Hg pressure and at 0°C weighs 1.42 g. If the pressure on the gas is increased to 6.00 atm at 0°C, what is the density of the O_2 in grams/liter?

B. CHARLES' LAW

For a given mass of gas at constant pressure the volume is directly proportional to the absolute temperature.

Expressed mathematically, $V = $ constant $\times T$ or $V/T = $ constant. Charles' law may also be stated in the form $V_1/T_1 = V_2/T_2$, where V_1 is the volume at absolute temperature T_1, and V_2 is the volume at absolute temperature T_2.

Example 6-c. If 2.00 liters of a gas are heated from 0°C to 91°C at constant pressure, what is the volume of gas at the higher temperature?

Solution.

Original conditions	Final conditions
$V_1 = 2.00$ liters	$V_2 = $ new volume
$T_1 = 273°K$	$T_2 = 91° + 273° = 364°K$

(a) $V_2 = 2.00$ liters $\times$ correction factor for temperature.

The correction factor will be a ratio of the absolute temperatures, that is, $364°K/273°K$ or $273°K/364°K$. Since the temperature is increasing, the volume must increase (Charles' law); hence we must use the larger of the two ratios, $364°K/273°K$, as the correction factor. Consequently,

$$V_2 = 2.00 \text{ liters} \times \frac{364°K}{273°K} = 2.67 \text{ liters}$$

(b) Or, by substituting in the equation $V_1/T_1 = V_2/T_2$, we obtain

$$\frac{2.00 \text{ liters}}{273°K} = \frac{V_2}{364°K}$$

$$V_2 = \frac{2.00 \text{ liters} \times 364°K}{273°K} = 2.67 \text{ liters}$$

Example 6-d. To what temperature (°C) must 1.00 liter of gas at 0°C be heated to expand its volume to 2.75 liters? Assume that there is no change in pressure.

> *Solution.*
>
> | Original conditions | Final conditions |
> |---|---|
> | $V_1 = 1.00$ liter | $V_2 = 2.75$ liters |
> | $T_1 = 273°K$ | $T_2 = $ new temperature |
>
> $T_2 = $ initial temperature × correction for volume change.
>
> $$T_2 = 273°K \times \text{correction factor}$$
>
> Since the volume increases from 1.00 to 2.75 liters, the temperature must increase, hence the ratio of volumes greater than unity must be used.
>
> $$\text{New temperature} = 273°K \times \frac{2.75 \text{ liters}}{1.00 \text{ liter}} = 750°K$$
>
> Centigrade temperature $= 750°K - 273°K = 477°C$

PROBLEMS (Charles' Law)

6-5. A sample of gas whose volume at 27°C is 272 ml is heated at constant pressure until the volume becomes 680 ml. What is the final centigrade temperature of the gas?

6-6. 495 ml of gas at 24°C will occupy what volume at −21°C (pressure constant)?

6-7. The temperature of 640 ml of a gas is changed from 47°C to 7°C at constant pressure. What is the new volmue in milliliters?

6-8. If 125 liters of Cl_2 is heated at standard pressure from 0°C to 273°C, what will the final volume be?

6-9. What volume will 91 ml of gas at 0°C occupy when cooled to −60°C at constant pressure?

6-10. If 450 ml of hydrogen gas at 27°C is heated to 337°C, what will the new volume be if the pressure does not change?

C. COMBINED GAS LAWS

By combining Boyle's and Charles' laws, we may say: *For a given mass of gas, the volume is inversely proportional to pressure and directly proportional to absolute temperature.*

Expressed mathematically, $V = $ constant $\times T/P$, or $PV/T = $ constant. This relationship may be stated in the form $P_1V_1/T_1 = P_2V_2/T_2$, where V_1 is the volume at pressure P_1 and absolute temperature T_1, and V_2 is the volume at pressure P_2 and absolute temperature T_2.

It follows from the above that *at constant volume, the pressure of a gas is directly proportional to the absolute temperature*, that is,

$$\frac{P}{T} = \text{constant} \quad \text{or} \quad \frac{P_1}{T_1} = \frac{P_2}{T_2}$$

Example 6-e. 50.0 ft³ of O_2 gas at a pressure of 3.00 atm and a temperature of 20°C will occupy what volume at a pressure of 1.00 atm and a temperature of 50°C?

Solution.

Original conditions	Final conditions
$P_1 = 3.00$ atm	$P_2 = 1.00$ atm
$V_1 = 50.0$ ft³	$V_2 = $ new volume
$T_1 = 20° + 273° = 293°K$	$T_2 = 50° + 273° = 323°K$

$V_2 = 50.0$ ft³ $\times$ correction for pressure $\times$ correction for temperature.

Since pressure is decreasing from 3.00 atm to 1.00 atm, the volume must increase; thus the correction for pressure change must be 3.00 atm/1.00 atm. Since the temperature is increasing from 293 to 323°K, volume must increase; hence the correction factor for temperature change must be 323°K/293°K. Then

$$V_2 = 50.0 \text{ ft}^3 \times \frac{3.00 \text{ atm}}{1.00 \text{ atm}} \times \frac{323°K}{293°K} = 165 \text{ ft}^3$$

Or, by substituting in the formula

$$\frac{P_1V_1}{T_1} = \frac{P_2V_2}{T_2}$$

we obtain

$$\frac{(3.00 \text{ atm})(50.0 \text{ ft}^3)}{293°K} = \frac{(1 \text{ atm})(V_2)}{323°K}$$

$$V_2 = \frac{(50.0 \text{ ft}^3)(3.00 \text{ atm})(323°K)}{(1.00 \text{ atm})(293°K)} = 165 \text{ ft}^3$$

Example 6-f. A given weight of a certain gas occupies a volume of 10.0 liters at standard conditions. To what temperature must the gas be heated to change the volume to 50.0 liters at a pressure of 0.300 atm?

   Solution.

Original conditions	Final conditions
$P_1 = 1.00$ atm	$P_2 = 0.300$ atm
$V_1 = 10.0$ liters	$V_2 = 50.0$ liters
$T_1 = 273°K$	$T_2 =$ new temperature

$T_2 = 273°K \times$ correction for volume $\times$ correction for pressure.

$$T_2 = 273°K \times \frac{50.0 \text{ liters}}{10.0 \text{ liters}} \times \frac{0.300 \text{ atm}}{1.00 \text{ atm}} = 410°K$$

Centigrade temperature $= 410 - 273 = 137°C$

PROBLEMS (Combined Gas Laws)

6-11. A 500 ml sample of gas is collected at 380 mm of Hg pressure and a temperature of 273°C. What volume would this sample of gas occupy at standard conditions?

6-12. 640 ml of gas at a pressure of 57.0 cm of Hg and a temperature of 91°C will occupy what volume at standard conditions?

6-13. 380 ml of N_2 gas is collected over Hg at −23°C and 500 mm of Hg pressure. What is the volume at STP?

6-14. A 5 ft³ sample of He, initially at standard conditions, was heated to a final temperature of 273°C. What was the final pressure if the final volume was 10 ft³.

6-15. A 4.0 liter sample of O_2 gas was collected at a temperature of −91°C at a pressure of 380 mm of Hg. Calculate its volume at standard conditions.

6-16. A 450 ml portion of a gas, collected at 27°C and 600 mm of Hg, was stored at −73°C and 900 mm of Hg. What was the storage volume?

6-17. A mixture of gases is sealed into a rigid vessel at standard conditions. What is the pressure of the gas when room temperature is 68°F?

6-18. A 100-ml sample of gas is enclosed in a cylinder with a movable piston at standard conditions. If the pressure on the piston is doubled and the absolute temperature is tripled, what is the volume of the gas?

6-19. Calculate the volume of a gas at 855 mm of Hg and 47°C if its volume is 9.0 liters under standard conditions.

6-20. What pressure does a gas exert if 200 ml of a gas at a pressure of 20 atm and at 27°C is heated to 1227°C at constant volume?

6-21. A large steel gas cylinder contained 50.0 liters of O_2 under a pressure of 40 atm of Hg at a temperature of 95°F. What would the pressure in the cylinder have been during a storeroom fire when the temperature of the cylinder reached 530°F.

6-22. A gas occupied 5 m³ at a certain temperature and pressure. What will the volume of the gas be when its absolute temperature is doubled and the pressure over it is increased tenfold?

6-23. If a 36 ml portion of O_2 at 500 mm of Hg pressure and −73°C is heated to 227°C at constant volume, what will the new pressure be?

6-24. A 6.00 liter sample of Kr gas at 0.500 atm and 400°K is compressed until the pressure is 1.500 atm; the temperature is then changed at constant pressure until the volume is 9.00 liters. What is the new temperature?

D. DALTON'S LAW OF PARTIAL PRESSURES

Each gas in a gaseous mixture acts independently of other gases present, and the total pressure of the mixture is the sum of the pressures of the gases present. The pressure exerted by each gas in the mixture is termed its partial pressure.

$$P_t = p_1 + p_2 + p_3 + \cdots$$

where P_t is the total pressure and p_1, p_2, and p_3, etc., represent the respective pressures exerted by each gas in the mixture.

A practical application of this law is in the collection of gases in the laboratory by the displacement of H_2O. When a gas is collected in this way some H_2O is vaporized, and this vapor exerts a part of the total pressure. The vapor pressure due to H_2O is a function of temperature only and not of volume; hence it should be subtracted from the total pressure before other corrections are made.*

$$P_{\text{total}} = p_{\text{gas}} + p_{H_2O}$$
or
$$p_{\text{gas}} = P_{\text{total}} - p_{H_2O}$$

Example 6-g. 362 ml of O_2 gas is collected in the laboratory by the displacement of liquid H_2O at a temperature of 20°C and a pressure of 689.5 mm of Hg. Determine the volume of the dry gas at standard conditions. The vapor pressure of H_2O at 20°C is 17.5 mm of Hg.

---

* A table of vapor pressures of water at various temperatures is recorded in Appendix VI. The student will need to make use of this table to solve problems when a correction for the vapor pressure of water is necessary.

Solution. In collecting a gas over H_2O, the O_2 gas becomes saturated with H_2O vapor. According to Dalton's law

Total pressure = pressure of O_2 gas + pressure of H_2O vapor

or

$$p_{O_2} = P_t - p_{H_2O}$$

In this problem $p_{O_2} = 689.5 - 17.5 = 672$ mm of Hg $= P_1$

Original conditions	Final conditions
$P_1 = 672$ mm of Hg	$P_2 = 760$ mm of Hg
$V_1 = 362$ ml	$V_2 = $ new volume
$T_1 = 293°K$	$T_2 = 273°K$

$$V_2 = 362 \text{ ml} \times \frac{672 \text{ mm}}{760 \text{ mm}} \times \frac{273°K}{293°K} = 298 \text{ ml}$$

Example 6-h. A metal tank was filled with a mixture of four gases. The total pressure in the container was 1480 mm of Hg. The pressure exerted by O_2 was 350 mm of Hg, by N_2, 520 mm of Hg, and by CO_2, 150 mm of Hg. What is the partial pressure of the fourth gas in the tank?

Solution.

$$P_t = 1480 \text{ mm of Hg} = p_1 + p_2 + p_3 + p_4$$
$$p_1 = 350 \text{ mm of Hg}$$
$$p_2 = 520 \text{ mm of Hg}$$
$$p_3 = 150 \text{ mm of Hg}$$
$$p_4 = ?$$
$$p_4 = 1480 \text{ mm} - (350 \text{ mm} + 520 \text{ mm} + 150 \text{ mm})$$
$$= 460 \text{ mm of Hg}$$

Partial Pressure and Mole Fraction. In a mixture of gases, the partial pressure of each gas is proportional to its *mole fraction.* For each gas the latter is simply the ratio of the moles of given parts to the total moles of all gases present; for example, if a mixture of gases contained 2 moles of A, 3 moles of B, and 5 moles of C, the mole fraction of A would be $\frac{2}{2+3+5} = 0.2$, the mole fraction of B would be $\frac{3}{10} = 0.3$, and the mole fraction of C would be 0.5.

Example 6-i. The total pressure of a mixture of 4 moles of A and 1 mole of B is 7 atm. What is the partial pressure of each gas?

Solution.

$$\text{Mole fraction of } A = \frac{4}{4+1} = 0.8$$

$$\text{Mole fraction of } B = \frac{1}{4+1} = 0.2$$

Partial pressure of $A = (0.8)(7.0 \text{ atm}) = 5.6 \text{ atm}$

Partial pressure of $B = (0.2)(7.0 \text{ atm}) = 1.4 \text{ atm}$

Example 6-j. A mixture of gases containing 60% CO_2 and 40% O_2 by weight has a total pressure of 740 mm Hg. What is the partial pressure of each gas?

Solution. Assume 100 g of gas which would contain $(0.6 \times 100) =$ 60 g CO_2 and $(0.4 \times 100) = 40$ g O_2

$$\text{Moles } CO_2 = 60 \text{ g}/44 \text{ g/mol} = 1.36$$

$$\text{Moles } O_2 = 40 \text{ g}/32 \text{ g/mole} = 1.25$$

$$\text{Mole fraction } CO_2 = 1.36/2.61 = 0.52$$

$$\text{Mole fraction } O_2 = 1.25/2.61 = 0.48$$

$$p_{CO_2} = (0.52)(740 \text{ mm}) = 380 \text{ mm}$$

$$p_{O_2} = (0.48)(740 \text{ mm}) = 360 \text{ mm}$$

PROBLEMS (Dalton's Law)

6-25. One liter of O_2, 1 liter of N_2, and 1 liter of H_2 are collected at a pressure of 1 atm, each in a different container. The three gases are then forced into a single vessel of 1-liter capacity with the temperature kept constant. What is the resulting pressure in units of atmospheres?

6-26. Hydrogen gas was prepared in the laboratory by the reaction Zn + $H_2SO_4 \rightarrow ZnSO_4 + H_2$. A 400-ml sample of H_2 was collected over liquid H_2O at 29°C at a pressure of 793 mm of Hg. What volume would the dry gas occupy at 2.00 atm and 152°K?

6-27. A 2-liter container was filled with O_2, and a 3-liter container with N_2, at standard conditions. Both gases were then pumped from their containers into a 5-liter flask. What was the pressure of each gas in the 5-liter flask and what was the total gas pressure?

6-28. Dry air contains, by volume, 78.03% N_2, 20.9% O_2, 0·9% Ar, and 0.040% CO_2. Calculate the pressures of O_2 and CO_2 in millimeters of Hg if the total pressure is equal to standard pressure.

6-29. The volume of a dry gas at 370 mm of Hg and $-51°C$ was exactly 72 ft³. What volume would this gas occupy if stored over water at 23°C and a total pressure of 761 mm?

6-30. If 4 liters of methane gas, CH_4, at 1 atm and 25°C, together with 10 liters of ethane gas, C_2H_6, at 1 atm and 25°C, are placed in an empty container that has an 8-liter capacity, what will be the final total pressure at 25°C?

6-31. If a 0.100-liter sample of Ar gas is collected over H_2O at 25°C at a total pressure of 53.0 cm of Hg, what would be the corresponding volume of dry Ar gas at 65.0 cm and 25°C?

6-32. What is the partial pressure of each gas in a mixture with a total pressure of 100 atm which contains 1.5 moles A, 2.5 moles B, and 6.0 moles of C?

6-33. A mixture of gases contains 8.8 g CO_2, 1.0 g of H_2 and 12.8 g of O_2. If the total pressure of the mixture is 550 mm Hg, what is the partial pressure of each gas?

6-34. Air, by weight, is essentially a mixture of 79% N_2 and 21% O_2. Calculate the partial pressure of each gas at a total pressure of 700 mm Hg.

E.   GRAHAM'S LAW OF EFFUSION

The rate of effusion of a gas is inversely proportional to the square root of its density.

$$\text{Rate} = \frac{\text{constant}}{\sqrt{d}}$$

Since the density of a gas is proportional to its molecular weight, the latter may be substituted for density. In comparing the relative rates of effusion of two gases, we may use the formulation:

$$\frac{r_1}{r_2} = \sqrt{\frac{M_2}{M_1}}$$

where r_1 is the rate of effusion of the gas with molecular weight M_1, and r_2 is the rate of effusion of the gas with molecular weight M_2.

Example 6-k. What are the rates of effusion of the gas He (MW = 4) relative to the gases, CH_4, HCl, SO_2, and HBr, with the molecular weights 16, 36, 64, and 81, respectively?

> *Solution.* According to Graham's law of effusion, rates are inversely proportional to the square roots of molecular weights. Therefore, He will effuse $\sqrt{16}/\sqrt{4} = \frac{4}{2} = 2$ times as fast as CH_4; $\sqrt{36}/\sqrt{4} = \frac{6}{2} = 3$ times as fast as HCl; $\sqrt{64}/\sqrt{4} = 4$ times as fast as SO_2; $\frac{9}{2}$ as fast as HBr.

Example 6-1. If gas A effuses one-third as fast as methane, CH_4, what is the molecular weight of A?

Solution.

$$\frac{\text{Rate for } A}{\text{Rate for } CH_4} = \sqrt{\frac{\text{MW of } CH_4}{\text{MW of } A}}$$

$$\frac{1}{3} = \sqrt{\frac{16}{\text{MW of } A}}$$

$$\left(\frac{1}{3}\right)^2 = \frac{16}{\text{MW of } A}$$

$$\text{MW of } A = 9 \times 16 = 144$$

PROBLEMS (Effusion and Molecular Weight)

6-35. Two gases, CH_3OF and H_2, are allowed to effuse through the porous walls of a container in two separate experiments. (a) Which one will effuse faster? (b) How many times as fast?

6-36. The two gases HBr and SO_2 have molecular weights of 81 and 64 respectively. What is the *ratio* of the rate of the effusion of SO_2 to that of HBr?

6-37. Five liters of SO_2 pass through a small hole in 1 hr. Another gas effuses out through this same hole at a rate of 10 liters/hr. What is the molecular weight of the second gas?

6-38. Nitrosyl fluoride gas, NOF, effuses through a small opening at a rate of 1 micromole/hr. At the same conditions of temperature and pressure, what will be the rate of effusion of tetrafluoroethylene, C_2F_4, expressed in the same units?

6-39. A certain gas effuses through a small opening at a rate one-third as great as He. What is the molecular weight of the gas?

6-40. If 66 liters of SO_2 effuse through a porous partition in 22 min, what volume of CH_4 will effuse through the same partition at the same temperature and pressure in 30 min?

6-41. Gas A (MW = 81) effuses into one end of a tube 120 cm long, and gas B (MW = 36) effuses into the other end. The point at which the gases meet is marked by a white deposit on the walls. Where will the two gases meet?

F. GAY-LUSSAC'S LAW OF COMBINING VOLUMES

Whenever gases react or are formed in a reaction, they do so in the ratio of small whole numbers by volume, provided the gases are under the same conditions of temperature and pressure.

In a balanced equation involving gases, the coefficients give the volume relations; for example,

$$2\ CO(g) + O_2(g) \rightarrow 2\ CO_2(g)$$

Two volumes of CO react with *one* volume of O_2 to yield *two* volumes of CO_2. Any units of volume may be used, as long as the same units are employed for all gases.

Example 6-m. (a) What volume of oxygen is required to burn 100 liters of H_2S according to the equation

$$2\ H_2S(g) + 3\ O_2(g) \rightarrow 2\ H_2O(g) + 2\ SO_2(g)$$

with all gases under the same conditions of temperature and pressure? (b) What volume of SO_2 will be formed?

Solution. (a) According to Gay-Lussac's law, H_2S and O_2 react in the ratio of two or three by volume; hence the volume of oxygen required is

$$100 \text{ liters of } H_2S \times \frac{3 \text{ liters of } O_2}{2 \text{ liters of } H_2S} = 150 \text{ liters of } O_2$$

(b) Since two volumes of H_2S produce two volumes of SO_2, the volume of SO_2 produced is

$$100 \text{ liters of } H_2S \times \frac{2 \text{ liters of } SO_2}{2 \text{ liters of } H_2S} = 100 \text{ liters of } SO_2$$

PROBLEMS (Gay-Lussac's Law)

6-42. Zirconium metal, Zr, and Cl_2 gas react to form zirconium tetrachloride, $ZrCl_4$. At a temperature of 300°C and a pressure of $\frac{1}{2}$ atm, 600 ml of $ZrCl_4$ gas are obtained. What volume of Cl_2 gas is necessary at this same temperature and pressure?

6-43. At 200°C and 1.0 atm, 10 liters of H_2 and 5 liters of O_2 are mixed and the reaction is induced with a spark. Assuming the apparatus does not shatter, what will the volume be when the resulting gases are cooled to 200°C and 1.0 atm?

6-44. Calculate (a) the volume of O_2 required and (b) the total volume of all products when 50 ft³ of acetylene, C_2H_2, is burned according to the equation $2\ C_2H_2(g) + 5\ O_2(g) \rightarrow 4\ CO_2(g) + 2\ H_2O(g)$. Assume the products are cooled to the same temperature and pressure as the reactants.

G. AVOGADRO'S LAW AND THE MOLAR VOLUME OF A GAS

Equal volumes of gases under the same conditions of temperature and pressure contain the same number of molecules. The volume occupied by 1 gram mole of any gaseous substance at STP is 22.41 *liters*. The latter is termed the *molar volume. The number of molecules in 1 gram mole of any substance is* 6.023 × 10^{23}. This number is *Avogadro's number.*

The fact that 22.4 liters of any gas at STP has a mass equal to its gram molecular weight allows the conversion of gas volumes to masses and vice versa.

Example 6-n. (a) What is the mass of exactly 100 liters of O_2 gas at STP? (b) What volume at STP would be occupied by 48.0 g of O_2?

Solution. (a) Since the molecular weight of O_2 is 32.0, there are 32.0 g of O_2 per 22.4 liters at STP. The mass of 100 liters of O_2 at STP is obtained by multiplying by 32.0 g/22.4 liters.

$$\text{Mass of 100 liters at STP} = 100 \text{ liters} \times \frac{32.0 \text{ g}}{22.4 \text{ liters}} = 143 \text{ g}$$

(b) Number of moles of $O_2 = \dfrac{48.0 \text{ g}}{32.0 \text{ g/mole}} = 1.50 \text{ moles}$

Volume of 48.0 g of O_2 = (22.4 liters/mole) × 1.50 moles = 33.6 liters.

Example 6-o. 3.25 g of a certain gas occupies a volume of 750 ml at a pressure of 2.00 atm and a temperature of 22°C. (a) Calculate the mass of 1.00 liter, or the density, of the gas at STP. (b) What is the molecular weight of the gas?

Solution. (a) First, correct the volume to STP.

$$\text{Volume at STP} = 750 \text{ ml} \times \frac{2.00 \text{ atm}}{1.00 \text{ atm}} \times \frac{273°\text{K}}{295°\text{K}} = 1388 \text{ or } 1390 \text{ ml}$$

Since there is no change in mass, 1390 ml at STP weighs 3.25 g. The mass of 1.00 liter is calculated as follows:

$$\text{Density} = \frac{3.25 \text{ g}}{1390 \text{ ml}} \times \frac{1000 \text{ ml}}{1 \text{ liter}} = \frac{2.34 \text{ g}}{\text{liter}}$$

(b) Since 22.4 liters at STP is the volume of 1 mole, the molecular weight is numerically equal to the mass in grams of 22.4 liters of the gas at STP.

$$\text{Gram molecular weight} = 2.34 \text{ g/liter} \times \frac{22.4 \text{ liters}}{\text{mole}} = 52.4 \text{ g/mole}$$

Hence the molecular weight is 52.4

Example 6-p. How many formaldehyde molecules, CH_2O, are there in a sample of this gas which occupies 11.2 liters at STP?

Solution.

$$\text{Number of moles of } CH_2O = 11.2 \text{ liters} \times \frac{1 \text{ mole}}{22.4 \text{ liters}} = 0.500 \text{ mole}$$

Since 1 mole of any substance contains 6.02×10^{23} molecules, the number of CH_2O molecules

$$= 0.500 \text{ mole} \times \frac{6.02 \times 10^{23} \text{ molecules}}{1 \text{ mole}}$$

$$= 3.01 \times 10^{23} \text{ molecules}$$

PROBLEMS (Molar Volume and Avogadro's Number)

6-45. If 120 g of a gas occupied a volume of 16.8 liters at standard conditions, what was the molecular weight of the gas?

6-46. At standard conditions, 2.50 liters of a gas weighs 12.6 g. What is the molecular weight of the gas?

6-47. A 250-ml flask contains 0.65 g of a certain gas at 0°C and 760 mm of Hg pressure. What is the molecular weight of the gas?

6-48. How many molecules of H_2 will there be in 22.4 liters of the gas at 0°C and a pressure of 0.0100 atm?

6-49. The density of a gas measured at standard conditions was 6.25 g/liter. What was its molecular weight?

6-50. 403 mg of H_2 gas occupies what volume at standard conditions?

6-51. What is the weight in grams of 1.00 liter of O_2 at STP?

6-52. If 280 ml of a gas at STP weighs 1.50 g, what is the molecular weight of the gas?

6-53. If 8.96 liters of an unknown gas at standard conditions weighs 5.35 g, what is the molecular weight of the gas?

6-54. How many molecules are present in 0.224 liter of N_2 gas at standard conditions?

6-55. What is the volume in liters occupied by 3.00 moles of Ar gas at STP?

6-56. How many atoms are there in 5.60 liters of ozone gas, O_3, at standard conditions?

H. THE GENERAL OR IDEAL GAS LAW (EQUATION OF STATE)

According to Boyle's and Charles' laws, the volume of a given mass of gas is inversely proportional to the pressure and directly proportional to the absolute temperature, that is,

$$V \alpha \frac{T}{P}, \quad \text{or} \quad PV = \text{constant} \times T$$

Since 1 mole of any gas at 273.1°K and 1 atm pressure occupies a volume of 22.41 liters, we may substitute in the equation above and evaluate the constant for 1 mole.

$$\text{Constant} = \frac{PV}{T} = \frac{(1.000 \text{ atm})(22.41 \text{ liters/mole})}{273.1°K}$$

$$= 0.08206 \text{ (liters-atm) per (mole-°K)}$$

This value $\left(\dfrac{0.08206 \text{ liter-atm}}{\text{mole-°K}}\right)$ is termed the *molar gas constant* and is usually designated by the letter R.

$$\text{For 1 mole of gas } PV = RT$$
$$\text{For } n \text{ moles of gas } PV = n\,RT$$

The latter equation is very useful in calculating the number of moles of gas present under any conditions of volume, temperature, and pressure. Many of the problems of the preceding sections are easily solved by means of this equation.

Example 6-q. Exactly 500 ml of a gaseous compound at 70.0 cm of Hg pressure and a temperature of 27°C has a mass of 1.85 g. Determine the molecular weight of the compound.

Solution. (a) Substituting in the formula $PV = n\,RT$

$$P = 70.0 \text{ cm}/76.0 \text{ cm} = 0.921 \text{ atm}$$
$$V = 500 \text{ ml} = 0.500 \text{ liter}$$
$$R = \text{gas constant} = 0.0821 \text{ liter-atm per mole-°K}$$
$$T = 27°C = 300°K$$
$$n = \text{number of moles of gas} = ?$$

$$(0.921 \text{ atm})(0.500 \text{ liter}) = n\left(\frac{0.0821 \text{ liter-atm}}{\text{mole-°K}}\right)(300°K)$$

$$n = 0.0187 \text{ mole}$$
$$0.0187 \text{ mole weighs } 1.85 \text{ g}$$

Therefore, since $n = \dfrac{\text{Grams}}{\text{MW}}$,

MW of gas = 1.85 g/0.0187 mole = 99 g/mole

(b) An alternate solution to the problem would be

Volume of gas at STP = 500 ml × 700 mm/760 mm

$$\times\ 273°K/300°K = 419\ ml$$

$$\text{MW of gas} = \frac{1.85\ g}{419\ ml} \times \frac{22{,}400\ ml}{1\ mole} = 99\ g/mole$$

Example 6-r. What volume will be occupied by 33.0 g of CO_2 at 706 mm of Hg pressure and 27°C?

Solution. First, tabulating data,

$$P = \frac{706\ mm}{760\ mm} = 0.929\ atm$$

$$V = ?$$

$$n = \frac{33.0\ g}{44.0\ g/mole} = 0.750\ mole$$

$$R = 0.0821\ (\text{liter-atm})/(\text{mole-}°K)$$

$$T = 27°C = 300°K$$

Substituting in the equation $PV = n\,RT$

$$(0.929\ \text{atm})(V) = (0.750\ \text{mole})\left(\frac{0.0821\ \text{liter-atm}}{\text{mole-}°K}\right)(300°K)$$

$$V = 19.9\ \text{liters}$$

PROBLEMS (Ideal Gas Law)

6-57. Calculate the volume of 3.50 g of CO at −20°C and 2.00 atm pressure.
6-58. (a) How many moles of gas are present in 2.8 liters of a gas at STP?
(b) If the gas in part (a) has a mass of 24 g, what is its molecular weight?
6-59. To what temperature must 0.360 g of water vapor in a 0.500-liter flask be heated to have a pressure of 1.50 atm?
6-60. Exactly 250 ml of a gaseous compound at 78.0 cm of Hg pressure and at 100°C has a mass of 0.820 g. What is the molecular weight of the compound?

6-61. How many molecules are contained in 4.50 liters of NH_3 gas at 4.00 atm and $-10°C$?

6-62. A 100 liter cylinder containing oxygen gas at a pressure of 50 atm at 0°C becomes heated during a fire to a temperature of 1015°F. (a) At this temperature what was the pressure in the cylinder assuming no change in volume? (b) How many moles of oxygen gas were contained in the cylinder?

6-63. What weight of Cl_2 gas will occupy the same volume at 10°C as is occupied by 55 g of CO_2 gas at 40°C if the pressure of the two gases is the same?

6-64. Determine the pressure required to compress 5.0 moles of N_2 gas to a volume of 0.25 liters at a temperature of 27°C.

6-65. A steel cylinder with a volume of 20.0 liters contains N_2 gas at a pressure of 75.0 atm at a temperature of 0°C. What is the mass of N_2 in the cylinder?

6-66. A 5.0-liter cylinder contains 6.00×10^2 g of Ar gas at 27°C. Calculate the pressure of the gas in the cylinder.

GENERAL PROBLEMS (Gas Laws)

6-67. Five liters of O_2 gas at 1 atm pressure will occupy what volume at 10 atm pressure if the temperature remains constant?

6-68. 500 ft³ of air at a pressure of 1 atm is forced into a steel cylinder that has a volume of 5 ft³. What is the pressure of the air in the cylinder if a constant temperature was maintained?

6-69. 600 ml of a gas at a temperature of 27°C will occupy what volume at standard temperature, assuming there is no change in pressure?

6-70. In the laboratory mercuric oxide was decomposed as shown by the equation $2 HgO \rightarrow 2 Hg + O_2$. If 453 ml of O_2 gas was collected over water at a temperature of 18°C and a pressure of 695.5 mm of Hg, what would the volume have been if dry O_2 had been collected at standard conditions?

6-71. A 0.313-g sample of a metal reacts with excess HCl to produce 0.0250 g of dry hydrogen at 27°C and 57.0 cm of Hg pressure. (a) What is the volume of dry hydrogen produced at STP? (b) What is the equivalent weight of the metal?

6-72. 10.0 liters of He gas at STP will occupy what volume at 4.00 atm and 546°C?

6-73. Toy balloons are filled with H_2 gas at 0°C from a 10.0-liter cylinder of the gas. The initial pressure of the gas in the cylinder is exactly 100 atm. Assuming that each balloon is filled to a volume of 1.00 liter at standard pressure and that there is no change in temperature, how many balloons could be filled? (*Note:* Not quite all the gas in the cylinder can be used. Why?)

6-74. To what temperature must 560 ml of gas at 690 mm of Hg and 27°C be heated to change its volume to 1400 ml at 920 mm of Hg pressure?

6-75. What would be the mass of 1.00 liter of O_2 gas at 570 mm of Hg pressure and a temperature of 27°C?

6-76. Compare the relative rates of effusion of CH_4, O_2, HCl, HBr, SO_2, He, and NOF with that of H_2.

6-77. Exactly 10 ml of gas A (MW = 15.0) effuses through an opening in 2.0 sec. Exactly 10 ml of gas B effuses through the same opening in 8.0 sec. What is the molecular weight of gas B?

6-78. If you had a sample of hydrogen gas at STP which occupied 44.8 liters, how many hydrogen molecules would there be in the sample?

6-79. Given 200 ml of dry gas at 27°C and a pressure of 2 atmospheres, find the temperature at which the *volume* of the gas will be *doubled* at *constant* pressure.

6-80. Given a 500-ml sample of gas collected over water at 27°C and 595 mm, the vapor pressure of water at 27°C is 25 mm. Find the volume occupied by the dry gas at STP.

6-81. 5.6 liters of a certain gas at STP has a mass of 13 g. What is the molecular weight of the gas?

6-82. What volume at STP will be occupied by 176 g of CO_2?

6-83. How many molecules are there in 0.025 mole of H_2 gas?

6-84. Octane, C_8H_{18}, is a component of gasoline, and when it undergoes complete combustion the equation for the reaction is $2\ C_8H_{18}\ (g) + 25\ O_2\ (g) \longrightarrow 16\ CO_2(g) + 18\ H_2O(g)$. (a) What volume of O_2 at 200°C and 1 atm pressure is required to produce 3.2 liters of CO_2 at the same temperature and pressure? (b) What volume of water vapor would be produced when 3.2 liters of CO_2 are produced at the same temperature and pressure?

6-85. A 2-liter sample of krypton gas at 0.6 atm and 600°K is compressed until the pressure is 1.0 atm, then it is heated at constant pressure until the volume is 3 liters. What is the new temperature?

6-86. What is the volume of 4.5×10^{18} molecules of a gas at STP?

6-87. Determine the pressure necessary to compress 525 ft^3 of a gas at 3.50 atm pressure to a volume of 25.0 ft^3 at constant temperature.

6-88. 35.0 g of a gas occupies 10.0 liters at 0°C and 2.00 atm pressure. What is the molecular weight of the gas?

6-89. In the laboratory O_2 gas is collected by the displacement of H_2O. The gas so collected had a volume of 0.760 liter at a temperature of 27°C and a pressure of 627 mm of Hg. Calculate the volume which this gas (dry) would occupy at STP.

6-90. Exactly 0.500 liter of a certain gas was collected at 25°C and a pressure of $\frac{1}{3}$ atm. If the gas is allowed to expand at that temperature to 2.000 litres, what would the final pressure in the system be?

6-91. A sample of H_2 and Ar was collected over H_2O when the barometric pressure was 70.0 cm of Hg and the temperature was 26°C. It was determined that the pressure of Ar was 15 mm of Hg. What is the pressure of H_2?

6-92. Find the density of a gas in grams per liter at STP when 9.00 g of it occupies 20.0 liters at 57.0 cm of Hg at 182°C.

6-93. A sample of 0.286 g of a certain gas occupies 50.0 ml at a temperature of 0°C and 76.0 cm of Hg. What is the molecular weight of the gas?

6-94. A sample of O_2 was collected over water at 24°C when the barometric pressure was 692 mm of Hg. The volume of the gas as it was collected was 3.75 liters. (a) What would the volume of dry O_2 be at standard conditions? (b) How many grams of dry O_2 were collected?

6-95. Exactly 10 liters of a H_2—CO_2 mixture was collected in the laboratory at 0°C and a pressure of 1.5 atm wherein the partial pressure of CO_2 was determined to be 0.5 atm. The CO_2 was removed and the remaining gas was compressed to 1.0 liter at 273°C. What was the final pressure of H_2 gas?

6-96. How many moles of acetylene are contained in a 10-liter cylinder at a pressure of 30.0 atmospheres and a temperature of 25°C?

6-97. 20.0 g of hydrogen gas is pumped into a 50-liter cylinder at 27°C. What is the pressure in the cylinder?

6-98. Determine the temperature at which 5 moles of neon gas will occupy 8.20 liters at a pressure of 25 atm.

6-99. 1.60 g of a gas occupies a volume of 485 ml at 21°C and 700-mm pressure. What is the molecular weight of the gas?

6-100. The density of air at STP is 1.293 g/liter. What is the density of air when the barometer exhibits a pressure of 700 mm of Hg and the temperature is 21°C?

6-101. Exactly 1 liter of O_2 at 400 mm of Hg pressure and exactly 1 liter of H_2 at 360 mm of Hg pressure were mixed in a 1.0-liter evacuated flask. What is the pressure of the mixture of gases?

6-102. 0.20 mole of O_2 gas and 0.30 mole of H_2 gas are mixed in a 1.0-liter container at a temperature of 0°C. What is the total pressure of the mixture?

6-103. A quantity of gas is sealed in a vessel at 87°C and 60.0 cm of Hg pressure. Determine the pressure in the vessel if it is cooled to −183°C in liquid air. Assume no change in volume of vessel.

6-104. What is the volume of 6.02×10^{20} molecules of H_2S at 1.00×10^{-3} atm and 0°C?

6-105. How many molecules of H_2 would be contained in a volume of 4.48 mm³ at STP?

6-106. If 1.0 ft³ of H_2 gas effuses through an orifice in 5.0 min, what time will be required for 1.0 ft³ of O_2 gas to effuse through the same orifice?

6-107. One-fifth mole of a compound containing 10.0% H and 90.0% C is burned in oxygen. If 26.88 liters of CO_2 is obtained at STP, (a) what is the empirical formula for the compound and (b) what is the molecular formula?

6-108. If 12.0 g of O_2 is required to inflate a balloon to a certain size at 27°C, what weight of O_2 must be used to inflate it to the same size (and pressure) at 127°C?

6-109. A balloon is partly filled with H_2 gas at sea level where the pressure is exactly 1 atm and the temperature is 27°C. The volume of H_2 under these conditions is 24.0 liters. The balloon rises into the stratosphere where the temperature is −48°C and the internal pressure in the balloon is 0.200 atm. What volume in liters does the H_2 occupy under these new conditions?

6-110. 0.464 g of a sample of a metal reacted with excess HCl to produce 0.305 liter of H_2 at 100°C and 76.0 cm of Hg pressure. (a) What is the equivalent weight of the metal? (b) If its valence is 3, what is its atomic weight?

6-111. What volume will be occupied at 20°C and 750 mm of Hg pressure by exactly 21 g of N_2 and 20 g of Ar?

6-112. A gas cylinder having a 150-liter capacity contains He at a pressure of 67 atm at a temperature of 27°C. All the gas is used to fill a weather balloon. At a height of about 20 miles, the volume of the balloon is 2100 m³ and its temperature is −13°C. What is the pressure of the gas inside the balloon?

6-113. A mixture of 5 moles of O_2 and one mole of He is allowed to pass through a small orifice into a vacuum. What is the approximate composition of the mixture which passes through first?

6-114. A mixture of 6 g of He and 20 g of Ar is stored in a one liter cylinder at 25°C. What is the pressure of the gas mixture in the cylinder?

6-115. A 0.2-liter steel container is filled with 0.05 g hydrogen and 0.32 g oxygen gases at 25°C. (a) What is the total pressure of the gaseous mixture? (b) If, at 227°C, a spark brings about reaction to form water vapor, what substances are present and what is the partial pressure of each?

6-116. A student wished to determine the molar volume of oxygen gas (MW = 32), and to do so he needed to measure the volume of a given mass of the gas. He decided to heat a sample of solid HgO, which decomposes to produce Hg and O_2. He could determine the mass of the O_2 liberated by weighing the sample tube before and after heating. By collecting the O_2 by H_2O displacement from an inverted graduated cylinder he could determine the volume of O_2 at atmospheric pressure when the cylinder is adjusted so that the water level is the same inside and out. The following data were obtained:

Weight of tube plus HgO before heating . . . . . 26.220 g
Weight of the tube plus Hg plus unreacted HgO . . 25.830 g
Volume of oxygen collected . . . . 325 ml
Pressure of atmosphere . . . . 694.5 mm of Hg
Temperature. . . . . 17°C

What is the molar volume at STP?

6-117. (a) Calculate the average distance between centers of atoms of Kr gas at 0°C and 1 atm pressure. (b) If Kr has a radius of 1.69 Å, what fraction of the space occupied by the gas is actually occupied by atoms? Assume the atoms are spherical.

6-118. A student found in the laboratory that when a 0.0653 g sample of Fe was dissolved in dilute H_2SO_4, 30.3 ml of H_2 gas was collected over H_2O at a pressure of 757 mm of Hg at 27°C. (a) What equivalent weight of Fe did he find? (b) If the atomic weight of Fe is 55.85, what is its valence?

6-119. In 1928 Professor G. P. Baxter and coworkers at Harvard University determined the atomic weight of Ne from gas density measurements. At 0°C, the average densities for a large number of measurements at three different pressures are,

(a) Density = 0.89990 g/liter at a pressure of 760.000 mm of Hg.
(b) Density = 0.60004 g/liter at a pressure of 506.667 mm of Hg.
(c) Density = 0.30009 g/liter at a pressure of 253.333 mm of Hg.

Calculate the atomic weight of Ne as determined under each set of conditions.

6-120. A sample of dry air on analysis showed the following composition by weight: 75.6% N_2, 23.1% O_2, and 1.3% Ar. Neglecting the other minor constituents present in air (a) calculate the density of air in grams/liter at STP. (b) Calculate the density of air in grams/liter at 20°C and 0.900 atm pressure.

6-121. In one method of determining molecular weight, a sample of liquid is introduced into a weighed evacuated flask of known volume at a temperature sufficiently high to volatilize the liquid. Its pressure is measured at a constant temperature and then the flask is weighed again to determine the mass of the enclosed gas. A student obtained the following data on a compound containing only nitrogen and hydrogen:

Mass of evacuated flask	65.575 g
Mass of flask plus compound	65.649 g
Pressure of the gas	57.5 cm of Hg
Temperature of the gas	125°C
Volume of the flask	100.0 ml

(a) What is the molecular weight of the compound? (b) A chemical analysis of the compound indicated that it consisted of 12.5% H and 87.5% N. What is the true molecular formula of the compound?

6-122. A student was asked to determine the molecular weight of a substance containing only carbon and sulfur by the Victor Meyer method. A weighed sample of a volatile liquid was introduced into a system at a temperature such that all of the liquid was volatilized. The volume was then measured at a known pressure. (a) Calculate the molecular weight of the gas from his experimental data:

Mass of sample	0.0869 g
Volume of gaseous sample	35.0 ml
Temperature	100°C
Pressure	76.0 cm of Hg

(b) If the compound is composed of 15.8% C and 84.2% S, what is the true molecular formula of the substance?

6-123. Radium metal emits alpha particles which are in fact positively charged helium nuclei. These particles pick up electrons from surrounding matter to become electrically neutral helium atoms. From a particular sample of radium, alpha particles were produced at a rate of 7.02×10^{10} particles per second. Over a period of 121 days all the helium atoms from the radium decay were collected and, at 29°C and 1.00 atm pressure, they occupied a volume of 0.0302 ml. Calculate Avogadro's number from this data, assuming the rate of alpha particle emission is constant over the period of time involved and that there is no other source of alpha particles.

CHAPTER 7

Stoichiometry

Stoichiometry is a study of the relationship between amounts of reactants and products of a chemical change as shown by the chemical equation. The balanced equation gives at once the relationship between *moles* of reactants and products, because the coefficients in the balanced equation represent the number of moles of substances. For example, in the reaction

$$2 \text{ H}_2\text{S}(g) + 3 \text{ O}_2(g) \rightarrow 2 \text{ SO}_2(g) + 2 \text{ H}_2\text{O}(g)$$

two moles of H_2S gas react with *three* moles of O_2 gas to yield *two* moles of SO_2 gas and *two* moles of H_2O vapor. These amounts of reactants and products are termed *stoichiometric* amounts, that is, the amounts undergoing reaction as given by the chemical equation. Although amounts of reactants in other than stoichiometric ratios may be present, reaction occurs *only* in the stoichiometric ratios.

Amounts of chemicals are usually measured by their masses, although in the case of gases, the measurement may be by volume. Since we now know the relationship between *moles* and mass or volume, it is a simple matter to convert grams or liters (for a gas) to moles or vice versa. Careful attention to the units used is essential in working problems of the type described below.

The three principal types of problems based on equations are: (a) weight-weight, (b) weight-volume, and (c) volume-volume. All three types follow the schematic process outlined below. The weight (or volume) of gas *A* is converted to moles of *A*. From the balanced equation the number of moles of *B* produced (or reacted with *A*) is calculated. Finally, the weight of *B* (or the volume of gaseous *B*) is calculated.

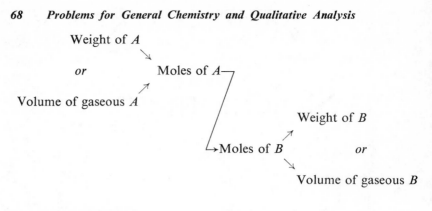

A. WEIGHT–WEIGHT CALCULATIONS

Problems of this nature involve the determination of an unknown weight of a reactant or product from a given weight of some substance in a chemical change. Interconversions between grams and moles were discussed in Chapter 3.

Example 7-a. According to the following equation, $4\ NH_3(g) + 5\ O_2(g) \rightarrow 4\ NO(g) + 6\ H_2O(g)$ what weight of O_2 in grams will be required to react with exactly 100 g of NH_3? The molecular weight of NH_3 is 17.0 and that of O_2 is 32.

Solution.

Number of moles of NH_3 used

$$= 100 \text{ g } NH_3 \times \frac{1 \text{ mole } NH_3}{17.0 \text{ g } NH_3} = 5.88 \text{ moles } NH_3$$

According to the equation, 4 moles of NH_3 require 5 moles of O_2.

$$\text{Number of moles of } O_2 \text{ required} = 5.88 \text{ moles } NH_3 \times \frac{5 \text{ moles } O_2}{4 \text{ moles } NH_3}$$

$$= 7.35 \text{ moles } O_2$$

$$\text{Mass of } O_2 \text{ required} = 5.35 \text{ moles } O_2 \times \frac{32.0 \text{ g } O_2}{1 \text{ mole } O_2} = 235 \text{ g } O_2$$

Example 7-b. In the reaction

$$2\ C_2H_2(g) + 5\ O_2(g) \rightarrow 4\ CO_2(g) + 2\ H_2O(g)$$

what weight of O_2 in grams will be required to burn 10 moles of acetlyene, C_2H_2?

Solution. Two moles of C_2H_2 require 5 moles of O_2. Hence the number of moles of O_2 required is

$$10 \text{ moles } C_2H_2 \times \frac{5 \text{ moles } O_2}{2 \text{ moles } C_2H_2} = 25 \text{ moles } O_2$$

$$\text{Mass of } O_2 \text{ required} = 25 \text{ moles } O_2 \times \frac{32 \text{ g } O_2}{1 \text{ mole } O_2} = 800 \text{ g } O_2$$

PROBLEMS (Weight–Weight)

7-1. From the balanced equation

$$2 \text{ H}_2S(g) + 3 \text{ O}_2(g) \rightarrow 2 \text{ SO}_2(g) + 2 \text{ H}_2O(g)$$

determine the following: (a) How many moles of O_2 are necessary to react with 0.40 mole of H_2S? (b) How many moles of SO_2 will be produced from 0.40 mole of H_2S? (c) How many grams of O_2 are required for 0.40 mole of H_2S? (d) What weight of SO_2 would be produced if 272 g of H_2S are used?

7-2. Benzene, C_6H_6, burns in O_2 according to the equation

$$2 \text{ C}_6H_6(g) + 15 \text{ O}_2(g) \rightarrow 12 \text{ CO}_2(g) + 6 \text{ H}_2O(g)$$

(a) Calculate the number of moles of O_2 necessary to burn 6 moles of C_6H_6. (b) What number of grams of O_2 is necessary in (a)? (c) How many moles of products are produced in (a)? (d) How many grams of C_6H_6 must be burned to produce 2.64 grams of CO_2?

7-3. Balance the following equation for the combustion of acetylene, C_2H_2, in oxygen.

$$C_2H_2(g) + O_2(g) \rightarrow CO_2(g) + H_2O(g)$$

(a) How many moles of O_2 will be required to burn 50 moles of C_2H_2? (b) How many moles of CO_2 will be formed? (c) If 52 g of C_2H_2 are burned, how many moles of O_2 are required? (d) How many moles of CO_2 will be formed from 52 g of C_2H_2? (e) How many grams of O_2 will be required to react with 52 g of C_2H_2? (f) How many grams of CO_2 will be formed from 52 g of C_2H_2? (g) How many moles of C_2H_2 could be burned by 160 g of O_2? (h) How many grams of C_2H_2 could be burned by 160 g of O_2?

7-4. Given the balanced equation

$$4 \text{ NH}_3(g) + 5 \text{ O}_2(g) \rightarrow 4 \text{ NO}(g) + 6 \text{ H}_2O(g)$$

(a) how many moles of NH_3 will be required to form 25 moles of NO? (b) How many grams of NH_3 will be required in (a)? (c) How many grams of O_2 will be required in (a)? (d) How many grams of NH_3 will be required to react with 160 g of O_2?

7-5. Complete combustion of PH_3 is shown by the equation

$$PH_3(g) + 2\,O_2(g) \rightarrow H_3PO_4(s)$$

For 34 g of PH_3 determine (a) the moles of O_2 required, (b) grams of O_2 required, (c) the moles of H_3PO_4 formed, and (d) the grams of H_3PO_4 formed.

7-6. Silver nitrate reacts with $CaCl_2$ as shown by the equation

$$2\,AgNO_3 + CaCl_2 \rightarrow Ca(NO_3)_2 + 2\,AgCl(s)$$

(a) What weight of $AgNO_3$ is required to produce 28.7 g of AgCl?
(b) What weight of $CaCl_2$ is required to produce 28.7 g of AgCl?
(c) What weight of $Ca(NO_3)_2$ is produced when 28.7 g of AgCl are produced?

7-7. In the reaction

$$5\,KI + KIO_3 + 6\,HNO_3 \rightarrow 6\,KNO_3 + 3\,I_2 + 3\,H_2O$$

76.2 g of I_2 is produced. (a) How many grams of KI are required? (b) How many grams of KIO_3 are required? (c) How many grams of HNO_3 are required? (d) How many grams of KNO_3 are produced?

7-8. Exactly 40 g of CuS are treated with excess dilute HNO_3 and the reaction

$$3\,CuS + 8\,HNO_3 \rightarrow 3\,Cu(NO_3)_2 + 2\,NO + 4\,H_2O + 3\,S$$

occurs. (a) How many grams of $Cu(NO_3)_2$ are produced? (b) How many grams of S are produced? (c) What is the minimum number of grams of HNO_3 required?

7-9. Silicon carbide, SiC, is made by the reaction

$$SiO_2 + 3\,C \rightarrow SiC + 2\,CO$$

(a) What weight of carbon is required to produce 50 kg of SiC?
(b) What weight of CO is produced?

7-10. Two of the several reactions in a blast furnace are

$$2\,C(s) + O_2(g) \rightarrow 2\,CO(g)$$
$$Fe_2O_3(s) + 3\,CO(g) \rightarrow 2\,Fe + 3\,CO_2(g)$$

Assume that these are the reactions which are primarily responsible for the production of Fe. (a) How many kilograms of C are required to reduce 200 kg of Fe_2O_3? (b) How many kilograms of Fe are produced from 200 kg of Fe_2O_3? (c) How many kilograms of CO_2 are produced? (d) How many kilograms of O_2 are required?

B. WEIGHT–VOLUME CALCULATIONS

Weight–volume calculations refer to stoichiometric calculations where one of the reactants or products is a gas. As in weight–weight calculations, the balanced chemical equation is used to relate the number of moles of a substance to the number of moles of other reactants or products. The conversion of moles or grams of a gas to volume at some given temperature and pressure is most readily accomplished by use of the ideal gas equation. Alternatively, since 1 mole of gas occupies 22.4 liters at STP, the volume of a given number of moles under these conditions may be calculated and subsequently converted to the desired conditions by use of the P, V, T relations. The conversion of the volume of a gas to moles may be accomplished by the reverse of the process just described. (See Chapter 6.)

Example 7-c. Using the equation in Example 7-a, determine the *volume* of O_2 at 27°C and 1 atm which would be required to react with 100 g of NH_3.

Solution.

$$\text{Number of moles of } NH_3 = \frac{100 \text{ g } NH_3}{17 \text{ g } NH_3/\text{mole } NH_3} = 5.88 \text{ moles } NH_3$$

Since 4 moles of NH_3 requires 5 moles of O_2,

$$\text{Number of moles of } O_2 = 5.88 \text{ moles } NH_3 \times \frac{5 \text{ moles } O_2}{4 \text{ moles } NH_3}$$

$$= 7.35 \text{ moles } O_2$$

The volume of O_2 required may be calculated from the ideal gas equation.

$$\text{Volume of } O_2 = 7.35 \text{ moles } O_2 \times 0.0821 \frac{\text{liter-atm}}{\text{mole-°K}} \times \frac{300°K}{1 \text{ atm}}$$

$$= 181 \text{ liters } O_2$$

Example 7-d. 1.00×10^2 liters of H_2S at STP is bubbled into an aqueous solution of $SbCl_3$. Sb_2S_3 is precipitated according to the equation

$$2 SbCl_3 + 3 H_2S(g) \rightarrow Sb_2S_3 + 6 HCl$$

Calculate the weight of precipitated Sb_2S_3 in grams if an excess of $SbCl_3$ is present. The formula weight of Sb_2S_3 is 340.

Solution.

$$\text{Number of moles of } H_2S = \frac{100 \text{ liters } H_2S}{22.4 \text{ liters } H_2S/\text{mole } H_2S}$$

$$= 4.46 \text{ moles } H_2S$$

The chemical equation requires that 1 mole of Sb_2S_3 precipitates for each 3 moles of H_2S gas; therefore

$$\text{Number of moles of } Sb_2S_3 = 4.46 \text{ moles } H_2S \times \frac{1 \text{ mole } Sb_2S_3}{3 \text{ moles } H_2S}$$

$$= 1.49 \text{ moles } Sb_2S_3$$

$$\text{Number of grams of } Sb_2S_3 = 1.49 \text{ moles } Sb_2S_3 \times \frac{340 \text{ g } Sb_2S_3}{1 \text{ mole } Sb_2S_3}$$

$$= 507 \text{ g } Sb_2S_3$$

PROBLEMS (Weight–Volume)

7-11. In the dissolution of excess Hg by HNO_3, NO gas is produced.

$$6 \text{ Hg} + 8 \text{ HNO}_3 \rightarrow 3 \text{ Hg}_2(NO_3)_2 + 2 \text{ NO}(g) + 4 \text{ H}_2O$$

What volume of NO is produced at STP if 12.6 g of HNO_3 is allowed to react with excess Hg?

7-12. Ammonium chloride is a by-product of the Solvay process, and ammonia is recovered from NH_4Cl by the reaction

$$CaO + 2 \text{ NH}_4Cl \rightarrow CaCl_2 + H_2O(l) + 2 \text{ NH}_3(g)$$

What volume of NH_3 gas at STP can be prepared from 21.4 g of NH_4Cl?

7-13. In the Solvay process for the production of $NaHCO_3$, a solution saturated with both NaCl and NH_3 is treated with CO_2 gas and the following reaction occurs:

$$NaCl + NH_3(g) + H_2O(l) + CO_2(g) \rightarrow NaHCO_3(s) + NH_4Cl$$

(a) What volume of $NH_3(g)$ at STP is required to produce exactly 300 g of $NaHCO_3$? (b) What volume of $CO_2(g)$ at 20°C and 1.2 atm is required to produce exactly 600 g of $NaHCO_3$?

7-14. The gas phosphine, PH_3, is prepared from P_4 by the reaction

$$P_4(s) + 3 \text{ NaOH} + 3 \text{ H}_2O(l) \rightarrow 3 \text{ NaH}_2PO_2 + PH_3(g)$$

(a) What volume of PH_3 at STP can be prepared from 62 g of phosphorus? (b) What volume of PH_3 at STP can be prepared from 40 g of NaOH? (c) What volume of PH_3 at 37°C and 0.90 atm is produced if 284 g of NaH_2PO_2 is produced?

7-15. In one step of the commercial production of Cu metal, the reaction

$$Cu_2S + 2 \text{ Cu}_2O \rightarrow 6 \text{ Cu} + SO_2(g)$$

is very important. How many liters of SO_2 at 327°C and 740 mm are produced from 3.18 kg of Cu_2S?

C. VOLUME–VOLUME CALCULATIONS

It was pointed out in Chapter 6, Section F, that Gay-Lussac's law of combining volumes states "Whenever gases react or are formed in a reaction, they do so in the ratio of small whole numbers by volume, provided the gases are under the same conditions of temperature and pressure." The coefficients in the balanced equation give the volume relations for the gaseous substances. This follows from the facts that the coefficients give the number of moles of each gas involved, and that 1 mole of each gas occupies the same volume at the same temperature and pressure.

When two or more of the gases are at different temperatures and pressures, one of the following procedures must be employed. All gases can be treated *as if* the reaction was carried out at the particular temperature and pressure of one of the gases; volume corrections are then made for the differences in conditions of the two gases involved by use of the P, V, T relations. *Alternatively*, the volume of one of the gases can be converted to moles by use of the ideal gas equation; then the number of moles of the second gas can be calculated; and finally, the volume of the second gas is calculated under the desired conditions. Such a problem is illustrated in Example 7-f.

Example 7-e. What volume of oxygen would be required to burn exactly 500 liters of acetylene gas, C_2H_2, according to the equation $2\,C_2H_2(g) + 5\,O_2(g) \rightarrow 4\,CO_2(g) + 2\,H_2O(g)$, if all gases are at the same temperature and pressure?

> **Solution.** According to the equation, two volumes of C_2H_2 require five volumes of O_2. Therefore, the volume of O_2 required is

$$500 \text{ liters } C_2H_2 \times \frac{5 \text{ liters } O_2}{2 \text{ liters } C_2H_2} = 1250 \text{ liters } O_2$$

Example 7-f. H_2S burns in O_2 according to the equation

$$2\,H_2S(g) + 3\,O_2(g) \rightarrow 2\,H_2O(g) + 2\,SO_2(g)$$

Determine (a) the volume of O_2 at STP required to burn 20.0 liters of H_2S at STP, and (b) the volume of SO_2 obtained at a pressure of 70.0 cm and a temperature of 773°K.

Solution. (a) Two volumes of H_2S require three volumes of O_2; therefore, the volume of oxygen required for 20.0 liters of H_2S is

$$20.0 \text{ liters } H_2S \times \frac{3 \text{ liters } O_2}{2 \text{ liters } H_2S} = 30.0 \text{ liters } O_2 \text{ at STP}$$

(b) Two volumes of H_2S yield two volumes of SO_2; therefore 20.0 liters of H_2S at STP will yield 20.0 liters of SO_2 at STP. At 70.0 cm Hg pressure and 773°K the volume of SO_2 will be

$$20.0 \text{ liters } \times \frac{76.0 \text{ cm}}{70.0 \text{ cm}} \times \frac{773°K}{273°K} = 61.5 \text{ liters } SO_2$$

Alternate Solution. From the ideal gas equation find the number of moles of H_2S to be burned; determine the moles of SO_2 produced; calculate the volume of SO_2 produced at 70.0 cm and 773°K.

PROBLEMS (Volume–Volume)

7-16. The compound C_2H_6 reacts in a limited amount of O_2 to form CO and H_2O. (a) Balance the equation for the reaction

$$C_2H_6(g) + O_2(g) \rightarrow CO(g) + H_2O(l)$$

(b) What volume of O_2 reacts with 5.0 liters of C_2H_6 at STP? (c) What volume of CO measured at STP is produced from 5.0 liters of C_2H_6 at STP?

7-17. $P_4(g)$ reacts with $Cl_2(g)$ to form $PCl_3(g)$. (a) Balance the equation for the reaction

$$P_4(g) + Cl_2(g) \rightarrow PCl_3(g)$$

(b) What volume of Cl_2 is required to react with 20 liters of gaseous P_4 at a given temperature? (c) What volume of PCl_3 is produced from 20 liters of P_4 if the absolute temperature of the PCl_3 is twice that of the reactants and the pressure is constant?

7-18. $SiCl_4(g)$ reacts with $H_2O(g)$ at elevated temperatures as shown in the equation

$$SiCl_4(g) + 2 H_2O(g) \rightarrow SiO_2(s) + 4 HCl(g)$$

If 40 liters of $SiCl_4$ at 300°C and $\frac{1}{2}$ atm pressure react with H_2O, (a) what volume of H_2O at this temperature and pressure will be consumed, (b) what volume of HCl will be produced, (c) what is the volume of that amount of HCl gas at STP?

GENERAL PROBLEMS

7-19. Nitric acid, HNO_3, may be prepared by distillation from a mixture of solid $NaNO_3$ and concentrated H_2SO_4.

$$NaNO_3(s) + H_2SO_4(l) \rightarrow NaHSO_4(s) + HNO_3(g)$$

(a) How many grams of H_2SO_4 are required to produce 5 moles of HNO_3?
(b) How many grams of $NaHSO_4$ are produced when 5 moles of HNO_3 are produced? (c) How many moles of HNO_3 are produced from 170 g of $NaNO_3$? (d) How many grams of HNO_3 are produced from 170 g of $NaNO_3$?

7-20. Sodium carbonate, washing soda, is produced commercially by heating baking soda, $NaHCO_3$. This salt decomposes on heating as indicated by the equation

$$2\,NaHCO_3(s) \xrightarrow{\Delta} Na_2CO_3(s) + CO_2(g) + H_2O(g)$$

(a) How many moles of $NaHCO_3$ are required for each mole of CO_2 produced? (b) How many grams of CO_2 are produced for each mole of $NaHCO_3$ decomposed? (c) How many moles of each product are obtained for each mole of $NaHCO_3$ decomposed? (d) What is the total weight of products for each kilogram of $NaHCO_3$ used? (e) How many grams of $NaHCO_3$ will be required to produce 318 g of Na_2CO_3? (f) How many moles of CO_2 would be produced simultaneously in (e)?

7-21. What quantity (in pounds) of C (coke) will be necessary to produce 128 pounds of CaC_2 according to the equation:

$$CaO(s) + 3\,C(s) \rightarrow CaC_2(s) + CO(g)$$

7-22. Acetylene, C_2H_2, may be produced from the reaction

$$CaC_2(s) + 2\,H_2O(l) \rightarrow Ca(OH)_2(s) + C_2H_2(g)$$

What volume of C_2H_2 (at STP) will be produced for each gram of CaC_2 used?

7-23. Calcium carbide, CaC_2, is produced commercially by heating lime, CaO, and carbon, C, together in an electric furnace. The reaction taking place is

$$CaO(s) + 3\,C(s) \rightarrow CaC_2(s) + CO(g)$$

For each kilogram of CaC_2 produced, (a) how many kilograms of CaO and, (b) how many kilograms of C are required?

7-24. (a) Calculate the number of grams of O_2 obtainable from the decomposition of exactly 4 moles of $KClO_3$ according to the equation

$$2\,KClO_3(s) \rightarrow 2\,KCl(s) + 3\,O_2(g)$$

(b) How many grams of KCl will be obtained in (a)?

7-25. Chlorine gas is produced commercially by the electrolysis of an aqueous NaCl solution as shown by the equation

$$2\,NaCl + 2\,H_2O \rightarrow 2\,NaOH + Cl_2(g) + H_2(g)$$

(a) How many grams of Cl_2 are obtained for each kilogram of NaCl used?
(b) What volume of Cl_2 gas is obtained at STP in (a)? (c) What volume of H_2 gas is obtained at STP in (a)? (d) What mass of H_2 gas is obtained in (a)?

(e) How many kilograms of NaOH (lye) are obtained for each kilogram of Cl_2 produced?

7-26. Gasoline, with an average composition of C_8H_{18}, burns in oxygen or air according to the equation

$$2 C_8H_{18}(l) + 25 O_2(g) \rightarrow 16 CO_2(g) + 18 H_2O(g)$$

(a) Starting with exactly 1 liter of liquid C_8H_{18} (density 0.800 g/ml), what volume of O_2 at STP is required for complete combustion? (b) What volume of CO_2 at STP is produced? (c) How much air (20% oxygen) will be required in (a) at STP?

7-27. Consider the reaction

$$2 H_2S(g) + 3 O_2(g) \rightarrow 2 H_2O(g) + 2 SO_2(g)$$

(a) Calculate the number of grams of O_2 required to react with 554 g of H_2S. (b) What volume of O_2 at STP is required in (a)? (c) Assuming all gases are under the same conditions of temperature and pressure, what volume of O_2 is required to burn 500 ft^3 of H_2S?

7-28. In the commercial production of nitric acid from ammonia, the first step is the oxidation of ammonia according to the equation

$$4 NH_3(g) + 5 O_2(g) \rightarrow 4 NO(g) + 6 H_2O(g)$$

(a) What volume of O_2 at STP will be required to burn exactly 6 moles of NH_3? (b) If all gases are under the same conditions of temperature and pressure, what volume of NO will be formed for each liter of NH_3 burned? (c) How many grams of O_2 are required in (a)? (d) What volume of NO at 327°C and 1.0 atm pressure will be produced from 8 liters of NH_3 at 27°C and 1.0 atm?

7-29. All heavy metal carbonates are unstable toward heat and yield CO_2 at elevated temperatures.

$$CuCO_3(s) \xrightarrow{\Delta} CuO(s) + CO_2(g)$$

$$BaCO_3(s) \xrightarrow{\Delta} BaO(s) + CO_2(g)$$

Calculate the mass of $BaCO_3$ which yields the same mass of CO_2 as 247 g of $CuCO_3$.

7-30. (a) What volume of CO_2 at STP is obtained from 247 g of $CuCO_3$? (b) What volume of CO_2 is produced at 400°C and 0.90 atm from 1.00 kg of $CuCO_3$?

7-31. Magnesium may be determined quantitatively by precipitation as $MgNH_4PO_4$,

$$Mg^{++} + NH_4^+ + PO_4^{---} \rightarrow MgNH_4PO_4$$

followed by ignition to magnesium pyrophosphate ($Mg_2P_2O_7$)

$$2 MgNH_4PO_4(s) \xrightarrow{\Delta} Mg_2P_2O_7(s) + 2 NH_3(g) + H_2O(g)$$

Assuming a 1.000-g sample of a magnesium mineral yields 0.890 g of $Mg_2P_2O_7$, what is the percentage of magnesium in the sample?

7-32. Phosphorus may also be determined quantitatively by conversion to $Mg_2P_2O_7$. A 2.000-g sample of phosphate rock [mainly $Ca_3(PO_4)_2$] yields 1.328 g of $Mg_2P_2O_7$. Assuming that all of the phosphorus is present as $Ca_3(PO_4)_2$, what is the percentage of $Ca_3(PO_4)_2$ in the sample?

7-33. The determination of the percentage composition of hydrocarbons involves burning in O_2 to form CO_2 and H_2O. 5.80 g of a certain compound containing only C and H yields 17.6 g of CO_2 and 9.00 g of H_2O on combustion in O_2. (a) How many grams of O_2 are used for the combustion? (b) What is the empirical formula of the hydrocarbon?

7-34. A chemist found that 0.460 g of an organic compound containing C, H, and O when burned in O_2 yielded 0.88 g of CO_2 and 0.54 g of H_2O. Determine the empirical formula for the compound.

7-35. If 0.5 mole of $FeCl_3$ is mixed with 0.3 mole of Na_2CO_3, what is the maximum number of moles of $Fe_2(CO_3)_3$ that can be formed?

7-36. One gram of sulfur-containing compound is analyzed for sulfur by precipitating $BaSO_4$. If 0.73 g of $BaSO_4$ is obtained what is the percentage of sulfur in the sample?

7-37. Assuming 100% conversion, what volume of NH_3 at STP could be produced from 7 kg of nitrogen gas?

7-38. HBr(*g*) may be prepared in the laboratory by the following reaction

$$2 \ NaBr(s) + H_3PO_4 \rightarrow Na_2HPO_4 + 2 \ HBr(g)$$

How many grams of NaBr and H_3PO_4 are required to produce 1238 ml of HBr(*g*) at 27°C and 70.0 cm of Hg pressure?

7-39. What volume of H_2 gas measured at 72.0 cm of Hg pressure and 22°C is obtained by the action of excess dilute H_2SO_4 on 21.69 g of Al?

$$2 \ Al(s) + 3 \ H_2SO_4 \rightarrow Al_2(SO_4)_3 + 3 \ H_2(g)$$

7-40. Oxygen is prepared commercially by electrolysis of water. A 4.0 liter steel cylinder is to be filled with oxygen gas. To attain a final pressure of 100 atm at a temperature of 17°C, how many grams of H_2O will have to be electrolyzed to furnish the necessary O_2 for filling the cylinder?

7-41. Assume you are going to prepare $KClO_3$ in the laboratory using the following sequence of reactions:

$$2 \ KMnO_4 + 16 \ HCl \rightarrow 2 \ KCl + 2 \ MnCl_2 + 5 \ Cl_2(g) + 8 \ H_2O$$
$$3 \ Cl_2 + 6 \ KOH \rightarrow KClO_3 + 5 \ KCl + 3 \ H_2O$$

(a) How many moles of $KClO_3$ are produced per mole of $KMnO_4$ used? (b) Calculate the masses of $KMnO_4$ and KOH necessary to produce 10.2 g of $KClO_3$. (c) What volume of Cl_2, in the first reaction, measured at STP is produced per mole of HCl used? (d) What volume of 40% HCl, density 1.20 g/ml, would be required in (b)?

7-42. Phosphorus may be produced commercially by heating together in an electric furnace the following: phosphate rock, $Ca_3(PO_4)_2$; sand, SiO_2; and coke, C. Phosphorus, P_4, is distilled away from the reaction mixture The overall reaction is

$$2\ Ca_3(PO_4)_2(s) + 6\ SiO_2(s) + 5\ C(s) \rightarrow P_4(g) + 6\ CaSiO_3(s) + 5\ CO_2(g)$$

Determine the mass of each reactant required to yield 1.0 kg of phosphorus.

7-43. SO_2 is a by-product of the roasting of sulfide ores, for example,

$$2\ PbS(s) + 3\ O_2(g) \rightarrow 2\ PbO(s) + 2\ SO_2(g)$$

What volume of SO_2 at 546°C and 1.0 atm pressure is obtained on roasting 1.0×10^3 kg of ore which contains 23.9% PbS?

7-44. In a certain metallurgical operation, the mineral "pyrites," FeS_2, is roasted in air according to the equation

$$4\ FeS_2(s) + 11\ O_2(g) \rightarrow 2\ Fe_2O_3(s) + 8\ SO_2(g)$$

SO_2 is then converted into H_2SO_4 by the reactions

$$2\ SO_2(g) + O_2(g) \rightarrow 2\ SO_3(g)$$
$$SO_3(g) + H_2SO_4(l) \rightarrow H_2S_2O_7(l)$$
$$H_2S_2O_7(l) + H_2O(l) \rightarrow 2\ H_2SO_4(l)$$

Assuming that a mineral is 12.0% FeS_2 and the remainder is inert, what weight of H_2SO_4 is produced per 1000 kg of mineral used if the processes are 100% efficient?

7-45. "Red lead" is an oxide of Pb which is useful as a paint pigment. When treated with H_2 gas, it is reduced to metallic Pb and water vapor. Starting with 1.714 g of the oxide, it was found that 1.554 g of Pb remained after complete reduction. (a) What is the formula for "red lead"? (b) Write an equation for the reaction with H_2.

7-46. To determine the composition of an oxide of Fe, a student passed H_2 gas over a weighed quantity of the heated oxide in a glass tube. Under these conditions the oxide was reduced to elemental Fe. The following data were collected:

> Weight of oxide taken   11.18 g
>
> Weight of Fe produced   7.82 g

(a) Determine the empirical formula of the oxide. (b) What is the percentage of Fe in the compound? (c) What weight of H_2O is produced?

7-47. Black gunpowder consists of a mixture of S, C, and KNO_3. An analytical method used for the determination of nitrates in such mixtures depends upon the reduction of the nitrate radical to $NO(g)$ and the subsequent measurement of the volume of NO gas under known conditions in a *nitrometer.* Metallic Hg is used as the reducing agent in H_2SO_4 solution. The equation for the reaction is

$$6 \text{ Hg} + 2 \text{ HNO}_3 + 3 \text{ H}_2\text{SO}_4 \rightarrow 3 \text{ Hg}_2\text{SO}_4 + 2 \text{ NO}(g) + 4 \text{ H}_2\text{O}$$

In a typical analysis, the following data were obtained:

Weight of sample . . . . 0.1308 g
Volume of NO collected over H_2O . . . . . 25.0 ml
Temperature . . . . 20°C
Pressure of gas in collecting bulb. . . . . . 69.8 cm of Hg

Calculate the percentage of KNO_3 in the mixture assuming this compound to be the only source of nitrate in the mixture.

7-48. When Mg is burned in air, a mixture of MgO and Mg_3N_2 (magnesium nitride) is formed. If 3.597 g of Mg produces 5.688 g of the mixture, what is the percentage of each compound in the mixture?

7-49. An alloy of Zn and Cd was analyzed by dissolving in dilute HCl and collecting the liberated H_2 gas by displacement of H_2O. Equations for the reactions are

$$\text{Zn} + 2 \text{ HCl} \rightarrow \text{ZnCl}_2 + \text{H}_2(g)$$
$$\text{Cd} + 2 \text{ HCl} \rightarrow \text{CdCl}_2 + \text{H}_2(g)$$

The following data were obtained:

Weight of alloy taken for analysis . . . . . 4.275 g
Volume of gas (wet) . . . . 1403 ml
Temperature . . . . 20°C
Pressure in collecting vessel . . . . 697.5 mm of Hg

What is the weight percentage of each element in the mixture?

7-50. 2.704 g of a Mg–Al alloy was burned in O_2 to produce 4.762 g of a mixture of the oxides of the two metals. Determine the weight percentage and atom percentage of Mg in the alloy.

7-51. In 1905, the atomic weight of chlorine was determined by Professor T. W. Richards and his coworkers at Harvard University. A known weight of Ag was dissolved in dilute HNO_3, and HCl was then added to precipitate AgCl. The ratio by weight of AgCl to Ag was determined to be 1.328667. Previously, the atomic weight of Ag had been shown to be 107.870. Calculate the atomic weight of chlorine.

7-52. Professors Richards and Willard computed the atomic weight of lithium in 1910 by determining the weight of AgCl which could be prepared from a known weight of LiCl. They found the ratio of LiCl used to AgCl produced was 0.295786. Assuming that the atomic weights of Ag and Cl are known to be 107.870 and 35.453, respectively, calculate the atomic weight of Li.

7-53. A compound of xenon and fluorine was prepared when a mixture of Xe and F_2 was heated to 400°C for 1 hr. A reaction mixture was prepared at 0°C by introduction of F_2 gas at 0.60 atm and Xe gas at 0.12 atm into a 0.25-liter nickel container. After the reaction was completed, the mixture was cooled to 0°C, where it was found that all the Xe had been removed from the gas phase as a solid Xe–F compound. The pressure remaining in the nickel container was 0.36 atm. What is the formula for the compound prepared?

7-54. 0.20 moles of H_2 and 0.20 moles of O_2 are placed in a 5.0-liter container at 327°C. (a) What is the pressure of the mixture? (b) A reaction between H_2 and O_2 to produce H_2O is induced by an electric spark. What is the composition of the final mixture in terms of moles of each of the components? (c) What is the total pressure of the final mixture at 327°C? Assume all substances are gases under the conditions of this experiment.

7-55. On being heated in excess oxygen a 4.000 gram mixture of the three sulfides, PbS, CdS, and ZnS, yielded 3.506 g of mixed oxides (PbO, CdO, and ZnO) plus SO_2. (a) If the weight percent of CdS and ZnS is the same in the original mixture, calculate the percentage of each sulfide in the mixture. (b) What weight of SO_2 was produced?

CHAPTER 8

Solids

Solids are characterized by rigidity and definite crystalline or geometric structures corresponding to relatively simple polyhedra. X-ray studies of metals have shown that these structures result because the atoms of the metal occupy certain fixed geometrical positions with respect to each other in a three-dimensional network or space lattice. The smallest portion of a crystal which has the structure characteristic of the space lattice is called the *unit cell*. The space lattice may, in principle, be generated by extensive repetition of the unit cell in three dimensions. The lengths of the edges of the unit cell, or the unit cell dimensions, are the distances between the nearest identical positions along the axes of the space lattice.

The preceding discussion has dealt with atoms occupying sites in the unit cells of metals. Unit cells of other substances may be considered to have atoms, ions, or molecules occupying the lattice sites as may be appropriate.

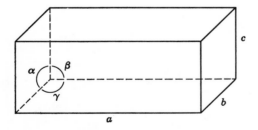

Fig. 8-1 Generalized unit cell.

The unit cells for the recognized crystal types may be characterized in terms of a parallelepiped as shown in Fig. 8-1. The lengths of the three axes are a, b, and c, and the angles between axes are α, β, and γ.

81

Unit cell	Axis length	Angles between axes
cubic	$a = b = c$	$\alpha = \beta = \gamma = 90°$
tetragonal	$a = b \neq c$	$\alpha = \beta = \gamma = 90°$
orthorhombic	$a \neq b \neq c$	$\alpha = \beta = \gamma = 90°$
monoclinic	$a \neq b \neq c$	$\alpha = \gamma = 90°$ $\beta \neq 90°$
triclinic	$a \neq b \neq c$	$\alpha \neq \beta \neq \gamma \neq 90°$
hexagonal	$a = b \neq c$	$\alpha = \beta = 90°$ $\gamma = 120°$
rhombohedral	$a = b = c$	$\alpha = \beta = \gamma \neq 90°$

Crystals belonging to the cubic system may be further classified as belonging to one of the unit cells shown in Fig. 8-2. The simplest of these is shown in Fig. 8-2a where there are atoms only at the corners of each cube. The structure shown in part b is the body-centered cubic unit cell in which a cube contains an atom at its center as is implied by its name as well as an atom at each corner. The third type shown in part c is the face-centered unit cell which has an atom located at the center of each face as well as at the corners of the cube.

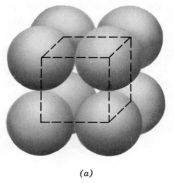

(a)

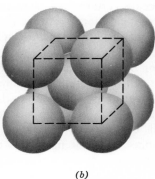

(b)

Fig. 8-2   Unit cells of the cubic system: (a) simple cubic; (b) body-centered cubic; and (c) face-centered cubic.

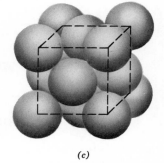

(c)

To determine the *number of atoms per unit cell*, we must consider how atoms are shared between adjacent unit cells. The following rules apply.

1. Atoms on a face are shared equally between two adjacent unit cells and hence are counted one-half for each.

2. Atoms on an edge are shared equally among four adjacent unit cells and hence are counted one-fourth for each.

3. Atoms at a corner of a unit cell are shared among eight adjacent unit cells and hence are counted one-eighth for each.

4. Atoms completely within the unit cell are not shared with others and hence are counted only for the cell in which located.

Atoms which surround a particular atom at the shortest distance between atoms are called *nearest neighbors*, and the number of nearest neighbors is termed the *crystallographic coordination number*. For example, in a body-centered cubic system, a metal atom at the center of the cube is surrounded by eight other metal atoms at the same distance; hence it has eight nearest neighbors or a coordination number of 8.

Crystals are often classified on the basis of the units which occupy the various sites in the space lattice and the bonding between these units. On this basis, there are four types of crystals: metallic or atomic, ionic, molecular, and covalent. Only the first two types will be discussed here.

A. METALLIC OR ATOMIC CRYSTALS

The preceding discussion is directly applicable to this type of crystal. Some representative problems follow.

Example 8-a. At 20°C, iron metal crystallizes in a form known as α-Fe having a body-centered cubic structure with a unit cell dimension of 2.86 Å. (a) How many atoms are there per unit cell? (b) How many nearest neighbors does a given iron atom have? (c) What is the volume in cubic centimeters of the unit cell? (d) What is the density of α-Fe in grams per cubic centimeter?

> *Solution.* (a) In a body-centered cubic unit cell there are eight corner sites and one center site. The contribution of each of these occupied sites to the unit cube must be calculated.
>
> 8 corner sites × $\frac{1}{8}$ (atom/corner) = 1 atom
> 1 center site × 1 atom/site   = 1 atom
>   Total = 2 atoms per unit cell

(b) Each center atom has eight nearest neighbor atoms at the corners of the cube. Each corner atom has eight nearest neighbor atoms at the centers of each of the eight adjacent cubes. Hence each iron atom has eight nearest neighbors.

(c) Volume $= a \times b \times c = (2.86 \text{ Å})^3 = 23.4 \text{ Å}^3/\text{unit cell}$

$$\text{Volume} = 23.4 \text{ Å}^3 \times \left(\frac{10^{-8} \text{ cm}}{1 \text{ Å}}\right)^3 = 2.34 \times 10^{-23} \text{ cm}^3/\text{unit cell}$$

(d) Since density is the ratio of mass to volume and since the volume of the unit cell represents the volume occupied by two atoms, the *mass* of two atoms must be determined before the density can be calculated.

$$\text{Mass} = \frac{55.9 \text{ g}}{\text{g atom Fe}} \times \frac{1 \text{ g atom Fe}}{6.02 \times 10^{23} \text{ atoms}} \times \frac{2 \text{ atoms}}{\text{unit cell}}$$

$$= 1.86 \times 10^{-22} \text{ g/unit cell}$$

$$\text{Density} = \frac{\text{Mass}}{\text{Volume}} = \frac{1.86 \times 10^{-22} \text{ g/unit cell}}{2.34 \times 10^{-23} \text{ cm}^3/\text{unit cell}}$$

$$= 7.95 \text{ g/cm}^3$$

Example 8-b. A unit cell of Pd metal is a face-centered cube with an edge of 3.89 Å. (a) What is the distance between centers of nearest neighbors? (b) What is the radius of a Pd atom? (c) What is the gram atomic volume of Pd?

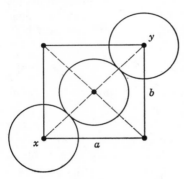

Solution. (a) First visualize the face of the unit cell as it is shown below. Each black dot represents the center of a Pd atom. The distance between nearest neighbors is one-half the distance, d, from atom x to atom y. We know from the theorem of Pythagoras, that the square of the hypotenuse of a right triangle is equal to the sum of the squares of the sides. Therefore

$$d^2 = a^2 + b^2 = (3.89 \text{ Å})^2 + (3.89 \text{ Å})^2 = 2(3.89 \text{ Å})^2$$

$$d = \sqrt{2} \ (3.89 \text{ Å}) = 5.50 \text{ Å}$$

The distance between centers of nearest neighbors is thus 5.50/2 or 2.75 Å.

(b) If atoms are assumed to be hard spheres which touch along the axis of closest approach, the radius is one-half the distance between centers of atoms or one-half of 2.75 Å = 1.38 Å.

(c) From the unit cell dimensions and the number of atoms per unit cell, the volume per atom may be calculated. By multiplying by Avogadro's number, the gram atomic volume may be obtained.

Volume $= (3.89 \text{ Å})^3/\text{unit cell} = 5.89 \times 10^{-23} \text{ cm}^3/\text{unit cell}$

In a face-centered cubic unit cell there are eight corner sites and six face sites.

8 corner sites × ($\frac{1}{8}$) atom/corner site = 1 atom
6 face sites × ($\frac{1}{2}$) atom/face site   = 3 atoms
                            Total = 4 atoms/unit cell

Gram atomic volume

$$= \frac{5.89 \times 10^{-23} \text{ cm}^3/\text{unit cell}}{4 \text{ atoms/unit cell}} \times \frac{6.02 \times 10^{23} \text{ atoms}}{\text{g atom}}$$

$$= 8.86 \text{ cm}^3/\text{g atom}$$

PROBLEMS (Metals)

8-1. How many atoms per unit cell are there in elements crystallizing in each of the following systems? (a) Simple cubic. (b) Face-centered cubic. (c) Simple tetragonal. (d) Face-centered tetragonal. (e) Body-centered tetragonal. (f) Simple orthorhombic.

8-2. For each of the cubic systems whose unit cells are shown in Fig. 8-2, how many nearest neighbors would an atom of a metal have?

8-3. Na metal crystallizes in a body-centered cubic system with a distance of 3.71 Å between nearest neighbors. (a) What is the radius of the Na atom? (b) What is the length of the edge of the unit cell? (c) What is the gram atomic volume of Na?

8-4. Ag metal crystallizes in the cubic system with a face-centered unit cell. (a) If the distance of closest approach of two Ag atoms is 2.89 Å, what is the length of the edge of the unit cell? (b) What is the density of Ag metal?

8-5. Rb metal crystallizes in a body-centered cubic system with a unit cell dimension of 5.72 Å. (a) What is the distance between the centers of nearest neighbors? (b) What is the volume of the unit cell? (c) What is the gram atomic volume? (d) What is the density of Rb metal?

8-6. Tin metal exists in two allotropic forms, α and β. β-Sn crystallizes with a simple tetragonal unit cell having dimensions $a = b = 3.02$ Å and $c = 3.18$ Å. (a) How many atoms are there per unit cell in β-Sn? (b) What is the density of β-Sn? (c) What is the gram atomic volume?

8-7. Indium metal crystallizes with a face-centered tetragonal unit cell with the dimensions $a = b = 4.59$ Å and $c = 4.94$ Å. (a) How many atoms are there per unit cell? (b) What is the volume of the unit cell? (c) What is the gram atomic volume? (d) What is the density of In metal?

B. IONIC CRYSTALS

Ionic crystals are those in which positive and negative ions occupy the various sites in the space lattice. Depending on the types of ions in the compound, the unit cells are more or less complicated. Two relatively simple types, shown by numerous compounds belonging to the cubic system, are those shown by sodium chloride and cesium chloride. The unit cells of each of these are shown in Fig. 8-3. In the sodium chloride

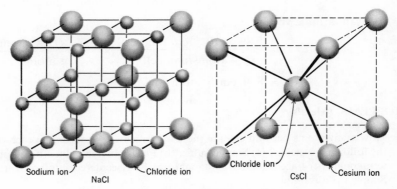

Sodium ion   Chloride ion
NaCl

Chloride ion   Cesium ion
CsCl

Fig. 8-3 Types of unit cells.

type, each positive ion is surrounded by six nearest neighbor chloride ions and each chloride ion is surrounded by six nearest neighbor sodium ions. In the cesium chloride type, each positive and negative ion has eight ions of opposite charge for nearest neighbors.

Example 8-c. The edge of the unit cell of CsCl is 4.11 Å. (a) What is the distance between the center of a Cs^+ ion and the center of the nearest Cl^- ion? (b) If the radius of Cl^- is 1.81 Å, what is the radius of Cs^+? (c) How many ions are there per unit cell?

> *Solution.* (a) The shortest distance between centers of a Cs^+ ion and a Cl^- ion is along the diagonal of the unit cube. Let the distance between opposite corners of the CsCl unit cell be d as shown below. From geometry,

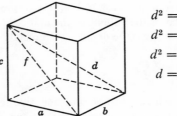

$$d^2 = f^2 + b^2$$
$$d^2 = a^2 + c^2 + b^2$$
$$d^2 = 3a^2 = 3(4.41 \text{ Å})^2 = 50.7 \text{ Å}^2$$
$$d = 7.12 \text{ Å}$$

The distance between centers of the Cs^+ and Cl^- is $7.12/2 = 3.56$ Å.

(b) Assume that the distance between the centers of Cs^+ and Cl^- is the sum of their radii. Then,

$$(r_{Cs^+}) + (r_{Cl^-}) = 3.56 \text{ Å}$$
$$(r_{Cs^+}) = 3.56 - 1.81 = 1.75 \text{ Å}$$

(c) The number of ions per unit cell is calculated as follows:

$$1 \text{ center } Cl^- = 1 \ Cl^-$$
$$8 \text{ corner sites} \times \tfrac{1}{8} \ (Cs^+/\text{corner site}) = 1 \ Cs^+$$

There is, therefore, one CsCl per unit cell.

Example 8-d. Crystals may have vacant or unoccupied sites in the crystal lattice. These are said to have *lattice defects*. If a crystal does have unoccupied sites, its density should be less than that calculated from unit cell dimensions with all sites filled. By comparing observed and theoretical densities, the fraction of sites unoccupied may be determined; for example, the observed density of NaCl is 2.165 and the unit cell of NaCl is 5.628 Å along each edge. Determine the theoretical density and then calculate the percentage of unoccupied sites.

Solution.

Volume of unit cell $= (5.628 \times 10^{-8} \text{ cm})^3 = 178.3 \times 10^{-24} \text{ cm}^3$

Since there are 4 Na^+ and 4 Cl^- per unit cell, the

$$\text{Mass per unit cell} = \frac{58.44 \times 4}{6.02 \times 10^{23}} = 38.83 \times 10^{-23} \text{ g}$$

$$\text{Density (theoretical)} = \frac{3.883 \times 10^{-22} \text{ g}}{1.783 \times 10^{-22} \text{ cm}^3} = 2.178 \text{ g/cm}^3$$

$$\text{Percentage of unoccupied sites} = \frac{2.178 - 2.165}{2.178} \times 100 = 0.60\%$$

PROBLEMS (Ionic Crystals)

8-8. The space lattice of NaCl is shown above as two interpenetrating face-centered cubes of Na^+ and Cl^-. The edge of the unit cell is 5.628 Å. (a) How many ions of each type are present in the unit cell? (b) What is the distance between the center of a Na^+ and the center of the nearest Cl^-? (c) What is the distance between centers of the nearest two Na^+ ions? (d) What is the volume of the unit cell?

8-9. KF crystallizes in the NaCl type structure. (a) If the radius of K^+ is 1.32 Å and F^- is 1.35 Å, what is the shortest K–F distance? (b) What is the length of the edge of the unit cell? (c) What is the shortest K–K distance? (d) What is the volume of the unit cell?

8-10. RbI crystallizes in the CsCl type structure. (a) If the radius of Rb^+ is 1.49 Å and I^- is 2.17 Å, what is the shortest Rb–I distance? (b) What is the shortest Rb–Rb distance? (c) How many Rb^+ ions are there at this distance around a given Rb^+ ion? (d) What is the volume of the unit cell?

8-11. RbCl crystallizes in the NaCl structure. (a) What are the unit cell dimensions if the Rb–Cl distance is 3.27 Å? (b) What is the density of RbCl?

8-12. CsI crystallizes in the CsCl type structure. (a) What is the length of the unit cell if the Cs–I distance is 3.95 Å? (b) What is the molar volume of CsI?

8-13. CsBr crystallizes in the CsCl structure. (a) If the density of CsBr is 4.438 g/cm^3, what is the length of the edge of the unit cube? (b) What is the shortest Cs–Br distance?

GENERAL PROBLEMS

8-14. In tetragonal β-Sn having a simple tetragonal unit cell, the dimensions are $a = b = 3.022$ Å and $c = 3.181$ Å. There are two different distances between the centers of Sn atoms. Sketch the unit cell. (a) What are these distances? (b) How many nearest neighbors does a particular atom have at each distance?

8-15. Cs metal crystallizes in a body-centered cubic system with a unit cell length $a = 6.08$ Å. What is the radius of the Cs atom?

8-16. Ni metal crystallizes in a face-centered cubic system with a unit cell length $a = 3.52$ Å. (a) What is the radius of the Ni atom? (b) How many close neighbors does a Ni atom have?

8-17. Indium crystals have a face-centered tetragonal unit cell of dimensions $a = b = 4.59$ Å and $c = 4.94$ Å, and there are two different distances between indium atoms. Sketch the unit cell. What are these distances?

8-18. The ionic radii of Ba^{++} and O^{--} ions are 1.36 and 1.40 Å, respectively. Barium oxide crystallizes in the NaCl type structure. (a) What is the volume of the unit cell? (b) What is the shortest distance between centers of adjacent Ba^{++} ions? (c) How many Ba^{++} ions are there around a particular Ba^{++} at this distance? (d) What is the density of BaO?

8-19. PbS crystallizes in the NaCl type structure. If the shortest distance between centers of Pb^{++} and S^{--} ions is 2.962 Å, (a) what is the length of the unit cell edge? (b) What is the volume of the unit cell? (c) What is the distance between nearest Pb^{++} ions? (d) What is the molar volume of PbS?

8-20. Calculate Avogadro's number from the fact that Au belongs to the face-centered cubic system and has a unit cell dimension $a = 4.078$ Å. Its density is 19.30 g/cm^3.

8-21. SrO has a density of 5.05 g/cm³ and there are 4 Sr⁺⁺ and 4 O⁻⁻ ions per unit cell. What is the length of the unit cell if it is cubic?

8-22. LiH crystallizes in a cubic structure and has a density of 0.77 g/cm³. If the edge of the unit cube is 4.086 Å, how many Li⁺ and H⁻ ions are there in a unit cell?

8-23. Calculate Avogadro's number from the fact that CsBr crystallizes in the CsCl type structure and has a unit cell edge length of 4.289 Å. The density of CsBr is 4.477 g/cm³.

8-24. (a) For each of the three unit cubes shown in Fig. 8-2, calculate the fraction of the cube actually occupied by atoms. Assume that the atoms are hard spheres and touch along the shortest interatomic distance. (b) Which of the three exhibits the most efficient packing of atoms?

8-25. Data are recorded in the table for a number of compounds which crystallize in the cubic system, either as NaCl or CsCl types. From the information given for each compound, carry out calculations to obtain data to fill in the blanks.

NaCl Type

	Compound	Edge of unit cell in Å	Dist. between centers of like ions (atoms) in Å	Dist. between centers of unlike ions (atoms) in Å	Molar volume (ml)	Density theoretical	g/ml observed
(a)	NaF	4.620	—	2.31	—	2.84	2.79
(b)	LiI	—	4.24	—	32.5	—	4.06
(c)	RbBr	6.854	—	—	—	3.41	3.35
(d)	CaO	—	3.40	2.405	16.75	—	3.35
(e)	KBr	6.6000	—	—	—	—	2.75
(f)	LiF	—	—	2.01	9.76	2.66	2.60
(g)	AgCl	5.547	—	—	—	5.57	5.56

CsCl Type

	Compound	Edge of unit cell in Å	Dist. between centers of like ions (atoms) in Å	Dist. between centers of unlike ions (atoms) in Å	Molar volume (ml)	Density theoretical	g/ml observed
(k)	CsBr	4.286	—	—	47.4	—	4.438
(l)	CsI	—	4.567	—	57.3	4.53	4.51
(m)	TlCl	—	3.834	—	—	7.07	7.00
(n)	AgCd	3.33	—	—	—	—	—
(o)	LiAg	3.168	3.168	—	16.2	—	—
(p)	MgSr	—	3.90	3.37	—	3.14	—

8-26. Use the table above in answering this question. By comparing the observed density with that calculated from unit cell dimensions, determine the extent of "lattice defect," that is, the percentage of unoccupied sites in the crystal lattice for the compounds NaF, AgCl, and CsI.

CHAPTER 9

Expressing Concentrations of Solutions

The concentration of a solution may be expressed in a number of ways, the more important of which are listed below.

A. PERCENTAGE (by Weight)

% Solute = Parts by weight of solute per 100 parts by weight of solution

or

$$\% \text{ Solute} = \frac{\text{Parts of solute}}{\text{Parts solute} + \text{Parts solvent}} \times 100$$

For example, a solution containing 10 g of solute in 100 g of solution (10 g solute, 90 g solvent) is a 10% solution; one containing 50 g solute in 50 g of solvent is a 50% solution, etc. Note that the denominator in this expression is the weight of *solution*, which includes both solute and solvent.

Example 9-a. Calculate the percentage of K_2CO_3 in a solution which is made by dissolving 15 g of K_2CO_3 in 60 g of H_2O.

Solution.

$$\text{Percentage } K_2CO_3 = \frac{\text{Grams } K_2CO_3}{\text{Grams } K_2CO_3 + \text{Grams } H_2O} \times 100$$

$$= \frac{15 \text{ g}}{15 \text{ g} + 60 \text{ g}} \times 100 = 20\%$$

Example 9-b. How many grams of NaCl would have to be dissolved in 54 g H_2O to give a 10% solution?

Solution.

$$\text{Percentage NaCl} = \frac{\text{Grams NaCl}}{\text{Grams NaCl} + \text{Grams } H_2O} \times 100$$

Let

$$X = \text{Grams of NaCl}$$

then

$$0.1 = \frac{X}{X + 54 \text{ g}}$$

$$0.1X + 5.4 \text{ g} = X$$

$$X = \frac{5.4 \text{ g}}{0.9} = 6 \text{ g}$$

PROBLEMS (Percentage)

9-1. How many grams of NaCl are contained in 0.25 kg of an 8% solution?

9-2. In making up a 20% aqueous solution of sucrose, $C_{12}H_{22}O_{11}$, how many grams of sugar and water would be used to prepare 1.0 kg of solution?

9-3. A certain solution is 10.0% $C_{12}H_{22}O_{11}$. How many grams of sugar must be added to 0.400 kg of this solution to bring the percentage of sugar to 20.0%?

B. MOLARITY (Formality*)

Molarity is defined as the number of moles of solute per liter of solution. A solution which contains 1 mole per liter is termed 1 molar.

$$\text{Molarity} = \frac{\text{Moles solute}}{\text{Liters of solution}} \tag{1}$$

Since Moles = Grams/Molecular weight, we may substitute for the

---

* A similar term, *formality*, is defined as the number of formula weights of solute per liter of solution. Formality may be used with substances where only the empirical formula is known, such as for ionized substances. In practice, however, there is often no distinction made between molecular and formula weights insofar as these types of calculations are concerned, and hence the two terms molarity and formality may be used interchangeably. In the discussion which follows, only the term molarity will be employed.

numerator and obtain

$$\text{Molarity} = \frac{\text{Grams of solute}}{\text{MW of solute} \times \text{Liters of solution}} \qquad (2)$$

Example 9-c. What is the molarity of a solution which contains 49 g of H_2SO_4 in 2.0 liters of solution? The molecular weight of H_2SO_4 is 98.

Solution.

$$\text{Number of moles } H_2SO_4 = \frac{49 \text{ g}}{98 \text{ g/mole}} = 0.50 \text{ mole}$$

$$\text{Molarity} = \frac{0.50 \text{ mole}}{2.0 \text{ liters}} = 0.25 \text{ moles/liter or } 0.25 \text{ } M \text{ (molar) or, by}$$
substituting in formula (2) above,

$$\text{Molarity} = \frac{49 \text{ g}}{98 \text{ g/mole} \times 2.0 \text{ liters}} = 0.25 \text{ } M$$

Example 9-d. How many grams of H_2SO_4 are contained in 0.050 liter of 0.50 M solution?

Solution. 0.50 M H_2SO_4 would contain 0.50 mole/liter $\times$ 98 g/mole = 49 g/liter.

0.050 liter would contain 0.050 liter $\times$ 49 g/liter = 2.45 g of H_2SO_4. *Or* substituting in formula (2),

$$\frac{0.50 \text{ mole}}{\text{liter}} = \frac{\text{Grams } H_2SO_4}{\frac{98 \text{ g}}{\text{mole}} \times 0.050 \text{ liter}}$$

$$\text{Grams } H_2SO_4 = \left(\frac{0.50 \text{ mole}}{\text{liter}}\right)\left(\frac{98 \text{ g}}{\text{mole}}\right)(0.050 \text{ liter}) = 2.45 \text{ g}$$

Example 9-e. Determine the molarity of a solution prepared by diluting 1 liter of 0.5 M solution to 2 liters.

Solution. Molarity = Moles/Liters or Moles = Molarity $\times$ Liters. Moles of solute in 1 liter of 0.5 M solution = 0.5 mole/liter $\times$ 1 liter = 0.5 mole. This will also be the number of moles in the diluted solution. Hence molarity of diluted solution = 0.5 mole/ 2 liters = 0.25 mole/liter.

Example 9-f. To what volume must 250 ml of 0.15 M H_2SO_4 be diluted to yield a 0.025 M solution?

Solution. 250 ml of 0.15 M solution contains (0.25 liter)(0.15 mole/liter) = 0.0375 mole of H_2SO_4. Since Molarity = Moles/Liter or Liters = Moles/Molarity

$$\text{Volume of diluted solution} = \frac{0.0375 \text{ mole}}{0.025 \text{ moles/liter}}$$

$$= 1.5 \text{ liters or } 1500 \text{ ml}$$

PROBLEMS (Molarity)

9-4. What is the molarity of a solution in which 9.80 g of H_2SO_4 is dissolved in 0.100 liter of solution?

9-5. Determine the molarity of solutions of each of the following solutes in which exactly 100 g of solute is present in 4.00 liters of solution.

(a) NaOH   (c) KNO_3
(b) H_2SO_4 (d) $C_{12}H_{22}O_{11}$

9-6. How many grams of H_3PO_4 must be weighed to prepare 3.000 liters of 0.3000 M solution?

9-7. Calculate the molarity of a solution of $Ca(OH)_2$ which contains 0.370 g of solute in 2.00×10^2 ml of solution.

9-8. How many grams of HNO_3 will be required to prepare 1.5 liters of 0.50 M solution?

9-9. Calculate the molarity of a solution in which 1.46 g of HCl is dissolved in toluene and then diluted to 0.500 liter.

9-10. How many grams of each of the following solutes would be needed to prepare 0.500 liter quantities of 0.200 M solution?

(a) H_3PO_4 (c) KOH
(b) $Al_2(SO_4)_3$ (d) $(NH_4)_2CO_3$

9-11. How many liters of 0.555 M KOH solution can be prepared from 224 g of KOH?

9-12. If 52 ml of 0.15 M H_2SO_4 is added to an excess of solid Na_2CO_3, what volume of CO_2 gas is evolved at STP?

9-13. If 50 ml of 16 M HNO_3 are diluted to 400 ml, what is the molarity of the diluted solution?

9-14. To what volume must 0.050 liter of 0.30 M H_3PO_4 be diluted to yield a 0.060 M solution?

9-15. A 5.85 g sample of NaCl is dissolved in enough water to yield 0.250 liter of solution. This solution is then diluted to exactly 1 liter.

(a) How many moles of NaCl are present in the undiluted solution?
(b) How many moles of NaCl are present in the diluted solution?
(c) Determine the molarity of each solution.

C. NORMALITY

Normality is defined as the number of equivalent weights of solute per liter of solution.

$$\text{Normality} = \frac{\text{Equivalent weights of solute}}{\text{Liters of solution}} \tag{3}$$

The equivalent weight of a compound depends upon the reaction which it undergoes, and further, an equivalent weight of one substance reacts exactly with the equivalent weight of a second substance. For acid-base reactions, the equivalent weight of an acid is the number of grams that furnishes one mole (1.008 g) or one Avogadro's number of H^+; an equivalent weight of a base is the number of grams of the compound that reacts with one mole of H^+ or with one equivalent weight of an acid. For example, the equivalent weight of H_2SO_4 is equal to the formula weight/2 since one mole of the acid contains two moles of H^+. For $Ba(OH)_2$, a formula weight reacts with two moles of H^+ or two equivalents of H^+, and therefore the equivalent weight is one-half the formula weight.

For salts in metathesis reactions, usually the equivalent weight is the number of grams furnishing one Avogadro's number of positive charges or one Avogadro's number of negative charges. For example, one mole of K_2CO_3 contains two moles or two Avogadro's numbers of K^+; hence the equivalent weight of K_2CO_3 is FW/2; the equivalent weight of $AlCl_3$ is FW/3; that for $Cr_2(SO_4)_3$ is FW/6; that for $Th_3(PO_4)_4$ is FW/12. Again, we must carefully consider the reaction involved. For example, if we are concerned with the reaction of either K^+ or H^+ in $KHSO_4$, the equivalent weight is equal to the formula weight, but if we are concerned with the reaction of SO_4^{--}, then the equivalent weight is equal to one-half of the formula weight.

Since the number of equivalents is the number of grams divided by the equivalent weight, it is found on substituting in formula (3) that

$$\text{Normality} = \frac{\text{Grams of solute}}{\text{EW of solute} \times \text{Liters of solution}} \tag{4}$$

The equivalent weight is always some simple fraction of the molecular weight (if not equal to it), and the normality of a given solution will always be some simple multiple of the molarity (if not equal to it).

Example 9-g. Determine the normality of a solution of H_2SO_4 which contains 49 g of H_2SO_4 in 2.0 liters of solution. The molecular weight of H_2SO_4 is 98; the equivalent weight is 49.

Solution.

$$\text{Number of equiv} = \frac{49 \text{ g}}{49 \text{ g/equiv}} = 1.0 \text{ equiv}$$

Since 2.0 liters of this solution contain 1.0 equiv, 1 liter contains 0.50 equiv and hence the normality is 0.50 equiv/liter or 0.50 N, *or*, substituting in (4),

$$\text{Normality} = \frac{49 \text{ g}}{49 \text{ g/equiv} \times 2.0 \text{ liters}} = 0.50 \text{ equiv/liter or } 0.50 \text{ } N$$

Since the equivalent weight of H_2SO_4 is one-half the molecular weight, the molarity of this solution will be $\frac{1}{2} \times 0.50 = 0.25$ M.

Example 9-h. How many grams of $Ca(OH)_2$ are contained in 0.250 liter of 0.01000 N solution? The formula weight of $Ca(OH)_2$ is 74.0; its equivalent weight is 74.0/2 or 37.0.

Solution. A 0.01000 N solution of $Ca(OH)_2$ would contain 0.01000 equiv/liter $\times$ 37.0 g/equiv = 0.370 g/liter.

0.250 liter or $\frac{1}{4}$ liter would contain $\frac{1}{4}$ liter $\times$ 0.370 g/liter = 0.0925 g, *or*, substituting in (4),

$$0.01000 \text{ equiv/liter} = \frac{\text{Grams Ca(OH)}_2}{\dfrac{37.0 \text{ g}}{\text{equiv}} \times 0.250 \text{ liter}}$$

$$\text{Grams Ca(OH)}_2 = \frac{0.01000 \text{ equiv}}{\text{liter}} \times 0.250 \text{ liter} \times \frac{37.0 \text{ g}}{\text{equiv}}$$

$$= 0.0925 \text{ g}$$

PROBLEMS (Normality)

9-16. (a) Determine the normality of a solution which contains 6.87 g of $Ba(OH)_2$ in exactly 500 ml of solution. (b) What is the molarity of the solution?

9-17. 25 ml of a 0.15 N solution of H_2SO_4 would contain how many grams of H_2SO_4?

9-18. How many grams of H_3PO_4 would be needed to prepare 2.0 liters of 0.15 N solution?

9-19. How many equivalents of H_2SO_4 are contained in 200 ml of 0.4 N solution?

9-20. (a) Determine the normality of a solution which contains 5.66 g of $Zr(SO_4)_2$ in 500 ml of solution. (b) What is its molarity?

9-21. How many grams of selenic acid, H_2SeO_4, are contained in 2.0 liters of 0.20 N solution?

D. MOLE FRACTION

The *mole fraction* is defined as the ratio of moles of a given component to the total moles of all components in a solution; that is,

Mole fraction of component $A = X_A = \dfrac{\text{Moles of } A}{\text{Total moles of all components}}$

For a solution containing a single solute in a solvent

$$\text{Mole fraction of solute} = \frac{\text{Moles of solute}}{\text{Moles of solute} + \text{Moles of solvent}} \tag{5}$$

$$\text{Mole fraction of solvent} = \frac{\text{Moles of solvent}}{\text{Moles of solute} + \text{Moles of solvent}} \tag{6}$$

Obviously, the sum of the two fractions (5) and (6) must be unity.

Example 9-i. What is the mole fraction of sugar, $C_{12}H_{22}O_{11}$, in a solution made by dissolving 17.1 g of $C_{12}H_{22}O_{11}$ in 89.0 g of H_2O?

Solution. The molecular weight of $C_{12}H_{22}O_{11}$ is 342, and that of H_2O is 18.

$$\text{Number of moles of } C_{12}H_{22}O_{11} = \frac{17.1 \text{ g}}{342 \text{ g/mole}} = 0.0500 \text{ mole}$$

$$\text{Number of moles of } H_2O = \frac{89.0 \text{ g}}{18.0 \text{ g/mole}} = 4.95 \text{ moles}$$

$$\text{Mole fraction of } C_{12}H_{22}O_{11} = \frac{0.0500}{0.050 + 4.95} = 0.0100$$

Example 9-j. What is the mole fraction of NaCl in a solution containing 1.00 mole of solute in 1.00 kg of H_2O?

Solution.

$$\text{Number of moles of } H_2O = \frac{1000 \text{ g}}{18.0 \text{ g/mole}} = 55.5 \text{ moles}$$

$$\text{Mole fraction NaCl} = \frac{1.00 \text{ mole}}{1.00 \text{ mole} + 55.5 \text{ moles}}$$

$$= \frac{1.00 \text{ mole}}{56.5 \text{ moles}} = 0.0177$$

PROBLEMS (Mole Fraction)

9-22. Calculate the mole fractions of solute and solvent in a solution prepared by dissolving 234 g of NaCl in 3.00 kg of H_2O.

9-23. Calculate the mole fraction of H_2SO_4 in a solution containing 4.50 moles in 1.00 kg of H_2O.

9-24. What is the mole fraction of H_2SO_4 in an 8.60% aqueous solution?

9-25. Calculate the mole fraction of H_2SO_4 in a 1.20 M aqueous solution which has a density of 1.070 g/ml.

9-26. Calculate the mole fraction of H_2SO_4 in 15.5 M H_2SO_4. The density of this solution is 1.760 g/ml.

E. MOLALITY

Molality is defined as moles of solute per kilogram of solvent. A 1 molal solution contains 1 mole of solute in 1 kg of solvent.

$$\text{Molality} = \frac{\text{Moles of solute}}{\text{Kg of solvent}}$$

or

$$\text{Molality} = \frac{\text{Grams of solute}}{\text{MW of solute} \times \text{Kg of solvent}} \tag{7}$$

Molarity varies with temperature since the volume of a solution changes with temperature; whereas molality, which is based on weight rather than volume, is independent of temperature. This method of expressing the concentration of a solution is ordinarily used when dealing with *colligative* properties; such properties of solutions are the boiling and freezing points which are dependent upon the *number* of dissolved particles and not upon their nature or kind. In dilute solutions, molarity and molality have virtually the same values, but they have wide differences in concentrated solutions.

Example 9-k. What is the molality of a solution in which exactly 100 g of NaOH is dissolved in 0.250 kg of H_2O?

   Solution.

$$\text{Number of moles of NaOH} = \frac{100 \text{ g NaOH}}{40.0 \text{ g NaOH/mole NaOH}}$$

$$= 2.50 \text{ moles NaOH}$$

Since the solution contains 2.50 moles in 0.250 kg of H_2O, there are 10.0 moles in 1.00 kg of H_2O. Therefore the molality is 10.0 moles/ kg H_2O, *or*, by substituting in (7),

$$\text{Molality} = \frac{100 \text{ g}}{40.0 \text{ g/mole} \times 0.250 \text{ kg } H_2O} = 10.0 \text{ moles/kg } H_2O$$

PROBLEMS (Molality)

9-27. What is the molality of a solution which contains 24.5 g of H_3PO_4 in 1.00×10^2 g of H_2O?

9-28. In how many kilograms of H_2O should 234 g of NaCl be dissolved to obtain a 0.25 molal solution?

9-29. 250 g of a 0.20 molal solution of K_2SO_4 is evaporated to dryness. What is the mass in grams of the K_2SO_4 remaining?

GENERAL PROBLEMS

9-30. What is the molarity of a solution which contains 24.5 g of H_2SO_4 in 1 liter of solution?

9-31. What is the molarity of a solution which contains exactly 8 g of NaOH in 250 ml of solution?

9-32. What weight of $KMnO_4$ will be required to prepare 1.000 liter of 0.25 M $KMnO_4$?

9-33. What weight of $MgCl_2 \cdot 6 H_2O$ is required to prepare 150 ml of 0.125 M solution?

9-34. What is the normality of a solution containing 2.45 g of H_2SO_4 in 1.000 liter of solution?

9-35. 24.5 g of H_2SO_4 is dissolved in water and the solution diluted to two liters. (a) What is the molarity of the solution? (b) What is the normality?

9-36. Calculate the weight of NaOH required to prepare 600 g of 0.40 molal solution of NaOH.

9-37. Determine the molarity of a solution containing 4.90 g of H_2SO_4 in exactly 100 ml of solution. What is the normality of the solution?

9-38. Calculate the molarity of a solution of NaOH, 0.200 liter of which contain 10.0 g of solute. What is the normality of the same solution?

9-39. A liter of 0.200 M solution of H_2SO_4 contains (a) how many grams of H_2SO_4, (b) how many moles of H_2SO_4, (c) how many equivalents of H_2SO_4?

9-40. Determine the molarity and normality of each of the following aqueous solutions in which 20.0 g of the solute are dissolved in exactly 500 ml of solution: (a) NaOH, (b) H_2SO_4, (c) $Ba(OH)_2$, (d) $Al_2(SO_4)_3$, (e) H_3PO_4.

9-41. Determine the number of grams of solute in 0.200 liter of each of the following solutions: (a) 0.216 M HCl, (b) 0.024 N Ca(OH)$_2$, (c) 0.418 N H$_2$SO$_4$, (d) 0.50 M MgSO$_4$·7 H$_2$O, (e) 0.75 N H$_3$PO$_4$.

9-42. To what total volume would each of the following solutions have to be diluted to give 0.2 N solutions? (a) 10 ml of 10 M HCl, (b) 50 ml of 6 N HNO$_3$, (c) 1.0 ml of 18 M H$_2$SO$_4$, (d) 0.5 ml of 12 M NH$_3$.

9-43. How many moles and equivalents are there in each of the following? (a) 100 ml of 0.1 N H$_2$SO$_4$, (b) 4 g of solid NaOH, (c) 50 ml of 12 M HCl, (d) 10 liters of 0.2 N H$_3$PO$_4$, (e) 24.5 g of H$_2$SO$_4$ in 2 liters of solution.

9-44. What are the mole fractions of solute and solvent in a solution made by dissolving 120.0 g of urea, CO(NH$_2$)$_2$, in 270 g of H$_2$O?

9-45. The density of a 1.245 M aqueous solution of ZnSO$_4$ is 1.193 g/ml at 15°C. (a) What is the percentage of ZnSO$_4$ in the solution? (b) What is the mole fraction of ZnSO$_4$?

9-46. (a) What is the mole fraction of H$_2$SO$_4$ in a 20.00% aqueous solution? (b) If the density of the solution in part (a) is 1.1384 g/ml, what is the molarity of the solution?

9-47. A solution made by dissolving 40.0 g of 100% H$_2$SO$_4$ in 60.0 g of water has a density of 1.303 g/ml. Express the concentration of H$_2$SO$_4$ in this solution as (a) weight per cent, (b) molarity, (c) normality, (d) molality, (e) mole fraction of H$_2$SO$_4$.

9-48. Determine the density of an aqueous solution of HBr which is 2.86 M and 3.09 molal.

9-49. What volume of 40.0% HCl (d = 1.20 g/ml) would be required to prepare each of the following solutions? (a) 0.100 liter of 2.50 M solution. (b) 0.100 kg of 2.50 molal solution. (c) 0.100 kg of 2.50% solution.

9-50. Determine the molality and mole fraction of solute in (a) 20 wt percent NaCl solution, (b) a solution of H$_2$SO$_4$ (28 wt percent) of sp gr 1.20, (c) a solution made by dissolving 320 g of CH$_3$OH in 540 g H$_2$O.

9-51. What weight of solute must be dissolved in 1 kg of 10 wt percent solution to yield a 25 wt percent solution?

9-52. 45 g of C$_6$H$_{12}$O$_6$ is dissolved in 500 g of a 0.2 molal solution. What is the (a) molality of the resultant solution and (b) weight percentage of solute?

9-53. The solubility of CaCrO$_4$ at 100°C is 3.0 wt percent and at 0°C is 12.0 wt percent. If 500 g of solution is saturated at 0°C, what weight of salt will precipitate on heating this solution to 100°C?

9-54. 372 g of ZnSO$_4$ is dissolved in 790 g H$_2$O at 15°C to give 821 ml of solution. Determine the following: (a) molality, (b) molarity, (c) mole fraction of ZnSO$_4$, (d) density of solution, (e) weight percent of ZnSO$_4$.

9-55. A saturated solution of KCl at 20°C contains 296 g per liter of solution. The density is 1.17. Calculate the following: (a) weight percentage of KCl, (b) molarity, (c) molality, (d) mole fraction of KCl.

9-56. The solubility of individual gases in a mixture of gases is directly proportional to their partial pressures. At 20°C one volume of water dissolves 0.03405 volumes of oxygen and 0.01696 volumes of nitrogen at one atm pressure. Assuming air is 20.0% oxygen and 80.0% nitrogen by volume, what is the percentage of oxygen and nitrogen by volume in dissolved air at 20°C and one atm?

CHAPTER 10

Colligative Properties of Solutions

Properties of solutions which depend upon the *number* of dissolved particles and not upon their kind are termed *colligative* properties. Such properties include vapor pressures, freezing and boiling points, and osmotic pressures. These properties are useful for the experimental determination of molecular weights of dissolved substances. Freezing points are particularly useful since they may be determined with considerable precision. Solutions of electrolytes exhibit abnormal colligative properties and will be discussed near the end of the chapter. The discussion which follows in Sections A, B, and C applies to solutions of nonelectrolytes.

A. VAPOR PRESSURES

The vapor pressure of a solution of a nonvolatile solute in a liquid is always less than the vapor pressure of pure solvent. In accordance with Raoult's law, the lowering of the vapor pressure, ΔP, is proportional to the mole fraction of solute:

$$\Delta P = P_0 - P = P_0 X_2 \tag{1}$$

where P_0 is the vapor pressure of pure solvent, P is the vapor pressure of the solution, and X_2 is the mole fraction of solute, *or alternatively*, the vapor pressure of the solution is proportional to the mole fraction of solvent.

$$P = P_0 X_1 \tag{2}$$

where X_1 is the mole fraction of solvent.

Example 10-a. The vapor pressure of water at 20°C is 17.54 mm of Hg. Determine the vapor pressure of a solution prepared by dissolving 12.0 g of urea, $CO(NH_2)_2$, in 0.500 kg of H_2O.

Solution.

$$\text{Mole fraction solvent} = X_1 = \frac{\text{Moles solvent}}{\text{Moles solvent} + \text{Moles solute}}$$

$$= \frac{500/18.0}{500/18.0 + 12.0/60} = 0.993$$

Substituting into formula (2) above, $P = P_0 X_1$

$$\text{Vapor pressure of the solution} = (17.54 \text{ mm})(0.993)$$
$$= 17.42 \text{ mm of Hg}$$

Example 10-b. The vapor pressure of benzene, C_6H_6, at 30°C = 121.8 mm of Hg. If 10.0 g of a nonvolatile solute dissolved in exactly 100 g of benzene lowers the vapor pressure 8.8 mm, what is the molecular weight of the solute?

Solution. Calculate the mole fraction of the solute by substituting in equation (1).

$$8.8 \text{ mm} = (121.8 \text{ mm})(X_2)$$

$$X_2 = \frac{8.8 \text{ mm}}{121.8 \text{ mm}} = 0.072$$

$$X_2 = \frac{\text{Moles solute}}{\text{Moles solute} + \text{Moles solvent}}$$

$$0.072 = \frac{\text{Moles solute}}{\text{Moles solute} + \dfrac{100 \text{ g}}{78 \text{ g/mole}}}$$

$$\text{Number of moles solute} = 0.099 \text{ mole}$$

$$\text{Molecular weight} = \frac{10.0 \text{ g}}{0.099 \text{ mole}} = 101 \text{ g/mole}$$

PROBLEMS (Vapor Pressure)

10-1. The vapor pressure of H_2O at 25°C is 23.76 mm of Hg. Determine the vapor pressure of a solution of 90.0 g of glucose, $C_6H_{12}O_6$, in 400.0 g of H_2O.

10-2. What is the vapor pressure of a 15.0% aqueous solution of urea, $CO(NH_2)_2$, at 25°C if the vapor pressure of pure water is 23.76 mm of Hg at this temperature?

10-3. The vapor pressure of benzene at 75°C is 640.0 mm of Hg. A solution of 7.36 g of a solute in 106 g of benzene has a vapor pressure of 615 mm of Hg at 75°C. Calculate the molecular weight of the solute.

10-4. The vapor pressure of water at 40°C is 55.32 mm of Hg. What weight of sucrose (MW = 342.3) must be dissolved in 100.0 g of H_2O at 40°C to lower the vapor pressure to 55.00 mm of Hg?

B. FREEZING POINTS AND BOILING POINTS OF SOLUTIONS OF NONELECTROLYTES

If 1 mole of a nonelectrolyte (nonionized substance) is dissolved in 1000 g of a solvent (this would be a 1 molal solution), the freezing point is depressed a constant amount below the freezing point of the pure solvent. It makes no difference what the nonelectrolyte is. The effect produced by 1 mole of solute in a kilogram of a given solvent is termed the molal freezing point depression constant for that particular solvent. For water the constant is 1.86°C/(mole of solute/kg of solvent); consequently a 1 molal aqueous solution of a nonionized solute freezes at −1.86°C and a 2 molal solution freezes at −3.72°C. In other words, ΔT = Molality × f.p. depression constant, where ΔT is the change in temperature.

In a similar manner a nonvolatile solute elevates the boiling point of a solution. A solution of 1 mole of a nonelectrolyte in 1000 g of water boils at a temperature of 100.52°C at standard pressure. The boiling point has been elevated a matter of 0.52°C for a 1 molal solution. The molal boiling point elevation constant for water is therefore 0.52°C/(mole of solute/kg of solvent). Other solvents have different numerical values for this constant. The boiling point elevation is a result of a lowering of the vapor pressure of the solvent by the dissolved substance. As a nonvolatile solute is added to a solvent, the vapor pressure of the solution decreases in accordance with Raoult's law. Therefore the temperature (B.P.) at which the vapor pressure equals the external pressure is increased.

What has been said about the effect of a nonvolatile solute on the boiling point does not hold for a volatile solute. It will be recalled that a liquid boils when the vapor pressure of the liquid becomes equal to the external pressure (usually atmospheric). For a solution of a volatile solute, both solute and solvent exert a pressure at a given temperature. Hence the solution boils when the combination of the two pressures equals the external pressure. The boiling point of such a solution may be either lower or higher than the boiling point of the solvent.

Since one molecular weight of a nonvolatile nonelectrolyte dissolved in 1 kg of solvent produces a known effect on the boiling and freezing point of any particular solvent, the molecular weight of the solute can be obtained by determining the boiling or freezing points of solutions with known quantities of solute and solvent. How this is done will be evident from a study of the problem examples to follow.

Example 10-c. If 20 g of the nonelectrolyte urea, $CO(NH_2)_2$, is dissolved in 250 g of H_2O, (a) what is the freezing point of the solution and (b) what is the boiling point at standard pressure? The molecular weight of urea is 60.

Solution. (a) First, determine the molality; next, obtain the change in temperature of the freezing point by multiplying the molality by the molal freezing point depression constant. Subtract the change in temperature from the normal freezing point of the solvent to obtain the freezing point of the solution.

$$\text{Number of moles of } CO(NH_2)_2 = 20 \text{ g} \times \frac{1 \text{ mole}}{60 \text{ g}} = \frac{1}{3} \text{ mole}$$

$$\text{Molality} = \frac{\frac{1}{3} \text{ mole}}{0.250 \text{ kg } H_2O} = \frac{4}{3} (\text{moles/kg } H_2O)$$

$$\Delta T = 4 (\text{moles/kg } H_2O) \times \frac{1.86°C}{(\text{mole/kg } H_2O)} = 2.48°C$$

$$\text{Freezing point} = 0.00 - 2.48 = -2.48°C$$

(b) The boiling point of the solution is obtained in a manner similar to the freezing point.

$$\Delta T = \frac{4}{3} (\text{mole/kg } H_2O) \times \frac{0.52°C}{(\text{mole/kg } H_2O)} = 0.69°C$$

$$\text{Boiling point} = 100.00 + 0.69 = 100.69°C$$

Example 10-d. A solution of exactly 10 g of A in 0.100 kg of H_2O boils at 100.78°C at standard pressure. What is the molecular weight of A if it is a nonvolatile nonelectrolyte?

Solution. First, calculate the molality from the boiling point elevation, and then, from the weights of solute and solvent, calculate the molecular weight.

$$\Delta T = 100.78 - 100.00 = 0.78°C$$

$$\Delta T = 0.78°C = \text{Molality} \times \frac{0.52°C}{(\text{mole/kg } H_2O)}$$

$$\text{Molality} = \frac{0.78°C/(\text{mole/kg } H_2O)}{0.52°C} = 1.50 \text{ mole/kg } H_2O$$

$$\text{Molality} = \frac{\text{Grams of solute}}{\text{MW} \times \text{Kg of solvent}}$$

and on rearranging

$$\text{MW} = \frac{\text{Grams of solute}}{\text{Molality} \times \text{Kg of solvent}}$$

$$= \frac{10 \text{ g}}{1.50 \text{ (mole/kg } H_2O) \times 0.100 \text{ (kg } H_2O)} = \frac{67 \text{ g}}{\text{mole}}$$

Hence the molecular weight is 67.

Example 10-e. What weight of sugar (MW = 342) must be dissolved in 4.00 kg of H_2O to yield a solution which will freeze at $-3.72°C$?

Solution. First, calculate the molality of a solution which will freeze at -3.72 C. Then calculate the weight of solute needed for 4.00 kg of solvent.

$$\Delta T = \text{Molality} \times \frac{1.86°C}{(\text{mole/kg } H_2O)}$$

$$\text{Molality} = \frac{3.72°C/(\text{mole/kg } H_2O)}{1.86°C} = 2.00 \text{ (mole/kg } H_2O)$$

$$\text{Molality} = \frac{\text{Grams of solute}}{\text{MW} \times \text{Kg of solvent}}$$

$$\text{Grams of sugar} = 2.00 \text{ (mole/kg } H_2O) \times \frac{342 \text{ grams}}{\text{mole}}$$

$$\times 4.00 \text{ kg of } H_2O = 2736 \text{ grams}$$

When rounded off, the number of grams required is 2740.

PROBLEMS (Solutions of Nonelectrolytes)

10-5. What is the freezing point of a solution prepared by dissolving 30.0 g of thioacetamide, CH_3CSNH_2, in 100 g of H_2O?

10-6. What is the boiling point at 760 mm of Hg pressure of the solution in Problem 10-5?

10-7. Calculate the boiling point at standard pressure of a solution made by dissolving 60 g of urea, $CO(NH_2)_2$, in 250 g of water.

10-8. Determine the freezing point of a solution which contains 186 g of the nonelectrolyte glycol, $C_2H_4(OH)_2$, dissolved in 250 g of H_2O.

10-9. 36.0 g of a nonelectrolyte is dissolved in 200 g of water. The resulting solution is found to freeze at $-3.72°C$. What is the molecular weight of the solute?

10-10. Calculate the freezing point of a solution which contains 18.4 g of the nonelectrolyte glycerine, $C_3H_8O_3$, dissolved in 125 g of water.

10-11. A solution of 54 g of a nonelectrolyte in 0.400 kg H_2O freezes at $-2.79°C$. What is the molecular weight of the solute?

10-12. A solution of $CO(NH_2)_2$ freezes at $-0.93°C$. What is the molality of the solution?

C. OSMOTIC PRESSURE

If a solution is separated from its pure solvent by a semipermeable membrane, solvent molecules may pass between the two sides of the membrane, but the solute molecules or ions may not (Fig. 10-1). Molecules

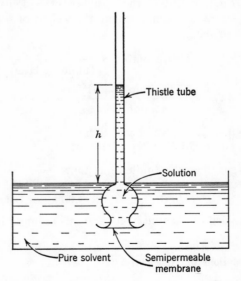

Fig. 10-1 An apparatus for the measurement of osmotic pressure.

from the pure solvent will diffuse through the membrane more rapidly than those from the solution can diffuse back. Because of this difference, there is a tendency for a net transfer of solvent molecules to the solution, thus diluting the solution. When the system reaches equilibrium, the pressure exerted downward by the column of solution of height h is just sufficient to prevent further dilution of the solution by the solvent. This pressure is termed the *osmotic* pressure, π, and the following relationships hold for dilute solutions:

$$\pi = \text{Molarity} \times RT \quad \text{or} \quad \pi = \text{Molality} \times RT$$

where R is the gas constant, 0.082 (liter-atm) per (mole-°K), and T is the absolute temperature. These relationships are particularly useful for the determination of molecular weights of large molecules, such as proteins.

Example 10-f. Determine the osmotic pressure of a 1.00×10^{-3} M aqueous solution at 0°C.

Solution.

$$\pi = \left(\frac{1.00 \times 10^{-3}\text{ mole}}{\text{liter}}\right)\left(\frac{0.0821\text{ liter-atm}}{\text{mole-°K}}\right)(273°\text{K}) = 0.0224\text{ atm}$$

Example 10-g. An aqueous solution containing 5.0 g of a protein in 0.20 liter of solution exerts an osmotic pressure of 0.0237 atm at 27°C. Determine the molecular weight of the protein.

Solution.

$$\text{Molarity of solution} = \frac{\pi}{RT} = \frac{0.0237\text{ atm}}{\left(\dfrac{0.0821\text{ liter-atm}}{\text{mole-°K}}\right)(300°\text{K})}$$

$$= 9.6 \times 10^{-4}\text{ mole/liter}$$

$$\text{Molarity} = \frac{\text{Grams of solute}}{\text{MW} \times \text{Liters of solution}}$$

Therefore,

$$\text{MW} = \frac{\text{Grams of solute}}{\text{Molarity} \times \text{Liters}} = \frac{5.0\text{ g}}{9.6 \times 10^{-4}\text{ (moles/liter)} \times 0.20\text{ liter}}$$

$$\text{MW} = 2.6 \times 10^4\text{ g/mole}$$

PROBLEMS (Osmotic Pressure)

10-13. Determine the osmotic pressure of a 0.020 M solution of sucrose at 20°C.

10-14. Determine the approximate molecular weight of a protein, 1 g of which dissolved in 0.100 liter of an aqueous solution exerts an osmotic pressure of 1.90 mm of Hg at 17°C.

10-15. The freezing point of human blood is −0.58°C. What is the osmotic pressure of blood at 98.6°F?

10-16. A 0.10% solution by weight of a certain protein exerts an osmotic pressure equivalent to a column of water 1.80-cm high at 20°C. Determine the approximate molecular weight of the protein.

D. SOLUTIONS OF ELECTROLYTES*

Solutions of electrolytes in aqueous solution are, to a certain extent, ionized or dissociated into positive and negative ions. Some electrolytes, such as NaCl, $CaBr_2$, and K_2SO_4 approach 100% dissociation and are called strong electrolytes. Others, such as acetic acid, ammonia, and hydrofluoric acid, are only partially dissociated and are called weak electrolytes. As the concentrations of the solutions of both weak and strong electrolytes decrease, the extents of dissociation increase. At infinite dilution, both types are presumed to be completely dissociated.

In a solution of an electrolyte, more particles are present per mole of dissolved substance than is indicated by the molality of the solution, and hence "abnormal" colligative properties, such as freezing points, are observed. Since the freezing point of a solution is dependent upon the number of dissolved particles, the extent to which a substance is ionized may be estimated from the freezing point.

Example 10-h. A solution of 20.0 g (1 mole) of HF in 1.00 kg of H_2O freezes at −1.92°C. What is the degree of ionization, that is, the fraction of the molecules present which are ionized? The ionization of HF may be represented as $HF \rightleftharpoons H^+ + F^-$ (note that whenever one HF is ionized, one H^+ and one F^- are produced).

> **Solution.**
> $$HF \rightleftharpoons H^+ + F^-$$

---

* This section may be omitted without loss of continuity or may be deferred until discussion of equilibrium and ionization. No problems of this type are included in the general problems set at the end of this chapter except 10-43, 10-45.

Let
$$X = \text{number of moles of } H^+ \text{ at equilibrium}$$

then
$$X = \text{number of moles of } F^- \text{ at equilibrium}$$

and
$$(1 - X) = \text{number of moles of HF at equilibrium}$$

Total number of moles of all species in solution
$$= X + X + (1 - X) = (X + 1)$$

$$\text{Total molality} = (X + 1) \text{ moles/kg } H_2O$$

Since

$$\Delta T = \text{Molality} \times \frac{1.86°C}{(\text{mole/kg } H_2O)}$$

$$1.92°C = (X + 1)(\text{mole/kg } H_2O) \times \frac{1.86°C}{(\text{mole/kg } H_2O)}$$

$$(X + 1) = \frac{1.92}{1.86} = 1.03$$

Therefore $X = 0.03$, and the degree (or fraction) of ionization is 0.03/1 or 3%.

Example 10-i. Determine the freezing point of a 2.0 molal aqueous solution of an organic acid, HA, which is 10% ionized.

Solution.
$$HA \rightleftharpoons H^+ + A^-$$

Moles of HA at equilibrium $= (0.90)(2.0) = 1.80$
Moles of H^+ at equilibrium $= (0.10)(2.0) = 0.20$
Moles of A^- at equilibrium $= (0.10)(2.0) = \underline{0.20}$
Total moles of all species $= \overline{2.20}$

Total molality of all species $= 2.20$ moles/kg H_2O

$$\Delta T = 2.20 \text{ (moles/kg } H_2O) \times \frac{1.86°C}{(\text{mole/kg } H_2O)} = 4.09°C$$

$$\text{Freezing point} = -4.09°C$$

PROBLEMS (Solutions of Electrolytes)

10-17. A 0.05 molal solution of NaCl may be considered to be completely ionized in aqueous solution. (a) What is the freezing point of the solution? (b) What is its boiling point?

10-18. A 0.200 molal solution of calcium perchlorate, $Ca(ClO_4)_2$, may be considered to be completely ionized in aqueous solution. (a) What is the freezing point of the solution? (b) What is its boiling point?

10-19. A solution of KNO_3 freezes at $-2.79°C$. What is the molality of the solution if the KNO_3 is considered to be completely ionized?

10-20. A 0.50 molal solution of $HC_2H_3O_2$ freezes at $-0.95°C$. Calculate the apparent degree of ionization.

10-21. 2.00×10^2 g of the acid, HX, of molecular weight 100, are dissolved in 5.000 kg of H_2O. If the acid is 25% ionized, what is the freezing point of the solution?

10-22. A 0.100 molal solution of NH_3 freezes at $-0.193°C$. Calculate the apparent degree of ionization. Consider the reaction to be $NH_3 + H_2O \rightleftharpoons NH_4^+ + OH^-$; ignore the amount of water consumed.

10-23. What is the freezing point of an aqueous 1.00 molal solution of HX which is 85% ionized?

10-24. A 0.15 molal solution of the coordination compound (or complex compound), $Co(NH_3)_4Cl_3$, freezes at $-0.56°C$. In this compound, some of the chloride ions and all the ammonia molecules are bound to the cobalt and are not free to dissociate. The balance of the chloride ions are free to form ions, and a solution of this compound consists of a single positive ion and one or more free chloride ions. From the data given above, determine the number of ions produced per formula unit of the compound.

GENERAL PROBLEMS

10-25. A certain sample of paraffin has a freezing point depression constant of $4.50°C/(mole/kg)$. 20.0 g of a solute is dissolved in 100 g of paraffin. The freezing point is found to be $6.00°C$ below the freezing point of pure paraffin. What is the molecular weight of the solute used in this case?

10-26. A solution of glucose, $C_6H_{12}O_6$, in water freezes at $-4.65°C$. What is the boiling point of this solution at standard pressure?

10-27. Ethyleneglycol, $C_2H_4(OH)_2$, is used as an antifreeze in automobile radiators. How much $C_2H_6O_2$ should be added to each quart of H_2O to give protection to $-28.0°C$? Assume that a quart of H_2O weighs 946 g.

10-28. How much ethyl alcohol, C_2H_5OH, should be added to 1 liter of H_2O to give a solution which freezes at approximately $-20.0°C$?

10-29. A solution of sucrose, $C_{12}H_{22}O_{11}$, with a density of 1.19 g/ml freezes at $-4.09°C$. What is (a) the molality, and (b) the molarity of the solution?

10-30. An ethyl alcohol solution of a substance has a vapor pressure of 31.76 mm of Hg. If the vapor pressure of pure alcohol, C_2H_5OH, is 32.16 mm of Hg, determine the molality of the solution.

10-31. Exactly 460 g of C_2H_5OH is dissolved in 1 quart of H_2O (946 ml). What is the approximate freezing point of the resulting solution?

10-32. 40.0 g of nonelectrolyte dissolved in 100.0 g of H_2O gives a solution that freezes at $-3.72°C$. What is its molecular weight?

10-33. A 20.0 wt percent aqueous solution of a nonelectrolyte freezes at $-18.6°C$. What is the molecular weight of the solute?

10-34. Calculate the boiling point of a solution of 0.05 mole of a nonvolatile solute in 100 g of chloroform. Chloroform boils at $61.3°C$ at 1 atm pressure and has a boiling point constant of 4.70.

10-35. A solution of 25 g of a nonvolatile solute in 250 g of water freezes at $-0.93°C$. What is the molecular weight of the solute?

10-36. A 5 wt percent aqueous solution freezes at $-0.372°C$. What is the molecular weight of the solute?

10-37. An organic compound analyzes $C = 40.7\%$, $H = 5.1\%$, $O = 54.2\%$. A solution of 23.6 g of this compound in 50 g H_2O freezes at $-7.44°C$. What is the molecular formula of the compound?

10-38. Calculate the freezing point depression constant for benzene, C_6H_6 from the following data: 32.0 g of naphthalene, $C_{10}H_8$, in exactly 500 g of C_6H_6 depress the freezing point $2.45°C$.

10-39. (a) Calculate the vapor pressure of a 4.00 wt percent aqueous solution of urea, $CO(NH_2)_2$, at $30°C$. The vapor pressure of H_2O at $30° = 31.82$ mm of Hg. (b) Calculate the osmotic pressure of the solution in (a).

10-40. A solution made by dissolving 16.0 g of an organic compound in 100.0 g of chloroform, $CHCl_3$, freezes at a temperature of $3.13°C$ below the freezing point of pure $CHCl_3$. If the freezing point depression constant for $CHCl_3$ is $4.70°C/(mole/kg)$, what is the molecular weight of the solute?

10-41. (a) If 15.0 g of $CO(NH_2)_2$ in 0.200 kg of a solvent lowers the freezing point of the latter 1.65 degrees, what is the freezing point depression constant for the solvent? (b) Calculate the freezing point depression for a 15.0 wt percent solution of $CO(NH_2)_2$ in this solvent.

10-42. When 5.70 g of an organic compound was dissolved in 100.0 ml of $CHCl_3$, it was found that the solution solidified at a temperature 2.35 degrees below that of pure $CHCl_3$. The density of $CHCl_3$ is 1.50 g/ml and the freezing point depression constant for $CHCl_3$ is $4.70°C/(mole/kg)$. Analysis of the compound gave the following results: 32.0% C; 6.7% H; 18.7% N; 42.7% O. From this data, determine the true molecular formula of the compound.

10-43. The blood of cold-blooded marine animals and fish has about the same osmotic pressure (isotonic) as sea water. If seawater freezes at $-2.30°C$, calculate the osmotic pressure of the blood of marine life at $17°C$.

10-44. The vapor pressure of an aqueous solution of sucrose, $C_{12}H_{22}O_{11}$, at $25°C$ is 95% that of pure H_2O. What is the osmotic pressure of this solution?

10-45. A method recently tested for the desalinization of seawater (removal of salt) is a reverse osmosis process. With a membrane permeable by water but not by dissolved salts, fresh water can be obtained by exerting a pressure on the salt-water side of the membrane. If seawater is considered to be composed of 3.0 wt percent NaCl by weight for the purposes of this calculation, what minimum pressure (in addition to atmospheric pressure) must be applied to the seawater side of a membrane to cause reverse osmosis to occur at 25°C?

CHAPTER 11

Thermal Changes in Chemical Processes

Energy changes accompany all chemical and physical changes. The energy is usually manifested as heat energy. If heat is evolved to the surroundings, the process is said to be exothermic; if heat is absorbed from the surroundings, it is endothermic. In a system insulated from its surroundings, an increase in temperature when a change occurs indicates that the process is exothermic; conversely, a decrease in temperature indicates that it is endothermic.

In considering heat changes, the following fundamental definitions* are important.

The *calorie* (cal) is the principal unit of heat energy employed in chemical problems. It is the energy required to raise the temperature of 1 g of water by 1°C. (from 14.5°C to 15.5°C.) A kilocalorie (kcal) is 1000 cal.

The *specific heat* is the number of calories required to raise the temperature of 1 g of substance 1°C. The specific heat of water is 1 cal/(g-°C), since 1 cal is required to raise the temperature of 1 g of water 1°C.

The *specific heat of vaporization* is the heat energy required to convert 1 g of liquid at its boiling point to 1 g of vapor at the same temperature. The heat of vaporization of water is about 540 cal/g.

The *specific heat of condensation* is the energy evolved when 1 g of vapor condenses to liquid at the boiling point. It is numerically equal but opposite in sign to the heat of vaporization.

The *specific heat of fusion* (melting) is the heat energy required to

---

* Many of the following terms are defined as specific heats or the heats per gram. The corresponding molar heats may be obtained by multiplying the specific heats by the number of grams per mole.

convert 1 g of solid to 1 g of liquid at the melting point. The heat of fusion for ice is about 80 cal/g.

The *specific heat of solidification* is the heat evolved when 1 g of liquid solidifies. It is numerically equal, but opposite in sign, to the heat of fusion.

The *molar heat of formation* is the energy change when 1 mole of a compound is formed from its elements. Heats of formation for a few compounds are shown in Table 11-1.

The *heat of reaction* (also called "enthalpy" of reaction, see Chapter 20) is the heat evolved or absorbed for a given chemical change when that number of moles shown in the balanced equation reacts. This heat may be considered to be the difference in the heats of formation of products and reactants at the same temperature and pressure. Two conventions for showing heat changes in reactions are employed. The student should consult his text to determine which is preferred by his instructor.

1. Energy changes may be shown as an adjunct to the chemical equation. For the reaction, $A + B \rightarrow C + D - \Delta H$, the symbol ΔH indicates the heat change accompanying the chemical reaction. ΔH has negative values for exothermic reactions and positive values for endothermic reactions. The exothermic reaction of CO and O_2 may be represented as

$$2\,CO + O_2 \rightarrow 2\,CO_2 - \Delta H$$

or as

$$2\,CO + O_2 \rightarrow 2\,CO_2 + 135.2 \text{ kcal}$$

It should be obvious that $\Delta H = -135.2$ kcal. Fractional coefficients are permitted where it is desirable to show the energy change per mole of some reactant or product, as in the equation

$$CO + \tfrac{1}{2} O_2 \rightarrow CO_2 + 67.6 \text{ kcal}$$

The endothermic process of formation of NO from its elements where $\Delta H = +43$ kcal may be represented

$$N_2 + O_2 \rightarrow 2\,NO - 43 \text{ kcal}$$

2. Alternatively, energy changes may be shown following the equation to which they apply. Thus for the equations above,

$$2\,CO + O_2 \rightarrow 2\,CO_2 \qquad \Delta H = -135.2 \text{ kcal}$$
$$CO + \tfrac{1}{2} O_2 \rightarrow CO_2 \qquad \Delta H = -67.6 \text{ kcal}$$
$$N_2 + O_2 \rightarrow 2\,NO \qquad \Delta H = 43 \text{ kcal}$$

This latter system will be used in the illustrations to follow.

Hess's law of constant heat summation states that the heat change in a chemical reaction is the same whether the reaction proceeds in one or more steps. In other words, the heat change is independent of the path. Heats of reaction may be calculated from heats of formation of reactants and products. The heat of reaction is simply the heats of formation of all reactants subtracted from heats of formation of all products, if all the heats have been measured at constant temperature. The heat of formation of an element is taken as zero.

The *Dulong-Petit* rule states that the product of the specific heat of a metallic element and its atomic weight is approximately a constant, which has a numerical value of 6.4. This is an approximate rule only, but it is useful in estimating approximate atomic weights.

$$\text{Approximate atomic weight} = \frac{6.4}{\text{specific heat}}$$

If the equivalent weight of the element is known, the precise atomic weight may be calculated from the relationship

$$\text{Atomic weight} = \text{Equivalent weight} \times \text{Valence}$$

The approximate atomic weight from the Dulong-Petit rule allows one to obtain the valence. How this is accomplished is explained in Example 11-e.

Example 11-a. Calculate the quantity of heat required to raise the temperature of 100 g of H_2O from 0°C to 100°C if the specific heat of H_2O is 1 cal/(g-°C).

Solution. Heat required = mass of water × temperature change × specific heat.

$$\text{Heat required} = 100 \text{ g} \times (100 - 0)°C \times \frac{1 \text{ cal}}{(g\text{-}°C)}$$

$$= 10,000 \text{ cal} = 10 \text{ kcal}$$

Example 11-b. The heat of combustion of a sample of coal is 5500 cal/g. What quantity of this coal must be burned to heat exactly 500 g of water from 20°C to 90°C?

Solution. Heat required to increase the temperature of water = heat liberated by burning coal.

Heat required = mass of water × temperature change × specific heat.

$$\text{Heat required} = 500 \text{ g} \times (90°C - 20°C) \times \frac{1 \text{ cal}}{(g\text{-}°C)} = 35,000 \text{ cal}$$

Heat liberated = mass of coal × heat of combustion

$$\text{Mass of coal required} = \frac{\text{heat liberated}}{\text{heat of combustion}} = \frac{35{,}000 \text{ cal}}{5500 \text{ cal/g}} = 6.4 \text{ g}$$

Example 11-c. Calculate the quantity of heat required to convert exactly 20 g of $H_2O(s)$ at 0°C to 20 g of $H_2O(g)$ at 100°C. Heat of fusion = 80.0 cal/g; heat of vaporization = 540 cal/g.

Solution. This requires three steps:

1. To melt 20 g of ice at 0°C requires 20 g × 80 cal/g = 1600 cal
2. To heat 20 g of liquid from 0°C to 100°C requires

$$20 \text{ g} \times 100°C \times \frac{1 \text{ cal}}{(g\text{-}°C)} = 2000 \text{ cal}$$

3. To vaporize 20 g liquid at 100°C requires 20 g × 540 cal/g = 10,800 cal

$$\text{Total heat required} = 14{,}400 \text{ cal} = 14.4 \text{ kcal.}$$

Example 11-d. 50 g of $H_2O(s)$ at 0°C is added to 100 g of $H_2O(l)$ at 80°C. What is the temperature of the $H_2O(l)$ when all the $H_2O(s)$ is melted? Assume that no heat is lost to the surroundings.

Solution. Let X = final temperature of liquid in °C.

The amount of heat needed to melt the ice and then to warm that liquid to the temperature X, must be obtained by the cooling of the water originally at 80° to the temperature X. Therefore, the heat gained by water originally in ice = heat lost by water originally at 80°C.

$$50 \text{ g} \times 80 \text{ cal/g} + 50 \text{ g} \times (X - 0)°C \times \frac{1 \text{ cal}}{(g\text{-}°C)}$$

$$= 100 \text{ g} \times (80 - X)°C \times \frac{1 \text{ cal}}{(g\text{-}°C)}$$

$$4000 \text{ cal} + 50X \text{ cal} = 8000 \text{ cal} - 100X \text{ cal}$$

$$150X \text{ cal} = 4000 \text{ cal}$$

$$X = \frac{4000 \text{ cal}}{150 \text{ cal}} = 26.7$$

The final temperature is therefore 26.7°C.

Example 11-e. The specific heat of zinc is 0.096 cal/(g-°C). (a) Determine the approximate atomic weight of zinc from the Dulong-Petit rule. (b) Determine a precise atomic weight for zinc from the fact that 80.34 g of zinc combine with 19.66 g of oxygen to form zinc oxide.

Solution. (a) Approximate atomic weight = $\dfrac{6.4}{0.096}$ = 67.

(b) The equivalent weight of zinc will be the weight which combines with 8.000 g of oxygen, hence

$$\text{Equiv weight} = 8.000 \text{ g O/equiv} \times \frac{80.34 \text{ g Zn}}{19.66 \text{ g O}} = 32.69 \text{ g Zn/equiv}$$

The whole number which multiplied by the equivalent weight most nearly gives 67 is 2; therefore

$$\text{Atomic weight} = 2 \times 32.69 = 65.38$$

Table 11-1   Heats of Formation

Compound	ΔH kcal/mole	Compound	ΔH kcal/mole
$H_2O(g)$	−57.8	$HCl(g)$	−22.0
$H_2O(l)$	−68.3	$HF(g)$	−64.5
$CO_2(g)$	−94.0	$H_2S(g)$	−5.3
$CH_3Cl(g)$	−19.6	$NO_2(g)$	8.1
$C_2H_2(g)$	54.3	$NH_3(g)$	−10.9
$C_2H_4(g)$	12.4	$CO(g)$	−26.4
$C_2H_6(g)$	−20.2		

Example 11-f. Calculate the heat of reaction for

$$CO(g) + \tfrac{1}{2} O_2(g) \rightarrow CO_2(g)$$

from the molar heats of formation of CO and CO_2.

$$C(s) + O_2(g) \rightarrow CO_2(g) \qquad \Delta H_1 = -94.0 \text{ kcal} \qquad (1)$$
$$C(s) + \tfrac{1}{2} O_2(g) \rightarrow CO(g) \qquad \Delta H_2 = -26.4 \text{ kcal} \qquad (2)$$

Solution. The heat of reaction may be obtained by subtracting the heats of formation *per mole* of reactants from products, that is, heat of reaction = ΔH = heat of formation of CO_2 − (heat of formation of CO + $\tfrac{1}{2}$ heat of formation of O_2) = −94.0 kcal − (−26.4 kcal + 0 kcal) = −67.6 kcal.

Or, we might arrange equations (1) and (2) such that on addition we obtain the desired reaction; for example, turn equation (2) around, as in equation (2a), and add it to equation (1). ΔH of the desired reaction is then the sum of the individual heats of reaction.

$$C(s) + O_2(g) \rightarrow CO_2(g) \qquad \Delta H_1 = -94.0 \text{ kcal} \qquad (1)$$

$$CO(g) \rightarrow \tfrac{1}{2} O_2(g) + C(s) \qquad \Delta H_{2a} = 26.4 \text{ kcal} \qquad (2a)$$

$$CO(g) + \tfrac{1}{2} O_2(g) \rightarrow CO_2(g) \qquad \Delta H = -67.6 \text{ kcal}$$

BOND ENERGIES

Chemical bonds are the forces which hold atoms together. Energy is required to break these bonds and overcome the forces of attraction. This energy is termed the *bond* energy. It is usually expressed as kcal/mole—the energy required to break Avogadro's number of bonds.

Bond energies may be calculated from heats of formation. Since the heat of formation is the heat change in forming a compound from its elements, it follows that this is also a measure of the difference in energies of bonds in the compound and bonds in the element. If the latter are known, bond energy in compounds may be calculated.

Example 11-g. Determine the energy of the H—Cl bond. Given

$$H—H + Cl—Cl \rightarrow 2 H—Cl \qquad \Delta H = -44 \text{ kcal}$$

$$H + H \rightarrow H—H \qquad \Delta H = -104 \text{ kcal}$$

and

$$Cl + Cl \rightarrow Cl—Cl \qquad \Delta H = -58 \text{ kcal}$$

Table 11-2   Bond Energies

Bond	kcal/mole	Bond	kcal/mole
H—H	104	C—H	99
Cl—Cl	58	C—Br	66
Br—Br	46	N—H	93
F—F	37	F—Cl	61
C—C	83	I—Cl	50
O—O	33	F—H	135
O=O	120	Br—H	88
O—H	111	Cl—H	103
C—O	84	N≡N	226
C=C	146	I—I	36
C≡C	194		
C—Cl	78		
C=O	173		

Adding the above

$$2\,H + 2\,Cl \rightarrow 2\,H\!-\!Cl \qquad \Delta H = -206\ \text{kcal}$$

For the H—Cl bond $\Delta H = \frac{206}{2} = 103$ kcal/mole
A few bond energies are tabulated in Table 11-2 opposite.
Where bond energies are known, heats of formation of compounds may be calculated.

Example 11-h. Calculate the heat of formation of HF (see Table 11-2).

Solution.

To break one H—H bond requires 104 kcal.
To break one F—F bond requires 37 kcal.
Energy released in forming 2 H—F bonds $= 2 \times 135 = 270$ kcal
Net energy released per mole of HF formed $= \Delta H$

$$= \frac{270 - (104 + 37)}{2} = 64.5\ \text{kcal}$$

or $\Delta H = -64.5$ kcal.

Example 11-i. Calculate the heat of formation of CH_4, i.e.,

$$C(s) + 2\,H_2(g) \longrightarrow CH_4(g) \qquad \Delta H = ?$$

using the table of bond energies and given

$$C(s) \longrightarrow C(g) \qquad \Delta H = 171.7\ \text{kcal}$$

Solution. From Table 11-2,

Energy released in forming 4 C—H bonds

$$= 4 \times (-99) = -396\ \text{kcal}$$

and

to break 2 H—H bonds requires $2 \times 104 = 208$ kcal,

or

1. $\qquad C(g) + 4\,H(g) \longrightarrow CH_4(g) \qquad \Delta H = -396$

2. $\qquad\qquad 2\,H_2(g) \longrightarrow 4\,H(g) \qquad \Delta H = 208$

and

3. $\qquad\qquad C(s) \longrightarrow C(g) \qquad \Delta H = 171.7$ kcal

On adding 1, 2, and 3

$$C(s) + 2\,H_2(g) \longrightarrow CH_4(g) \qquad \Delta H = -16.3\ \text{kcal}$$

GENERAL PROBLEMS

11-1. The specific heat of Bi metal is 0.0300 cal/(g-°C). What quantity of heat would be required to raise the temperature of exactly 1 kg of Bi from 20° to its melting point at 850°C?

11-2. Estimate the specific heat of Sn which has an atomic weight of 119.

11-3. Determine the precise atomic weight of Fe from the facts that 1.000 g of Fe combines with 0.4296 g of O_2 and that the specific heat of Fe is 0.12 cal/(g-°C). An equivalent weight of Fe is that weight which combines with 8.000 g of O_2.

11-4. Treatment of metallic Sn with concentrated HNO_3 results in the formation of an oxide of Sn which analyzes 78.76% Sn and the remainder O. If the specific heat of Sn is 0.054 cal/(g-°C), calculate a precise atomic weight for Sn.

11-5. From the following data, determine a precise atomic weight for lanthanum, La. The specific heat is 0.045 cal/(g-°C). When treated with excess sulfuric acid, 0.5000 g of La produces 0.01083 g of H_2.

11-6. Determine the quantity of heat in kilocalories necessary to convert 50 g of $H_2O(s)$ at 0°C to $H_2O(g)$ at 100°C.

11-7. The heat of combustion of a bituminous coal is 6000 cal/g. What quantity of this coal on burning would supply sufficient heat to convert 100.0 lb of $H_2O(s)$ at 0°C to $H_2O(g)$ at 100°C?

11-8. One hundred grams of $H_2O(s)$ at 0°C is added to 300 g of $H_2O(l)$ at a temperature of 60°C. Assuming no transfer of heat to surroundings, what is the temperature of the $H_2O(l)$ after the ice is melted?

11-9. The following are physical properties of Hg: f.p. −39°C, b.p. 357°C, heat of fusion 2.77 cal/g, heat of vaporization 70.80 cal/g, specific heat 0.300 cal/(g-°C). Calculate the number of calories required to convert 40.1 g of $Hg(s)$ at its freezing point to $Hg(g)$ at the boiling point.

11-10. The specific heat of Tl is 0.032 cal/(g-°C). Tl forms an oxide in which 34.06 g of thallium are combined with 4.000 g of oxygen. What is the exact atomic weight of thallium and its valence?

11-11. Neglect heat gained from or lost to the surroundings in this problem. 50 g of $H_2O(s)$ at 0°C is added to a quantity of $H_2O(l)$ which has a temperature of 70°C. The final temperature reached is 10°C. To what quantity of $H_2O(l)$ was the ice added?

11-12. Calculate the heat of reaction for

$$2\,PbS(s) + 3\,O_2(g) \rightarrow 2\,PbO(s) + 2\,SO_2(g)$$

from the following heats of formation (ΔH): PbS = −22.3 kcal/mole, PbO = −52.5 kcal/mole, SO_2 = −70.9 kcal/mole, O_2 = 0 kcal/mole.

11-13. Calculate the heat of reaction for

$$PbS(s) + PbSO_4(s) \rightarrow 2\,Pb(s) + 2\,SO_2(g)$$

The heat of formation of $PbSO_4$ is -218.5 kcal/mole. See exercise 11.12 for other heats of formation.

11-14. Calculate the heat of reaction for

$$2\,SO_2(g) + O_2(g) \rightarrow 2\,SO_3(g)$$

The heats of formation (ΔH) are: $SO_2 = -70.9$ kcal/mole, $SO_3 = -94.5$ kcal/mole, and $O_2 = 0$ kcal/mole.

11-15. From the following data, calculate the molar heat of formation of solid PCl_5.

$$\begin{array}{ll} P_4(s) + 6\,Cl_2(g) \rightarrow 4\,PCl_3(l) & \Delta H = -304.0\text{ kcal} \\ PCl_3(l) + Cl_2(g) \rightarrow PCl_5(s) & \Delta H = -32.8\text{ kcal} \end{array}$$

11-16. Calculate the molar heat of formation of methyl alcohol,

$$C(s) + 2\,H_2(g) + \tfrac{1}{2}\,O_2(g) \rightarrow CH_3OH(l)$$

from the following:

$$\begin{array}{ll} CH_3OH(l) + \tfrac{3}{2}\,O_2(g) \rightarrow CO_2(g) + 2\,H_2O(l) & \Delta H = -170.9\text{ kcal} \\ C(s) + O_2(g) \rightarrow CO_2(g) & \Delta H = -94.0\text{ kcal} \\ H_2(g) + \tfrac{1}{2}\,O_2(g) \rightarrow H_2O(l) & \Delta H = -68.3\text{ kcal} \end{array}$$

11-17. A piece of Fe with a mass of 500 g at a temperature of 300°C is added to a kilogram of $H_2O(l)$ at a temperature of 20°C. What is the final equilibrium temperature of the $H_2O(l)$ and the Fe? The specific heat of Fe is 0.12 cal/(g-°C) and that for $H_2O(l)$ is 1.0 cal/(g-°C).

11-18. 95 g of $H_2O(s)$ at 0°C is added to exactly 100 g of $H_2O(l)$ at 60°C. When the temperature of the mixture is 0°C, what mass of $H_2O(s)$ is still present?

11-19. How many grams of ice at 0°C is necessary to cool 1 gallon of lemonade (about 3.8 kg) from 20°C to 5°C? Assume that the specific heat of lemonade is the same as H_2O.

11-20. From the following equations, determine the molar heat of formation of $HNO_2(aq)$:

$$\begin{array}{ll} NH_4NO_2(aq) \rightarrow N_2(g) + 2\,H_2O(l) & \Delta H = -76.5\text{ kcal} \\ NH_3(aq) + HNO_2(aq) \rightarrow NH_4NO_2(aq) & \Delta H = -9.0\text{ kcal} \\ 2\,NH_3(aq) \rightarrow N_2(g) + 3\,H_2(g) & \Delta H = +40.6\text{ kcal} \\ 2\,H_2(g) + O_2(g) \rightarrow 2\,H_2O(l) & \Delta H = -136.6\text{ kcal} \end{array}$$

11-21. From the following equations, determine the heat of formation of HBr.

$$H_2(g) + Br_2(g) \rightarrow 2\,HBr(g)$$

$$\begin{array}{ll} 2\,KBr(aq) + Cl_2(g) \rightarrow 2\,KCl(aq) + Br_2(aq) & \Delta H = -23.0\text{ kcal} \\ H_2(g) + Cl_2(g) \rightarrow 2\,HCl(aq) & \Delta H = -78.4\text{ kcal} \\ KOH(aq) + HCl(aq) \rightarrow KCl(aq) + H_2O & \Delta H = -13.7\text{ kcal} \\ KOH(aq) + HBr(aq) \rightarrow KBr(aq) + H_2O & \Delta H = -13.7\text{ kcal} \\ Br_2(g) \rightarrow Br_2(aq) & \Delta H = -1.0\text{ kcal} \\ HBr(g) \rightarrow HBr(aq) & \Delta H = -20.0\text{ kcal} \end{array}$$

11-22. The heat of formation of $H_2O(g)$ is -57.8 kcal/mole. From Table 11-2 which gives the bond energies of H_2 and O_2, determine the bond energy of O—H.

11-23. From thermal data in Tables 11-1 and 11-2 and in the example problems, determine ΔH for the following processes:

(a) $2\ C(g) + 4\ H(g) \rightarrow C_2H_4(g)$

(b) $3\ C(s) + 4\ H_2(g) \rightarrow C_3H_8(g)$

(c) $\frac{1}{2}\ I_2(g) + \frac{1}{2}\ Cl_2(g) \rightarrow ICl(g)$

(d) $C_2H_2(g) + 2\ H_2(g) \rightarrow C_2H_6(g)$

(e) $\frac{1}{2}\ N_2(g) + 1\frac{1}{2}\ H_2(g) \rightarrow NH_3(g)$

(f) $\frac{1}{2}\ H_2(g) + \frac{1}{2}Br_2(g) \rightarrow HBr(g)$

(g) $F_2(g) + Cl_2(g) \rightarrow 2\ FCl(g)$

CHAPTER 12

Titrations

Frequently it is convenient to determine the concentration of a solution by allowing it to react with a second solution of known concentration. A given volume of the first solution is measured out and the second solution is added from a buret until equivalent amounts of the two solutions are present. The equivalence point or *end point* of the reaction may be determined by using an indicator. The process is termed *titration*.

Suppose solution A is being titrated with solution B. The end point of the titration occurs when stoichiometric or equivalent quantities of A and B are present; that is,

$$\text{Equiv of } A = \text{Equiv of } B \qquad (1)$$

Since Normality $= \dfrac{\text{Equiv}}{\text{Liters}}$ (see Chapter 9), we may substitute in equation (1) and obtain

$$\text{Normality of } A \times \text{Liters of } A = \text{Normality of } B \times \text{Liters of } B \qquad (2)$$

or

$$\text{Normality of } A \times \text{Volume of } A = \text{Normality of } B \times \text{Volume of } B$$

Any units of volume may be used as long as they are the same for both reagents. Equation (2) is very useful in solving titration problems for either unknown volumes or concentrations.

Since, by definition, Equiv of $A = \dfrac{\text{Grams of } A}{\text{EW of } A}$, problems in which a weight of reagent A is titrated with a solution of reagent B can easily be solved. Since at the equivalence point, Equiv of $A = $ Equiv of B, it follows that

$$\frac{\text{Grams of } A}{\text{EW of } A} = \text{Normality of } B \times \text{Liters of } B \qquad (3)$$

Calculations involving titrations are conveniently classified into two categories, (a) those which involve oxidation and reduction, and (b) those which do not. The need for this division is that the equivalent weight of a reactant or product depends upon the reaction. For reactions that do not involve oxidation-reduction, the equivalent weights of acids, bases, and salts are defined in the manner previously discussed in Chapter 9, Section C. For an oxidation-reduction process, the equivalent weight depends upon the oxidation number change or the number of electrons transferred per mole of substance. Such equivalent weights will be discussed in Section B of this chapter. Once the equivalent weights have been ascertained, the calculations for both types of reactions parallel one another.

The important thing to remember when dealing with equivalents is: "In any reaction where A reacts with B, whatever number of equivalents of A enter into the reaction, the same number of equivalents of B will be required." One equivalent of A reacts with one equivalent of B; five equivalents of A react with five equivalents of B; 0.235 equivalents of A react with 0.235 equivalents of B, etc.

A.   TITRATIONS (Nonoxidation-Reduction Reactions)

As stated in Chapter 9, Section C, the equivalent weight of an acid (or base) is that number of grams that provides (or reacts with) one Avogadro's number of H^+. The equivalent weight of a salt is that number of grams furnishing one Avogadro's number of positive charges or negative charges; it is also that number of grams furnishing ions equivalent to one Avogadro's number of H^+. The equivalent weights of some typical acids, bases, and salts are: HCl, 36.5/1; NaOH, 40/1; H_2SO_4, 98/2; NH_3, 17/1; $Ca(OH)_2$, 74/2; $BaBr_2$, 297/2; AgCl, 143/1; etc.

Example 12-a. What volume of 0.10 N H_2SO_4 would be required to neutralize exactly 50 ml of 0.050 N NaOH?

Solution.

Volume of H_2SO_4 × Normality of H_2SO_4

$$= \text{Volume of NaOH} \times \text{Normality of NaOH}$$

Volume of H_2SO_4 × 0.10 equiv/liter = 50 ml × 0.050 equiv/liter

$$\text{Volume of } H_2SO_4 = \frac{50 \text{ ml} \times 0.050 \text{ equiv/liter}}{0.10 \text{ equiv/liter}} = 25 \text{ ml}$$

Example 12-b. 35 ml of standard base (0.25 N) is used to neutralize 15 ml of a solution of an unknown acid. What is the concentration of the unknown acid?

> *Solution.*
>
> Normality of acid × 15 ml = 0.25 equiv/liter × 35 ml
>
> $$\text{Normality of acid} = \frac{(0.25 \text{ equiv/liter})(35 \text{ ml})}{15 \text{ ml}} = 0.58 \text{ equiv/liter}$$

Example 12-c. In the Kjeldahl method for determination of nitrogen, the sample is digested with hot concentrated H_2SO_4 to oxidize organic matter and convert nitrogen present to NH_4^+. NaOH is then added and (1) NH_3 is distilled from the solution, (2) NH_3 is absorbed in a measured excess of a standard solution of an acid, and (3) the acid in excess of that neutralized by NH_3 is then titrated with a standard basic solution. These reactions are illustrated as follows:

(1) $\quad NH_4^+ + OH^- \xrightarrow{\Delta} NH_3 \uparrow + H_2O$

(2) $\quad NH_3 + H^+ \rightarrow NH_4^+$

(3) $\quad H^+ \text{ (excess)} + OH^- \rightarrow H_2O$

From these data the weight of NH_3 can be calculated, and from this information the percentage of nitrogen in the original sample can be ascertained. Study the following problem and its solution.

The NH_3 obtained from a 1.40 g sample of an organic mixture is absorbed in exactly 70 ml of 0.10 N HNO_3. Exactly 20 ml of 0.15 N NaOH solution is required to neutralize the excess acid. What is the percentage of nitrogen in the sample?

> *Solution.* 20 ml of 0.15 N NaOH contains (0.020 liter) (0.15 equiv/liter) = 0.0030 equiv.
>
> 70 ml of 0.10 N HNO_3 contains (0.070 liter) (0.10 equiv/liter) = 0.0070 equiv.
>
> Equivalents of NH_3 distilled = 0.0070 − 0.0030 = 0.0040 equiv.
>
> Grams of NH_3 = (0.0040 equiv)(17 g/equiv) = 0.068 g.
>
> Grams of N = 0.068 g NH_3 × 14 g N/17 g NH_3 = 0.056 g.
>
> Percentage N in sample = (0.056 g/1.40 g) × 100 = 4.0%.

PROBLEMS (Nonoxidation-Reduction Reactions)

12-1. 50 ml of an unknown acid solution is titrated to neutrality with 20 ml of NaOH solution which has a normality of 0.40 equiv/liter. What is the normality of the unknown acid solution?

12-2. How many milliliters of 0.1 N NaOH is required to neutralize 60 ml of 0.2 N H_2SO_4?

12-3. How many milliliters of 0.1 N NaOH is required to neutralize 40 ml of 0.2 M H_2SO_4?

12-4. 25 ml of a solution of H_2SO_4 requires exactly 40 ml of 0.20 N NaOH for neutralization. (a) What is the normality of the H_2SO_4 solution? (b) What is its molarity? (c) How many grams of H_2SO_4 are contained in the 25 ml?

12-5. 40.0 ml of HCl is exactly neutralized by 20.0 ml of NaOH solution. The resulting neutral solution is evaporated to dryness and the residue is found to have a mass of 0.117 g. Calculate the normality of the HCl and NaOH solutions.

12-6. A series of Kjeldahl determinations gave the following data. In each case determine the percentage of nitrogen in the sample (see Example 12-c).

Sample	Weight of sample	Volume of 0.1000 N HNO_3 used	Volume of 0.1000 N NaOH to neutralize excess acid
A	3.50 g	100.0 ml	50.0 ml
B	0.35	50.0	40.0
C	0.936	67.00	22.00
D	0.500	25.00	13.00

12-7. The equivalent weights of organic acids are often determined by titration with a standard solution of base. Determine the equivalent weight of benzoic acid if 0.305 g of the latter requires 25.0 ml of 0.100 N NaOH for neutralization.

12-8. Determine the equivalent weight of succinic acid if 0.738 g requires 125 ml of 0.100 N base for neutralization.

12-9. Chloride ion in a sample may be determined by titration with a standard $AgNO_3$ solution, or alternatively, Ag^+ may be determined by titration with a standard Cl^- solution:

$$Cl^- + Ag^+ \rightarrow \underline{AgCl}$$

(a) If 20 ml of 0.1 N Ag NO_3 is required to react with 50 ml of a chloride ion solution, what is the normality of the latter? (b) 0.100 g of a silver coin is dissolved in HNO_3 and the resulting solution is titrated with 0.050 N NaCl solution. If 16.7 ml of the NaCl solution is required, what is the percentage of silver in the coin?

B.   TITRATIONS   (Oxidation-Reduction)

In oxidation-reduction reactions the gram equivalent weight is defined as that weight which reacts with or supplies one Avogadro's number of electrons (also called one equivalent or one mole of electrons). The equivalent weight is therefore determined by the loss or gain of electrons per formula unit; for example, when permanganate ion, MnO_4^-, is used as an oxidizing agent in acid solution, it is reduced to Mn^{++}, which requires 5 electrons per MnO_4^- ion. If $KMnO_4$ is used as the source of MnO_4^-, then 1 mole of $KMnO_4$ would accept 5 equivalents of electrons, and an equivalent weight would be $\frac{1}{5}$ of a formula weight. Furthermore, a 1 normal solution of $KMnO_4$ would contain $\frac{1}{5}$ mole per liter of solution. When MnO_4^- is used for oxidation in basic solution, it is reduced to MnO_2, and only 3 electrons are required per formula unit. The equivalent weight in this case is $\frac{1}{3}$ of the formula weight, and a 1 normal solution would contain $\frac{1}{3}$ mole per liter.

In a similar manner, the equivalent weight of a reducing agent is determined; for example, oxalate ion, $C_2O_4^{--}$, is oxidized to CO_2, according to the half-reaction, $C_2O_4^{--} \rightarrow 2\,CO_2 + 2\,e^-$. Since there are 2 electrons produced per $C_2O_4^{--}$, an equivalent weight is $\frac{1}{2}$ of the formula weight. A 1 normal solution of sodium oxalate will contain $\frac{1}{2}$ mole per liter of solution.

Example 12-d. Dichromate ion in acid solution will oxidize stannous ion according to the equation

$$3\,Sn^{++} + 14\,H^+ + Cr_2O_7^{--} \rightarrow 3\,Sn^{++++} + 2\,Cr^{+++} + 7\,H_2O$$

(a) If $SnCl_2$ is the source of Sn^{++}, how many grams of $SnCl_2$ would be contained in 2.00 liters of 0.100 N solution? (b) If $K_2Cr_2O_7$ is the source of $Cr_2O_7^{--}$, what is the normality of a solution containing 4.9 g of $K_2Cr_2O_7$ in 0.100 liter of solution?

Solution. (a) Since Sn^{++} is oxidized from $+2$ to $+4$ and gives up 2 electrons per Sn^{++}, an equivalent weight of $SnCl_2$ would be $\frac{1}{2}$ of the formula weight, that is, $\frac{1}{2}$ of 190 or 95.

Number of grams of $SnCl_2$ = Normality

$\times$ Equiv wt $SnCl_2$ $\times$ Liters $SnCl_2$

Grams of $SnCl_2$ = 0.100 (equiv/liter)

$\times$ 95 (g/equiv) $\times$ 2.00 liters = 19.0 g

(b) Since each chromium atom is reduced from $+6$ to $+3$, 6 electrons are added for every $Cr_2O_7^{--}$ ion. The equivalent weight of $K_2Cr_2O_7$ is one-sixth the formula weight; one-sixth of 294 or 49. Substituting in the equation

$$\text{Normality} = \frac{\text{Grams of } K_2Cr_2O_7}{\text{Equiv wt of } K_2Cr_2O_7 \times \text{Liters of solution}}$$

$$= \frac{4.9 \text{ g}}{49 \text{ (g/equiv)} \times 0.100 \text{ liter}} = 1.00 \text{ equiv/liter}$$

Example 12-e. Arsenites react with permanganates in acid solution as shown by

$$5 \text{ AsO}_3^{---} + 6 \text{ H}^+ + 2 \text{ MnO}_4^- \rightarrow 5 \text{ AsO}_4^{---} + 2 \text{ Mn}^{++} + 3 \text{ H}_2\text{O}$$

If 56 ml of 0.10 N $KMnO_4$ reacts with 28 ml of Na_3AsO_3 solution, (a) what is the normality of the sodium arsenite solution, (b) what weight of Na_3AsO_3 in grams is contained in the 28 ml of solution?

Solution. (a) Substitute in the relationship

Normality of $KMnO_4$ × Volume of $KMnO_4$
$$= \text{Normality of } Na_3AsO_3 \times \text{Volume of } Na_3AsO_3$$

Making the proper substitutions,

$$(0.10 \text{ equiv/liter})(56 \text{ ml}) = \text{Normality of } Na_3AsO_3 \ (28 \text{ ml})$$

$$\text{Normality of } Na_3AsO_3 \text{ solution} = \frac{(0.10 \text{ equiv/liter})(56 \text{ ml})}{28 \text{ ml}}$$

$$= 0.20 \text{ equiv/liter}$$

(b) The formulation,

$$\text{Normality} = \frac{\text{Grams of } Na_3AsO_3}{\text{Equiv wt of } Na_3AsO_3 \times \text{Liters of solution}}$$

relates the quantities involved. The formula weight of Na_3AsO_3 is 192 and since the oxidation number of arsenic increases from $+3$ to $+5$, the equivalent weight is $192/2 = 96$. Making the proper substitutions,

$$0.20 \text{ equiv/liter} = \frac{\text{Grams } Na_3AsO_3}{96 \text{ g/equiv} \times 0.028 \text{ liter}}$$

Grams of Na_3AsO_3
$$= (0.20 \text{ equiv/liter})(96 \text{ g/equiv})(0.028 \text{ liter}) = 0.54 \text{ g}$$

Example 12-f. Exactly 2 g of a sample of iron ore, after conversion of all Fe^{+++} to Fe^{++}, is titrated with potassium dichromate solution. If 53.5 ml of 0.400 N potassium dichromate solution is required, what is the percentage of Fe in the original sample? The equation for the reaction during titration is

$$6\ Fe^{++} + Cr_2O_7^{--} + 14\ H^+ \rightarrow 6\ Fe^{+++} + 2\ Cr^{+++} + 7\ H_2O$$

Solution. The equivalent weight of iron is 55.9, its atomic weight, since ferrous ion changes only 1 in oxidation number during the titration.

Number of equivalents of $K_2Cr_2O_7$ used = Normality × Liters

$$= 0.400\ \text{equiv/liter} \times 0.0535\ \text{liter} = 0.0214\ \text{equiv}$$

The number of equivalents of iron titrated is equal to the number of equivalents of $K_2Cr_2O_7$ used.

Weight in grams of Fe titrated

$$= 0.0214\ \text{equiv} \times 55.9\ \text{g/equiv} = 1.20\ \text{g}$$

$$\text{Percentage Fe in original sample} = \frac{\text{Weight of Fe} \times 100}{\text{Weight of sample}}$$

$$= \frac{1.20\ \text{g}}{2.00\ \text{g}} \times 100 = 60.0\%$$

PROBLEMS (Oxidation-Reduction)

12-10. Consider the reaction:

$$Sn^{++} + 2\ Ce^{++++} \rightarrow Sn^{++++} + 2\ Ce^{+++}$$

(a) How many grams of $SnCl_2$ would be contained in 0.125 liter of 0.250 N $SnCl_2$? (b) How many grams of $Ce(SO_4)_2$ would be contained in 0.350 liter of 1.25 N $Ce(SO_4)_2$? (c) What would be the normality of a solution containing 23.7 g of $SnCl_2$ in 375 ml of solution? (d) What would be the normality of a solution containing 37.4 g of $Ce(SO_4)_2$ in 0.450 liter?

12-11. Consider the reaction in aqueous solution:

$$H_2S + I_2 \rightarrow 2\ I^- + 2\ H^+ + S$$

(a) How many grams of I_2 would there be in 0.50 liter of a solution that is 0.20 N? (b) What volume of 0.20 N I_2 solution would be required to react with 0.170 g of H_2S?

12-12. Balance each of the following reactions. Then, for each oxidizing agent and each reducing agent, determine the number of equivalents per mole.
(a) $FeSO_4 + KMnO_4 + H_2SO_4 \rightarrow Fe_2(SO_4)_3 + K_2SO_4 + MnSO_4 + H_2O$
(b) $U(SO_4)_2 + KMnO_4 + KOH \rightarrow UO_2SO_4 + MnO_2 + K_2SO_4 + H_2O$

(c) $(NH_4)_4Fe(CN)_6 + Ce(SO_4)_2$
$$\rightarrow Ce_2(SO_4)_3 + (NH_4)_3Fe(CN)_6 + (NH_4)_2SO_4$$

(d) $H_2S + HNO_3 \rightarrow S + NO + H_2O$

(e) $KIO_3 + HCl + TiCl_3 \rightarrow KI + TiCl_4 + H_2O$

12-13. For each oxidizing agent and reducing agent in exercise 12-12, determine the weight in grams of each contained in 0.50 liter of 0.20 N solution.

12-14. For each oxidizing agent and reducing agent in exercise 12-12, determine the normality of a solution containing 0.100 g of solute in exactly 250 ml of solution.

12-15. Ceric ion, Ce^{++++}, may be used to titrate reducing solutions such as Fe^{++}.

$$Fe^{++} + Ce^{++++} \rightarrow Fe^{+++} + Ce^{+++}$$

A 0.50-g sample of an iron ore is dissolved in acid and all the Fe^{+++} reduced to the Fe^{++} state. Exactly 50 ml of 0.020 N Ce^{++++} solution is required to reoxidize all the Fe^{++}. What is the percentage of Fe in the sample?

GENERAL PROBLEMS

12-16. 10.0 ml of a solution of HBr neutralizes 15.0 ml of a solution of KOH. On evaporation the mixture yields 2.38 g of KBr. (a) How many equivalents of KBr are obtained? (b) What is the normality of the HBr solution? (c) The KOH solution?

12-17. 3.15 g of $Ba(OH)_2 \cdot 8 H_2O$ is neutralized by 100 ml of a solution of H_2SO_4. (a) What is the normality of the H_2SO_4 solution? (b) If the neutralized solution is evaporated to dryness, how many grams of $BaSO_4$ are obtained?

12-18. What volume of 0.500 N HCl is required to react completely with 2.12 g of solid Na_2CO_3 according to the equation

$$Na_2CO_3 + 2 HCl \rightarrow 2 NaCl + H_2O + CO_2$$

12-19. Calculate the volume of CO_2 produced at STP when 25 ml of 0.10 N H_2SO_4 acts on K_2CO_3 according to the equation

$$K_2CO_3 + H_2SO_4 \rightarrow K_2SO_4 + CO_2 + H_2O$$

12-20. A solution of KOH is standardized by titration with an aqueous solution containing a known weight of oxalic acid, $H_2C_2O_4 \cdot 2 H_2O$. If 30.0 ml of KOH solution is required to neutralize 0.408 g of $H_2C_2O_4 \cdot 2 H_2O$, what is the normality of the KOH solution?

12-21. An excess of $AgNO_3$ is added to 30.0 ml of 0.250 N HCl. The precipitated AgCl is filtered off, dried, and weighed. How many grams of AgCl are obtained?

12-22. 18.00 ml of a solution of 0.1500 M HCl is required to neutralize 24.00 ml of a solution of NaOH. What is the concentration of NaOH?

12-23. To what total volume (in milliliters) must 25 ml of 16 M HNO_3 solution be diluted to make a 0.8 M HNO_3 solution?

12-24. What volume of 0.20 N H_2SO_4 is required to neutralize 75 ml of 0.12 N KOH solution?

12-25. To what volume must exactly 30 ml of 0.50 N NaOH be diluted to yield a 0.20 N solution?

12-26. 85.0 ml of 0.150 N acid requires 17.0 ml of basic solution for neutralization. What is the normality of the base?

12-27. A solution of KOH was standardized by titration of solid oxalic acid, $H_2C_2O_4 \cdot 2\ H_2O$. If 50.0 ml of KOH solution was required to neutralize 1.26 g of the solid acid, what was the normality of the KOH?

12-28. If 40.40 ml of HCl solution is required to neutralize 0.224 g of pure KOH, what is the normality of the HCl solution?

12-29. 24.5 ml of H_2SO_4 solution requires 32.7 ml of KOH solution for neutralization. When the neutralized solution is evaporated to dryness, 0.348 g of K_2SO_4 remains. (a) How many equivalents of K_2SO_4 are present? (b) Calculate the normality of the H_2SO_4 solution, and (c) of the KOH solution.

12-30. If a 10.1-ml sample of vinegar requires 50.5 ml of 0.134 N base for neutralization (a) what is the normality of the acid in the vinegar? (b) Assuming that the acidity is due to acetic acid, $HC_2H_3O_2$, what is the percentage by weight of acetic acid if the density of the vinegar is 1.00 g/ml?

12-31. A 5.10-g sample of household ammonia, NH_3 in H_2O, requires 156 ml of 0.275 N H_2SO_4 for neutralization. (a) If the density of the NH_3 solution is 0.960 g/ml, what is the normality of the solution? (b) What is the weight per cent of NH_3 in the solution?

12-32. 0.300 g of tartaric acid is dissolved in 60.0 ml of 0.100 N NaOH. The resulting solution requires exactly 100 ml of 2.00×10^{-2} N HCl for neutralization. Determine the equivalent weight of tartaric acid without looking up the formula.

12-33. If $KMnO_4$ is used as the source of MnO_4^- in the following reaction (a) what weight (in grams) of $KMnO_4$ should be dissolved in one liter of solution to yield a 1.0 Normal solution?

$$2\ MnO_4^- + 5\ C_2O_4^{--} + 16\ H^+ \rightarrow 2\ Mn^{++} + 10\ CO_2 + 8\ H_2O$$

And (b) how many equivalents are there in one mole of $C_2O_4^{--}$?

12-34. 40.0 ml of an $H_2C_2O_4$ solution requires 24.2 ml of 0.105 N $KMnO_4$ in a titration.

$$5\ C_2O_4^{--} + 16\ H^+ + 2\ MnO_4^- \rightarrow 10\ CO_2 + 2\ Mn^{++} + 8\ H_2O$$

(a) How many equivalents of $KMnO_4$ are used? (b) How many equivalents of $H_2C_2O_4$ undergo reaction? (c) What is the normality of the $H_2C_2O_4$ solution?

12-35. In a certain Redox reaction 50 ml of 0.1 N oxidizing agent is used. (a) How many equivalents of oxidizing agent were used? (b) How many equivalents of reducing agent must be used?

12-36. I_2 may be titrated with sodium thiosulfate, $Na_2S_2O_3$, according to the equation

$$I_2 + 2\,Na_2S_2O_3 \rightarrow 2\,NaI + Na_2S_4O_6 \quad \text{(sodium tetrathionate)}$$

If 50.00 ml of $Na_2S_2O_3$ solution is required to react with 1.269 g of I_2, what is the normality of the $Na_2S_2O_3$ solution?

12-37. Given:

$$5\,C_2O_4^{--} + 2\,MnO_4^- + 16\,H^+ \rightarrow 10\,CO_2 + 2\,Mn^{++} + 8\,H_2O$$

(a) Assuming that the source of MnO_4^- is $KMnO_4$, what is the *equivalent weight* of $KMnO_4$ in this reaction? (b) If $H_2C_2O_4\cdot2\,H_2O$ is the source of $C_2O_4^{--}$, what weight of the former is *one equivalent*? (c) In titrating MnO_4^- with $C_2O_4^{--}$, it is found that 50 ml of $0.100\,N\,C_2O_4^{--}$ exactly reacts with 40 ml of MnO_4^- solution. What is the normality of the MnO_4^- solution?

12-38. A solution of $KMnO_4$ was standardized by weighing out 0.67 g of $Na_2C_2O_4$ (0.01 equiv.) and titrating. 10 ml of $KMnO_4$ was required. How many equivalents of $KMnO_4$ were used in the titration?

12-39. The copper in a 2.00-g sample of a copper mineral was determined quantitatively by the following reactions:

$$2\,Cu^{++} + 4\,I^- \rightarrow 2\,CuI + I_2$$
$$I_2 + 2\,S_2O_3^{--} \rightarrow S_4O_6^{--} + 2\,I^-$$

(a) How many grams of Cu^{++} are there in 1 g equiv. for the first reaction? (b) What is the equivalent weight of $Na_2S_2O_3$? (c) If 31.4 ml of $0.100\,N$ $Na_2S_2O_3$ were used in the titration, what is the percentage of Cu in the compound?

12-40. Arsenic in a sample may be determined through the following sequence of reactions:

$$As + 5\,HNO_3 \rightarrow H_3AsO_4 + 5\,NO_2 + H_2O$$
$$H_3AsO_4 + 2\,HI \rightarrow H_3AsO_3 + I_2 + H_2O$$
$$I_2 + 2\,Na_2S_2O_3 \rightarrow 2\,NaI + Na_2S_4O_6$$

If exactly 1 g of a sample of the As-containing substance ultimately required 50.0 ml of $0.100\,N\,Na_2S_2O_3$ to titrate to the equivalence point, what is the percentage of As in the sample?

12-41. In a procedure for its quantitative determination, Ca^{++} may be precipitated as CaC_2O_4, as shown by the equation, $Ca^{++} + C_2O_4^{--} \rightarrow CaC_2O_4$. The precipitate is filtered, suspended in H_2O, dissolved in acid, and titrated with $KMnO_4$ solution of known concentration. In a particular analysis, 5.00 g of a sample of limestone-bearing rock is dissolved in acid and then Ca^{++} is precipitated as CaC_2O_4. 50.0 ml of $0.200\,N\,KMnO_4$ is used to titrate the CaC_2O_4. The overall reaction may be represented as

$$5\,CaC_2O_4 + 2\,KMnO_4 + 8\,H_2SO_4$$
$$\rightarrow 5\,CaSO_4 + K_2SO_4 + 2\,MnSO_4 + 10\,CO_2 + 8\,H_2O$$

Calculate the percentage of calcium in the rock.

CHAPTER 13

Rates of Chemical Reactions*

A. FIRST-ORDER CHEMICAL REACTIONS

The rates at which reactions occur are questions of utmost importance to chemists. The term "rate of a reaction" refers to the amount of substance reacting to form products per unit time; for example, in the simple reaction

$$A \rightarrow B + C$$

the rate of the reaction R is equal to the rate of *decrease* in the concentration of species A with time. The symbol $d[A]/dt$ is used to represent the rate of change of the concentration of A with time. Assuming it has been shown experimentally that the rate of the above reaction is proportional to the concentration of A, the rate expression may be written

$$R = \frac{-d[A]}{dt} = k[A] \tag{1}$$

where k is the *specific reaction rate constant*. This equation is a differential equation and any reaction which obeys this law is called a *first-order* chemical reaction.

By the methods of the calculus, it is possible to integrate equation (1) so that it takes the form

$$2.303 \log \frac{[A]}{[A]_0} = -kt \tag{2}$$

* This chapter may be omitted without loss of continuity at the option of the instructor.

where [A] is the concentration of substance A remaining at time t, and [A]₀ is the amount originally present at zero time. The proportionality constant k is as defined in equation (1).

Equations (1) and (2) are useful in solving problems dealing with first-order reactions. The differential equation (1) can be used to obtain approximate solutions to problems as long as $d[A]$ is small compared to [A]. The integral form of the first-order rate equation [equation (2)] is valid under any conditions.

The time required for one-half of any given amount of material to decompose is known as the half-life of the substance in question. For a first-order reaction, the half-life is independent of the concentration or amount of starting material and is related to the specific reaction rate constant by the equation

$$t_{1/2} = \frac{0.693}{k} \tag{3}$$

Obviously, if either the rate constant or the half-life is known, the other may be calculated. A derivation of equation (3) is presented in Example 13-b.

Example 13-a. The decomposition of SO_2Cl_2 according to the equation $SO_2Cl_2(g) \rightarrow SO_2(g) + Cl_2(g)$, has been found to be a first-order chemical reaction. At 602°K, and at constant volume, 2.00% of the SO_2Cl_2 initially present had decomposed at the end of 6.72 min. What is the specific reaction rate constant at this temperature?

Solution. This problem may be solved exactly by employing the integrated form of the first-order rate equation. Since only 2.00% of the original amount has decomposed, 98.00% remains, or [A] = 0.9800 [SO_2Cl_2]₀ when t = 6.72 min. Substituting in equation (2)

$$2.303 \log \frac{[SO_2Cl_2]}{[SO_2Cl_2]_0} = 2.303 \log \frac{0.9800[SO_2Cl_2]_0}{[SO_2Cl_2]_0} = -k(6.72)$$

$$k = \frac{-2.303 \log 0.9800}{6.72 \text{ min}} = \frac{-2.303(-0.0080)}{6.72 \text{ min}} = 2.74 \times 10^{-3}/\text{min}$$

Alternate Solution. Using the differential form for the rate equation [equation (1)], an approximate solution to the problem is obtained. Since the decrease in concentration of SO_2Cl_2 is 2.00%, then

$-d[A] = 0.0200 \ [SO_2Cl_2]_0$ and the change in time is 6.72 min, hence $dt = 6.72$ min. Substituting into the expression

$$\frac{-d[A]}{dt} = k[A]$$

$$\frac{0.0200[SO_2Cl_2]_0}{6.72 \ min} = k[SO_2Cl_2]_0$$

$$k = \frac{0.0200}{6.72 \ min} = 2.97 \times 10^{-3}/min$$

This result is within 10% of the true reaction rate constant obtained from the integrated form of the equation. As the extent of the reaction increases, the differential form of the equation yields a less accurate solution to the problem and its use is to be discouraged.

Example 13-b. From the data and solution given in Example 13-a, what is the half-life of the reaction for the decomposition of SO_2Cl_2?

Solution. This is analogous to asking the question "What time is required for one-half of any given amount of material to decompose?" This time is usually denoted as $t_{1/2}$ and is substituted for t in the integrated equation when the amount of reactant has decreased to half its initial value, that is, $t = t_{1/2}$ when $[SO_2Cl_2] = [SO_2Cl_2]_0/2$. Substituting into equation (2),

$$-2.303 \log \frac{[SO_2Cl_2]_0}{2 \times [SO_2Cl_2]_0} = kt_{1/2}$$

$$-2.303 \log \tfrac{1}{2} = kt_{1/2}$$

$$-2.303(-0.301) = 0.693 = kt_{1/2}$$

Substituting the numerical value for k from Example 13-a, the half-life of SO_2Cl_2 at 602°K is

$$t_{1/2} = \frac{0.693}{k} = \frac{0.693}{2.74 \times 10^{-3}/min} = 2.53 \times 10^2 \ min$$

Note that since the reaction rate constant is in the units of reciprocal minutes, the half-life of the reaction is obtained in minutes.

Example 13-c. In 0.0400 M NaOH at 20°C, the decomposition of H_2O_2 by the reaction $2 \ H_2O_2 \rightarrow 2 \ H_2O + O_2$ has been shown to be first order. (a) If the half-life is found to be 654 min, what fraction of the original

H_2O_2 remains after exactly 100 min have elapsed? (b) What is the initial rate of reaction in a 0.0200 M solution of H_2O_2?

Solution. (a) To solve this problem, it is necessary to know k, the specific reaction rate constant. This can be obtained by substituting the preceding data into the integrated form of the rate expression. When $t_{1/2} = 654$ min, $[H_2O_2] = [H_2O_2]_0/2$, and hence

$$-2.303 \log \frac{[H_2O_2]_0}{2[H_2O_2]_0} = k\,(654 \text{ min})$$

$$k = \frac{-2.303 \times (-0.301)}{654 \text{ min}} = 1.060 \times 10^{-3}/\text{min}$$

Since $t_{1/2}$ is known, the specific reaction rate constant can also be evaluated from equation (3)

$$k = \frac{0.693}{t_{1/2}} = \frac{0.693}{654 \text{ min}} = 1.060 \times 10^{-3}/\text{min}$$

Knowing k, it is now possible to solve for the fraction of hydrogen peroxide, $[H_2O_2]/[H_2O_2]_0$, remaining after 100 min have elapsed. Making use of equation (2)

$$2.303 \log \frac{[H_2O_2]}{[H_2O_2]_0} = -(1.060 \times 10^{-3}/\text{min}) \times (100 \text{ min})$$

$$\log \frac{[H_2O_2]}{[H_2O_2]_0} = \frac{-1.060 \times 10^{-1}}{2.303} = -4.60 \times 10^{-2} = -0.0460$$

We obtain the antilog

$$\frac{[H_2O_2]}{[H_2O_2]_0} = 0.899$$

Hence 89.9% of the original H_2O_2 remains after the end of 100 min. (b) The initial rate of reaction is given by the equation $R = k[H_2O_2]_0$ and can easily be calculated if we know the rate constant and the concentration of H_2O_2. From the data given $[H_2O_2]_0 = 0.0200\,M$, and from the solution just presented, $k = 1.060 \times 10^{-3}/\text{min}$. Therefore, the rate $R = (1.060 \times 10^{-3}/\text{min}) \times (0.0200$ mole/liter$) = 2.12 \times 10^{-5}$ (mole/liter)/min.

PROBLEMS

13-1. The reaction for the isomerization of cyclopropane gas to propene gas proceeds as illustrated by the equation

$$
\begin{array}{c}
CH_2\text{——}CH_2 \\
\diagdown\;\;\diagup \\
CH_2
\end{array}
\rightarrow CH_3\text{—}CH\text{=}CH_2
$$

(cyclopropane)   (propene)

At 440°C and a pressure of 60.0 cm of Hg, the specific reaction rate constant for this first-order reaction is 5.40×10^{-2}/hr. (a) What is the half-life of the cyclopropane under these conditions? (b) What fraction of cyclopropane remains at the end of 24.0 hr?

13-2. The specific reaction rate constant for the *cis-trans* isomerization of liquid azobenzene is 2.22/hr at 77°C. The reaction is first order:

$$
\begin{array}{ccc}
C_6H_5 & \;\;\;\; C_6H_5 \;\;\; C_6H_5 & \\
\diagdown & \diagup \;\;\;\;\;\;\;\; \diagdown & \\
& N\text{=}N \;\;\;\;\;\;\; \rightarrow \;\;\;\;\;\;\; N\text{=}N & \\
\diagup & & \diagdown \\
& & \;\;\;\;\; C_6H_5 \\
& (cis) \;\;\;\;\;\;\;\;\;\;\; (trans) &
\end{array}
$$

(a) What fraction of the *cis* has been converted at the end of 30 minutes?
(b) What is the half-life of the *cis* form under these conditions?

13-3. Acetic anhydride reacts with water to form acetic acid by the reaction

$$
\begin{array}{ccccc}
O & O & & & O \\
\| & \| & & & \| \\
CH_3C\text{—}O\text{—}CCH_3 & + & H_2O & \rightarrow & 2\ CH_3C\text{—}OH
\end{array}
$$

At 5°C, the reaction follows first-order kinetics and it has been found that a solution initially containing 0.802 *M* acetic anhydride is only 0.527 *M* at the end of 10.7 min. (a) What is the specific reaction rate constant? (b) What is the half-life of acetic anhydride under these conditions? (c) What is the initial rate of reaction (hydrolysis) of the acetic anhydride with water? (d) What is the rate of reaction at the end of one half-life?

13-4. The decomposition of phosphine at 719°K proceeds according to the reaction

$$4\ PH_3(g) \rightarrow P_4(g) + 6\ H_2(g)$$

It is found that the reaction is first order and that the half-life is 37.9 sec. (a) What is the specific reaction rate constant? (b) What time is required for three-fourths of the material to decompose?

13-5. The inorganic complex ion, *trans*-dioxalatodiaquochromate (III), under-goes isomerization in aqueous solution at 17°C to form the *cis* complex by a first-order reaction as shown by Prof. Randall Hamm.

(*trans*)   (*cis*)

In an acidic aqueous solution, the half-life of the *trans* form is 1.07 hr. (a) What is the specific reaction rate constant for the reaction? (b) What fraction of the *trans* form remains at the end of 3.50 hr? (c) What is the initial rate of reaction in a 0.15 M solution? (d) What is the rate of reaction in a solution at the end of one half-life if the initial concentration of complex ion was 0.10 M?

13-6. At 25°C, the hydrolysis of succinic anhydride in aqueous solution has been shown to be first order. The reaction is

(a) If the half-life is 4.45 min, what is the specific reaction rate constant? (b) What fraction of material is left at the end of 30 min? (c) What fraction of material is left at the end of 13.3 min? (d) What is the initial rate of the reaction in a solution containing 0.141 mole of succinic anhydride per liter?

13-7. The experimentally determined initial rates of the aquation reaction

$$[Co(NH_3)_5Cl]^{++} + H_2O \rightarrow [Co(NH_3)_5(H_2O)]^{+++} + Cl^-$$

are given below. From these data (a) prove that the reaction is first order. (b) Calculate the specific reaction rate constant for each set of data.

Experiment No.	Initial $[Co(NH_3)_5Cl]^{++}$ [mole/liter]	Initial rate [(mole/liter)/min]
1	1×10^{-3}	1.3×10^{-7}
2	2×10^{-3}	2.6×10^{-7}
3	3×10^{-3}	3.9×10^{-7}
4	1×10^{-2}	1.3×10^{-6}

B. REACTIONS OTHER THAN FIRST ORDER

For the general reaction

$$aA + bB \rightarrow cC$$

it may be shown that the

$$\text{Rate of the reaction} = R = \frac{-d[A]}{dt} = k[A]^x[B]^y$$

The sum of the exponents x and y gives the "order of the reaction." The numerical values of x and y may be equal to the coefficients a and b in the chemical equation, but this must be demonstrated by experiment. It cannot be assumed that x and y are equal to a and b, respectively, without some form of experimental proof. It has even been found that for some reactions of the type $A \rightarrow B + C$, that the rate is independent of the concentration of A. Such a reaction is called a *zero-order* reaction.

Example 13-d. A reaction between gaseous substances A and B proceeds at a measurable rate. The following set of data has been accumulated.

Experiment No.	Initial [A] [mole/liter]	Initial [B] [mole/liter]	Initial rate [(mole/liter)/min]
1	1.00×10^{-3}	0.25×10^{-3}	0.26×10^{-9}
2	1.00×10^{-3}	0.50×10^{-3}	0.52×10^{-9}
3	1.00×10^{-3}	1.00×10^{-3}	1.04×10^{-9}
4	2.00×10^{-3}	1.00×10^{-3}	4.16×10^{-9}
5	3.00×10^{-3}	1.00×10^{-3}	9.36×10^{-9}
6	4.00×10^{-3}	1.00×10^{-3}	16.64×10^{-9}

(a) What is the rate expression? What is the order of the reaction? (b) What is the reaction rate constant? (c) What is the initial rate of reaction in a solution where each of the reactants is at a concentration of 4.0×10^{-4} M?

> **Solution.** (a) The questions which are really being asked are first, "What are the exponents x and y in the expression, Rate $= k[A]^x[B]^y$?" and second, "What is the sum of $(x + y)$?"
>
> Note in the data presented for experiments 1–3 that the concentration of A is held constant and in experiments 3–6, the concentration of B is held constant.
>
> Consider first the variation of rate at constant $[A]$ and varying $[B]$. The rate of the reaction doubles from experiment 1 to 2 as the

value of [B] is doubled, and likewise, in going from experiment 2 to 3, the rate doubles again as [B] is doubled. Therefore, there is a direct proportionality between the rate R and [B], or R ∝ [B].

Consider next the variation of rate at constant [B] and varying [A]. The rate of reaction increases by a factor of 4 as the [A] doubles, as is seen by comparing experiments 3 and 4, and experiments 4 and 6. In comparing experiments 3 and 5, the [A] triples and the rate increases by a factor of 9. When comparing experiments 3 and 6, the concentration increases by a factor of 4 and the rate increases by a factor of 16. Thus it is seen that the rate is increasing by the square of the factor of the increase of [A]. Hence the rate is proportional to the square of the concentration of A, or $R \propto [A]^2$.

The rate expression may now be written $R = k[A]^2[B]$ and the reaction is a third-order reaction because the sum of the exponents $(x + y)$ is 3.

(b) The reaction rate constant may be evaluated by substituting any set of data from the table above into the rate expression. For example, we may use the data from experiment number 5.

$$R = k[A]^2[B]$$
$$R = 9.36 \times 10^{-9} \text{ (mole/liter)/min}$$
$$[A] = 3.00 \times 10^{-3} \text{ mole/liter}$$
$$[B] = 1.00 \times 10^{-3} \text{ mole/liter}$$

Making the appropriate substitutions

9.36 × 10⁻⁹ (mole/liter)/min

$$= k(3.00 \times 10^{-3} \text{ mole/liter})^2(1.00 \times 10^{-3} \text{ mole/liter})$$

9.36 × 10⁻⁹ (mole/liter)/min

$$= k(9.00 \times 10^{-6})(1.00 \times 10^{-3})(\text{mole/liter})^3$$

$$k = \frac{9.36 \times 10^{-9}}{9.00 \times 10^{-9}} = 1.04/[(\text{mole/liter})^2\text{-min}]$$

(The student should verify the fact that each set of data gives the same specific reaction rate constant. This is required if the rate expression is valid.)

(c) The initial rate when $[A] = [B] = 4.0 \times 10^{-4} M$ is obtained using the rate expression and the constant evaluated in part (b).

$$R = k[A]^2[B] = (1.04/[(\text{mole/liter})^2\text{-min}])$$
$$\times [4.0 \times 10^{-4} \text{ (mole/liter)}]^2(4.0 \times 10^{-4} \text{ mole/liter})$$
$$= 1.04 \times 16 \times 10^{-8} \times 4.0 \times 10^{-4} \text{ (mole/liter)/min}$$
$$R = 6.7 \times 10^{-11} \text{ (mole/liter)/min}$$

PROBLEMS

13-8. The reaction $CO(g) + NO_2(g) \rightarrow CO_2(g) + NO(g)$ has been studied at a temperature of $540°K$. Some of the experimental data are listed in the table below.

Initial [CO] [mole/liter]	Initial [NO$_2$] [mole/liter]	Initial rate [(mole/liter)/hr]
5.1×10^{-4}	0.35×10^{-4}	3.4×10^{-8}
5.1×10^{-4}	0.70×10^{-4}	6.8×10^{-8}
5.1×10^{-4}	0.18×10^{-4}	1.7×10^{-8}
10.2×10^{-4}	0.35×10^{-4}	6.8×10^{-8}
15.3×10^{-4}	0.35×10^{-4}	10.2×10^{-8}

(a) Derive an expression for the rate of reaction in terms of concentration of CO and NO_2. (b) What is the order of the reaction? (c) What is the specific reaction rate constant? (d) What is the initial rate of reaction if each of the species is present at a concentration of $1.1 \times 10^{-4} M$?

13-9. The compound N_2O_5 is found to decompose by the reaction $2 N_2O_5(g) \rightarrow 4 NO_2(g) + O_2(g)$. At a certain temperature, the following data were obtained.

Initial [N$_2$O$_5$] [mole/liter]	Initial rate [(mole/liter)/sec]
2.00×10^{-4}	1.80×10^{-2}
3.00×10^{-4}	2.70×10^{-2}
4.00×10^{-4}	3.60×10^{-2}

(a) What is the rate expression? (b) What is the order of the reaction for the decomposition of N_2O_5? (c) What is the initial rate of the reaction when the initial concentration of N_2O_5 is $1.0 \times 10^{-3} M$?

13-10. Compound A, when enclosed in a glass vessel, spontaneously decomposes as shown by the equation $A \rightarrow B + C$. The following data were obtained for the rate of reaction as a function of initial concentration of A.

Initial [A] [mole/liter]	Initial rate [(mole/liter)/sec]
4.0×10^{-2}	1.02×10^{-9}
5.0×10^{-2}	1.02×10^{-9}
8.0×10^{-2}	1.02×10^{-9}

(a) What is the rate expression? (b) What is the order of the reaction?

13-11. Gaseous HI decomposes at $716°K$ by the reaction

$$HI(g) \rightarrow \tfrac{1}{2} H_2(g) + \tfrac{1}{2} I_2(g)$$

The experimental rate data are shown below.

Initial [HI] [mole/liter]	Initial rate [(mole/liter)/min]
1×10^{-3}	3×10^{-5}
2×10^{-3}	1.2×10^{-4}
3×10^{-3}	2.7×10^{-4}

(a) What is the rate expression? (b) What is the order of the reaction? (c) What would be the initial rate of reaction if the concentration of HI were $2 \times 10^{-4} M$?

GENERAL PROBLEMS

13-12. Derive the relationship between $t_{1/2}$ and the specific reaction rate constant for any first-order chemical reaction. Assume the equation $A \rightarrow B$, represents the reaction.

13-13. Acetyl fluoride hydrolyzes in mixed water-acetone solvent according to reaction

$$\overset{\overset{\displaystyle O}{\displaystyle \|}}{CH_3C}\!\!-\!\!F + HOH \rightarrow \overset{\overset{\displaystyle O}{\displaystyle \|}}{CH_3C}\!\!-\!\!OH + HF$$

The rate expression is first order with respect to acetyl fluoride and is independent of water concentration. The first-order specific reaction rate constant is 0.396/hr. Make a plot of the fraction of acetyl fluoride remaining unreacted versus t for a period of three half-lives.

13-14. Gaseous *cis*-dichloroethylene at $185°C$ is converted to the *trans* form by a process which was found experimentally to be first order.

If after 3.5 hr the amount of the *cis* form is 95% of that originally present, what is the specific reaction rate constant for the isomerization?

13-15. $H_2(g)$ and $I_2(g)$ react by means of a reaction which is first order with respect to each species. The specific reaction rate constant is $1.67 \times 10^{-2}/$ [(mole/liter)-min]. If the concentration of H_2 is 1.51×10^{-2} mole/liter and the concentration of I_2 is 3.02×10^{-2} mole/liter, what is the initial rate of the reaction?

13-16. $_{88}Ra^{226}$ decays by alpha emission to $_{86}Rn^{222}$. Such a process is first order (as are all radioactive decay processes) with a half-life of 1620 yr. (a) How long will it take for 10% of the Ra to decay? (b) If the original sample was 5.00 g, how much Ra will be left after 16,200 yr?

13-17. The carbon found in a sample of charcoal from a cave in the western United States has a C^{14}/C^{12} ratio roughly 19.0% as much as the C^{14}/C^{12} ratio of presently live wood. Assume that the only change in the C^{14}/C^{12} ratio is due to the first-order decay process $_6C^{14} \rightarrow {}_7N^{14} + {}_{-1}e^0$, and that the C^{14}/C^{12} in new wood today is the same as when the charcoal sample was part of a living tree. If the half-life of the C^{14} is 5730 yr, what is the age of the wood from which the charcoal was prepared?

13-18. The compound dimethyl ether decomposes as shown by the reaction

$$CH_3OCH_3(g) \rightarrow CH_4(g) + CO(g) + H_2(g)$$

by means of a first-order chemical reaction. At a certain temperature the reaction rate constant was 0.12/min. Make the plot of the fraction of dimethyl ether remaining in a sample versus the time since the reaction started. Plot the logarithm of this fraction versus the time since the reaction started. Which plot gives the most information?

13-19. The reaction of $2\,NO(g) + Cl_2(g) \rightarrow 2\,NOCl(g)$ appears to be a third-order reaction, and the rate determining step involves an intermediate with two molecules of NO and one of Cl_2. At a concentration of 0.20 M of each reactant, the specific reaction rate constant is found to be approximately $1.0 \times 10^{-3}/[(mole/liter)^2\text{-min}]$. (a) What is the rate expression, and (b) what is the initial rate of reaction under these conditions? (c) What is the initial rate of reaction if the concentrations of reactants are doubled?

13-20. In the aquation reaction

$$[Cr(NH_3)_5I]^{++} + H_2O \rightarrow [Cr(NH_3)_5(H_2O)]^{+++} + I^-$$

10.5% of the initial concentration of the iodo complex is aquated in 10.4 min. (a) If this is a first-order reaction, what time is required for 35% of the material to react? (b) What is the time required for the fraction of material reacted to change from 50% to 60%? (c) What is the initial rate of reaction in a 0.055 M solution of the complex?

CHAPTER 14

Molecular Equilibrium

Many chemical reactions are incomplete because the products of the reaction recombine to form the reactants. If the two processes are proceeding simultaneously, a state of equilibrium is eventually reached in which the rate of the forward reaction is equal to the rate of the reverse reaction. Consider the general equilibrium equation

$$mA + nB \rightleftharpoons pC + qD$$

where m, n, p, q are the coefficients of substances A, B, C, D, respectively, in the reaction. When equilibrium is established, the concentrations of reactants and products become fixed and are related through the mathematical relationship

$$\frac{[C]^p \times [D]^q}{[A]^m \times [B]^n} = K_c = \text{constant}$$

K_c is termed the *equilibrium constant*, and has a particular value for each chemical reaction and each temperature. The brackets denote concentrations expressed in moles per liter, and it should be observed that each concentration is raised to that power which corresponds to the coefficient in the balanced equation. For example, in the reaction

$$2\,NOCl(g) \rightleftharpoons 2\,NO(g) + Cl_2(g)$$

$$K_c = \frac{[NO]^2[Cl_2]}{[NOCl]^2}$$

By various methods of chemical analysis, we can evaluate the concentration of reactants and products at equilibrium. Once this information is available, the calculation of equilibrium constants* is a simple matter.

---

* Equilibrium constants may or may not be unitless quantities. If the equilibrium reaction involves the same number of moles of reactants as products, K_c will be just a number without units; where there is a difference in numbers of moles of reactants versus products, K_c would carry units of concentration, that is, moles/liter, (moles/liter)^2, (moles/liter)$^{-1}$, etc. In this presentation, the values of K_c will be expressed simply as *numbers*. The units in each case are easily obtainable from the equilibrium constant expression if desired.

144

Example 14-a. The equilibrium reaction

$$2 \, NOCl(g) \rightleftharpoons 2 \, NO(g) + Cl_2(g)$$

was studied at a temperature of 462°C and at a constant volume of 1.00 liter. Initially, 2.00 moles of NOCl were placed in the container, and when equilibrium was established, 33% of the NOCl was found to have dissociated. From these data, calculate the equilibrium constant.

Solution. Reaction:

$$2 \, NOCl(g) \rightleftharpoons 2 \, NO(g) + Cl_2(g)$$

Initial amounts: NOCl = 2.00 moles; NO = 0 mole; Cl_2 = 0 mole

Moles NOCl dissociating = $2.00 \times 0.33 = 0.66$ mole

At equilibrium:

Moles NOCl remaining = (Total − Moles dissociating)

$$= 2.00 - 0.66 = 1.34 \text{ moles}$$

Because 2 moles of NOCl form only 1 mole of Cl_2 for every 2 moles NO,

$$\text{Moles } Cl_2 = \frac{\text{Moles NO}}{2} = \frac{0.66}{2} = 0.33 \text{ mole}$$

$$[NOCl] = 1.34 \text{ moles/liter}$$

$$[NO] = 0.66 \text{ mole/liter}$$

$$[Cl_2] = 0.33 \text{ mole/liter}$$

$$K_c = \frac{[NO]^2[Cl_2]}{[NOCl]^2} = \frac{(0.66)^2(0.33)}{(1.34)^2}$$

$$K_c = \frac{0.436 \times 0.33}{1.80} = 8.0 \times 10^{-2}$$

Example 14-b. 1.00 mole of Br_2 gas is enclosed in a 1.00-liter container at 1483°C. It is found to be 1% dissociated under these conditions. Determine the equilibrium constant.

Solution. Reaction:

$$Br_2(g) \rightleftharpoons 2 \, Br(g)$$

$$K_c = \frac{[Br]^2}{[Br_2]}$$

Initial:

$$Br_2 = 1.00 \text{ mole}; \qquad Br = 0 \text{ mole}$$

Moles of Br_2 dissociating $= 1.00 \times 0.01 = 0.01$ mole

At equilibrium:

$$\text{Moles } Br_2 = \text{Total moles } Br_2 - \text{Moles } Br_2 \text{ dissociated}$$
$$= 1.00 - 0.01 = 0.99 \text{ mole}$$

Since Br_2 forms 2 Br atoms,

$$\text{Moles } Br = 2 \times \text{Moles } Br_2 \text{ dissociated} = 2 \times 0.01 = 0.02 \text{ mole}$$
$$[Br_2] = 0.99 \text{ mole/liter}$$
$$[Br] = 0.02 \text{ mole/liter}$$

$$K_c = \frac{(2 \times 10^{-2})^2}{0.99} = 4 \times 10^{-4}$$

Example 14-c. A sample containing 2.00 moles of HI is enclosed in a 1.00-liter flask and heated to 628°C. The equilibrium reaction $2\,HI \rightleftharpoons H_2 + I_2$ is established for which the equilibrium constant is 3.80×10^{-2}. What is the percentage of dissociation under these conditions? What is the concentration of each species in the mixture?

Solution. Reaction:

$$2\,HI(g) \rightleftharpoons H_2(g) + I_2(g)$$

$$K_c = \frac{[H_2][I_2]}{[HI]^2}$$

At equilibrium, let X be the number of moles of HI dissociated.

$$[HI] = \frac{\text{Total moles HI} - \text{Number of moles dissociated}}{\text{Number of liters}}$$

$$= \frac{2.00 - X}{1.00} = (2.00 - X) \text{ moles/liter}$$

Since two moles of HI form 1 mole of H_2 and 1 mole of I_2

$$[H_2] = \frac{X}{2} \text{ moles/liter}$$

and

$$[I_2] = \frac{X}{2} \text{ moles/liter}$$

Substituting in the equilibrium constant expression,

$$3.80 \times 10^{-2} = \frac{(X/2)(X/2)}{(2.00 - X)^2}$$

Taking the square root of both sides

$$\sqrt{3.80 \times 10^{-2}} = \frac{X/2}{2.00 - X}$$

$$1.95 \times 10^{-1} = \frac{X/2}{2.00 - X}$$

$$0.195 \times 2.00 - 0.195X = 0.500X$$

$$0.390 = 0.695X$$

$$X = \frac{0.390}{0.695} = 0.561 \text{ mole/liter}$$

$$[HI] = 2.00 - 0.561 = 1.44 \text{ moles/liter}$$

$$[H_2] = [I_2] = 0.28 \text{ mole/liter}$$

$$\text{Percentage dissociation} = \frac{(\text{Total [HI]} - \text{Equil [HI]}) \times 100}{\text{Total [HI]}}$$

$$= \frac{(2.00 - 1.44)100}{2.00} = 28\%$$

Example 14-d. Two and one-half moles of $COBr_2$ are introduced into a 2.00-liter flask at 0°C and heated to a temperature of 73°C. At equilibrium, the equilibrium constant K_c was found to be 0.190. What is the equilibrium concentration of each species? What is the percentage of dissociation? The chemical reaction involved is

$$COBr_2(g) \rightleftharpoons CO(g) + Br_2(g)$$

Solution. Initial concentrations:

$$[COBr_2] = \frac{2.50 \text{ moles}}{2.00 \text{ liters}} = 1.25 \text{ moles/liter}$$

$$[CO] = [Br_2] = 0 \text{ mole/liter}$$

Let

$$X = \text{amount of } COBr_2 \text{ dissociated in moles/liter}$$

At equilibrium:

$$[Br_2] = [CO] = X \text{ mole/liter}$$

and

$$[COBr_2] = (1.25 - X) \text{ mole/liter}$$

$$K_c = \frac{[CO][Br_2]}{[COBr_2]} = \frac{(X)(X)}{1.25 - X} = 0.190$$

By use of the quadratic,

$$X = 0.40 \text{ mole/liter}$$

Therefore, at equilibrium:

$$[CO] = [Br_2] = 0.40 \text{ mole/liter}$$

and

$$[COBr_2] = 1.25 - 0.40 = 0.85 \text{ mole/liter}$$

$$\text{Percentage dissociation} = \frac{(\text{Total } [COBr_2] - \text{Equil } [COBr_2]) \times 100}{\text{Total } [COBr_2]}$$

$$= \frac{(1.25 - 0.85)100}{1.25} = 32\%$$

Example 14-e. If we add 2.00 moles of carbon monoxide to the above equilibrium system, what is the concentration of each species when equilibrium is reestablished?

Solution. From LeChatelier's principle, we can predict that the equilibrium would be shifted toward the formation of more $COBr_2$ and that the concentration of Br_2 would be decreased.

Let X be the concentration decrease of Br_2 because of reaction with the added CO. At the new equilibrium

$$[Br_2] = (0.40 - X) \text{ mole/liter}$$

$$[CO] = (0.40 - X) \text{ mole/liter} + \text{Concentration added}$$

$$= (0.40 - X) \text{ mole/liter} + \frac{2.00 \text{ moles}}{2.00 \text{ liters}}$$

$$= (0.40 + 1.00 - X) \text{ moles/liter} = (1.40 - X) \text{ moles/liter}$$

$$[COBr_2] = \text{Equilibrium concentration} + \text{Concentration increase}$$

$$= (0.85 + X) \text{ mole/liter}$$

By substitution into the equilibrium constant expression,

$$K_c = 0.190 = \frac{(0.40 - X)(1.40 - X)}{0.85 + X}$$

and using the quadratic expression,

$$X = 0.23 \text{ mole/liter}$$

Hence

$$[Br_2] = 0.40 - 0.23 = 0.17 \text{ mole/liter}$$
$$[CO] = 1.40 - 0.23 = 1.17 \text{ moles/liter}$$
$$[COBr_2] = 0.85 + 0.23 = 1.08 \text{ moles/liter}$$

Example 14-f. It is often equally as convenient to express the equilibrium constant in terms of the partial pressures of the various gases present rather than as concentrations. This is shown in the following example where K_p is used to designate the equilibrium constant to distinguish it from the concentration constant K_c.

A sample of N_2 and H_2 is heated together at a temperature of 623°C and at a constant pressure of 30 atm. If a stoichiometric ratio of N_2 and H_2 is used, it is found that at equilibrium the pressure of N_2 is 4.0 atm. Using pressures rather than concentrations, calculate the equilibrium constant for the reaction

$$N_2(g) + 3 H_2(g) \rightleftharpoons 2 NH_3(g)$$

Solution. When equilibrium is finally established,

$$P_{total} = 30 \text{ atm} = p_{N_2} + p_{NH_3} + p_{H_2}$$

where p represents partial pressures. In a stoichiometric mixture, $p_{H_2} = 3 p_{N_2}$ because in the balanced chemical equation, the ratio (moles of H_2/moles of N_2) = 3. If

$$p_{N_2} = 4.0 \text{ atm} \quad \text{then} \quad p_{H_2} = 12.0 \text{ atm}$$
$$p_{NH_3} = 30 - p_{N_2} - p_{H_2} = 30 - 4 - 12 = 14 \text{ atm}$$

$$K_p = \frac{(p_{NH_3})^2}{(p_{N_2})(p_{H_2})^3} = \frac{(14)^2}{(4)(12)^3} = 0.028$$

Example 14-g. In the case of equilibrium in heterogeneous systems where a gaseous phase and one or more condensed phases are present, the equilibrium constant is written similarly. In the equilibrium reaction

$$C(s) + CO_2(g) \rightleftharpoons 2 CO(g)$$

$$K_p = \frac{(p_{CO})^2}{p_{CO_2} p_{C(s)}}$$

Since the vapor pressure of a pure solid (or a pure liquid) is constant at constant temperature, a new equilibrium constant K'_p may be defined as $K'_p = K_p \times p_{C(s)}$, and hence

$$K'_p = \frac{(p_{CO})^2}{p_{CO_2}}$$

At $1000°K$ when equilibrium between $CO_2(g)$, $CO(g)$, and $C(s)$ is established, the total pressure is found to be 4.70 atm. Calculate the equilibrium pressure of CO_2 and CO, if at $1000°K$ the value of K'_p is 1.72 and the vapor pressure of $C(s)$ is negligible.

Solution. Let X = pressure of CO_2 in atm at equilibrium,

$$p_{total} = 4.70 \text{ atm} = p_{CO_2} + p_{CO}$$

$$p_{CO} = (4.70 - X) \text{ atm}$$

On substitution into the equilibrium constant expression,

$$1.72 = \frac{(4.70 - X)^2}{X}$$

On solution of this equation, it is found that $X = 2.59$ atm, and hence

$$p_{CO_2} = 2.59 \text{ atm} \quad \text{and} \quad p_{CO} = 4.70 - 2.59 = 2.11 \text{ atm}$$

Example 14-h. The equilibrium constants K_c and K_p are related by the equation

$$K_p = K_c(RT)^{\Delta n}$$

where Δn is the difference in number of moles of gaseous products and reactants. If K_p for the equilibrium

$$PCl_5(g) \rightleftharpoons PCl_3(g) + Cl_2(g)$$

is 1.78 at $250°C$, calculate K_c.

Solution.

$$K_p = K_c(RT)^{\Delta n} \quad \text{or} \quad K_c = \frac{K_p}{(RT)^{\Delta n}}$$

$$\Delta n = 2 \text{ moles products} - 1 \text{ mole reactant} = 1$$

Substituting

$$K_c = \frac{1.78}{(0.082)(523)} = 4.15 \times 10^{-2}$$

THE DISTRIBUTION LAW

If a solute is added to a system of two immiscible liquids the solute will distribute itself between the two solvents in the same proportion as the ratio of the solubilities of the solute in the two solvents at any given temperature; for example, iodine dissolves in water to the extent of 0.029 g/100 g H_2O and dissolves in carbon tetrachloride to the extent of 2.60 g/100 g CCl_4 at 20°C. The ratio of solubilities in this case is 2.60/0.029 = 89.6. If lesser amounts of iodine are used, the ratio of concentrations of iodine dissolved in the two liquids at 20°C will be 89.6. Since an equilibrium exists between the two phases, an equilibrium constant, K_d, called a distribution coefficient may be determined—it is simply the ratio of the concentrations of solute in the two solutions. Concentrations are usually expressed in moles per liter. The distribution coefficients has practical application in the extraction of substances in certain industrial processes.

Example 14-i. The distribution coefficient at 20°C of iodine between carbon disulfide and water is 420. Calculate the weight of iodine which will be extracted from exactly one liter of 1.000×10^{-3} M aqueous solution by shaking with exactly 100 ml of CS_2.

Solution.

1 liter of 1.000×10^{-3} M I_2 solution contains 1.000×10^{-3} mole I_2

Let X = Moles of I_2 extracted by CS_2, then $1.000 \times 10^{-3} - X =$ Moles of I_2 remaining in the aqueous solution.

$$\text{Concentration in } CS_2 \text{ solution} = \frac{X}{0.1 \text{ liter}}$$

$$= 10X \text{ moles per liter}$$

$$\text{Concentration in } H_2O \text{ solution} = \frac{1.000 \times 10^{-3} - X}{1 \text{ liter}}$$

$$= 1.000 \times 10^{-3} - X \text{ moles/liter}$$

Since

$$\frac{\text{Conc. in } CS_2}{\text{Conc. in } H_2O} = 420, \quad \text{then} \quad \frac{10X}{1.000 \times 10^{-3} - X} = 420$$

$$10X = 0.420 - 420X$$

$$X = 0.000977 \text{ moles}$$

or $(0.000977 \text{ moles})(253.2 \text{ g/mole}) = 0.247 \text{ g}$

PROBLEMS

14-1. A reaction vessel with a capacity of 1 liter in which the reaction $SO_2(g) + NO_2(g) \rightleftharpoons SO_3(g) + NO(g)$ had reached a state of equilibrium, was found to contain 0.8 mole of SO_3, 0.4 mole of NO, 0.2 mole of NO_2, and 0.4 mole of SO_2. Calculate the equilibrium constant for the reaction.

14-2. Consider the equilibrium $H_2(g) + I_2(g) \rightleftharpoons 2\,HI(g)$. Set up the expression for the equilibrium constant and evaluate K_c if at $300°K$ the concentrations are $[H_2] = 0.3\,M$; $[I_2] = 0.6\,M$; $[HI] = 0.3\,M$.

14-3. Given the equilibrium: $2\,A(g) + B(g) \rightleftharpoons C(g) + 3\,D(g)$. The concentrations are, respectively: $[A] = 3\,M$; $[B] = 0.4\,M$; $[C] = 0.6\,M$; $[D] = 2\,M$. Calculate the value of the equilibrium constant.

14-4. The equilibrium system, $2\,SO_2(g) + O_2(g) \rightleftharpoons 2\,SO_3(g)$ was analyzed and found to contain the following concentrations: $[SO_2] = 0.40\,M$, $[O_2] = 0.20\,M$, $[SO_3] = 0.60\,M$. Calculate the equilibrium constant for this system.

14-5. The gaseous reaction $NO_2(g) + SO_2(g) \rightleftharpoons SO_3(g) + NO(g)$ is found to occur at elevated temperatures. In a given experiment, SO_2 and NO were mixed and allowed to come to equilibrium. Analysis of the mixture at equilibrium at a specific temperature demonstrated the presence of 0.40 mole of SO_3, 0.50 mole of NO, 0.30 mole of NO_2, and 0.10 mole of SO_2 in the 1.0-liter container. What is the equilibrium constant for the reaction?

14-6. The equilibrium constant for the formation of NH_3 by the reaction, $N_2(g) + 3\,H_2(g) \rightleftharpoons 2\,NH_3(g)$ is 4.0. If the equilibrium concentration of N_2 in a mixture is $2.0\,M$, and H_2 is $2.0\,M$, what is the concentration of ammonia?

14-7. At $2000°K$ a mixture of H_2, S_2, and H_2S rapidly reaches a state of equilibrium, represented by the equation

$$2\,H_2S(g) \rightleftharpoons 2\,H_2(g) + S_2(g)$$

An analysis of the equilibrium mixture shows that there are 2.00 mole of H_2S, 0.100 mole of H_2, and 0.64 mole of S_2 in the 2.00-liter flask. Calculate the concentration equilibrium constant.

14-8. The equilibrium constant is 10^{-4} for the dissociation of I_2 at elevated temperatures, as represented by the reaction $I_2(g) \rightleftharpoons 2\,I(g)$. If the concentration of atomic iodine is equal to $10^{-2}\,M$, what is the concentration of molecular iodine?

14-9. The equilibrium constant for the formation of CO_2 by the reaction $2\,CO(g) + O_2(g) \rightleftharpoons 2\,CO_2(g)$ is 1×10^4. If the concentration of CO_2 is $20\,M$ and the concentration of O_2 is $4.0\,M$, what is the concentration of CO?

14-10. A mixture of 1.00 mole of Cl_2 gas and 1.00 mole of Br_2 gas is enclosed in a 2.0-liter flask. At a certain temperature the reaction $Br_2(g) + Cl_2(g) \rightleftharpoons 2\,BrCl(g)$ occurs. When equilibrium is established, it is found that 9.8 % of the Br_2 has been used up. What is the equilibrium constant for the reaction?

14-11. The equilibrium constant for the reaction

$$N_2(g) + 3 H_2(g) \rightleftharpoons 2 NH_3(g)$$

is 0.50 at a certain temperature. If 1.0 mole of N_2 and 3.0 moles of H_2 are in equilibrium with NH_3 in a 1.0-liter vessel at this temperature, what is the equilibrium concentration of NH_3? What fraction of the *total* nitrogen in the system remains unreacted?

14-12. The equilibrium constant for the reaction

$$SO_2(g) + NO_2(g) \rightleftharpoons SO_3(g) + NO(g)$$

is 1.0 at a certain temperature. If 3 moles of SO_2 and 1 mole of NO_2 are placed together in a 1-liter vessel at this temperature, what will be the concentration of SO_3 at equilibrium?

14-13. A sample containing 1.50 mole of $POCl_3$ is enclosed in a 0.500-liter vessel at a certain temperature. When the equilibrium for the dissociation reaction $POCl_3(g) \rightleftharpoons POCl(g) + Cl_2(g)$ is attained, it is found that the compound is 25.0% dissociated. What is the equilibrium constant?

14-14. In the equilibrium system, $2 NO(g) + O_2(g) \rightleftharpoons 2 NO_2(g)$, at 1000°K, it is found that the concentration of NO is 2.00 M, and the concentration of O_2 is 0.075 M. If the equilibrium constant at this temperature is 1.20, (a) what is the concentration of NO_2, and (b) what fraction of the original NO has been converted to NO_2?

14-15. Exactly 4.00 moles of HBr is enclosed in a 1.00-liter flask and heated to 1297°K, whereupon the equilibrium

$$2 HBr(g) \rightleftharpoons H_2(g) + Br_2(g)$$

is rapidly established. The equilibrium constant is 7.32×10^{-6}. (a) What is the percentage of dissociation under these conditions? (b) What is the concentration of each species? (c) What is the pressure of each species?

14-16. The brown gas, NO_2, on cooling is converted to the colorless gas, dinitrogen tetroxide, N_2O_4, as described by the reaction $2 NO_2(g) \rightleftharpoons N_2O_4(g)$. If the original concentration of NO_2 is 0.2 M, and at equilibrium its concentration is only 0.1 M, what is the equilibrium constant for the reaction? What fraction of the NO_2 has reacted?

14-17. For the reaction $PCl_3(g) + PBr_3(g) \rightleftharpoons PCl_2Br(g) + PClBr_2(g)$, the equilibrium constant is 1.5 at a certain temperature. The pressure of PCl_2Br is equal to the pressure of $PClBr_2$ and is found to be 9.0 cm of Hg. If the pressure of PCl_3 is 3 cm of Hg, what is the pressure of PBr_3?

14-18. How many moles of PCl_5 must be added to a 10-liter flask at 250°C to obtain a concentration of 0.1 mole/liter of Cl_2? $K_c = 0.0415$.

14-19. At a certain temperature, the equilibrium constant for the reaction $CO_2(g) + H_2(g) \rightleftharpoons CO(g) + H_2O(g)$ is 16.0. Originally, equal numbers of moles of H_2 and CO_2 were placed in a flask. At equilibrium, the pressure of H_2 is 0.6 atm. What is the pressure of each of the other gases?

14-20. A sample, 0.500 mole, of $COBr_2$ was placed in a 1.00-liter reaction vessel and allowed to come to equilibrium at a temperature of 454°K. After the equilibrium $COBr_2(g) \rightleftharpoons CO(g) + Br_2(g)$ was established, it was found that 64.0% of the initial material had dissociated. (a) What is the concentration equilibrium constant of the reaction? (b) If 0.750 mole of Br_2 is added to the equilibrium mixture, what is the concentration of each species when equilibrium is reestablished?

14-21. At a certain temperature the equilibrium mixture

$$SO_2(g) + NO_2(g) \rightleftharpoons SO_3(g) + NO(g)$$

in a 1.0-liter flask is analyzed and found to contain 0.40 mole of SO_3, 0.25 mole of NO, 0.10 mole of SO_2, and 0.20 mole of NO_2. If 0.3 mole of SO_2 is added to the flask at constant temperature, calculate the new concentration of each gas in the mixture after establishment of equilibrium.

14-22. For the dissociation of $2\ HI(g) \rightleftharpoons H_2(g) + I_2(g)$, the equilibrium constant is 154 at 500°K. If 1.00 mole of HI is suddenly injected into a 2.00-liter flask at this temperature (a) what are the equilibrium concentrations of the various species? (b) What percentage of HI has dissociated?

14-23. The equation $CO_2(g) + H_2(g) \rightleftharpoons CO(g) + H_2O(g)$ represents the reaction between CO_2 and H_2 at elevated temperatures. (a) If at 1800°C there are 0.20 mole of CO_2, 0.20 mole of H_2, 0.40 mole of CO, and 0.40 mole of H_2O in a 1.00-liter flask, what is the concentration equilibrium constant? (b) what number of moles of CO_2 must be added to increase the concentration of CO to 0.50 M?

14-24. At 650°C, the equilibrium $N_2(g) + 3\ H_2(g) \rightleftharpoons 2\ NH_3(g)$ is established. If a stoichiometric mixture of 3.83 moles of N_2 and 11.49 moles of H_2 is heated to 650°K in a 1.000-liter container, it is found that the reaction is 74.0% complete when equilibrium is established. (a) What are the equilibrium concentrations of the various species? (b) What is the equilibrium constant? (c) How many moles of NH_3 must be added to raise the equilibrium concentration of H_2 to 3.50 M?

14-25. The equilibrium constant for the formation of HBr by the reaction $H_2(g) + Br_2(g) \rightleftharpoons 2\ HBr(g)$ at 1000°K is 2.20×10^6. If 0.0800 mole of HBr is suddenly introduced into a 4.00-liter flask at 1000°K, what fraction of the HBr undergoes dissociation?

14-26. At 690°K, solid $TiCl_3$ is in equilibrium with HCl, $TiCl_4$, and H_2 as illustrated by the equation

$$2\ TiCl_3(s) + 2\ HCl(g) \rightleftharpoons 2\ TiCl_4(g) + H_2(g)$$

At equilibrium, the pressure of HCl is 0.706 atm, the pressure of $TiCl_4$ is 1.500 atm, and the total pressure of the system is 4.206 atm. (a) What is the equilibrium constant? (b) What must the pressure of H_2 be for the pressure of HCl to equal the pressure of $TiCl_4$? Assume the vapor pressure of $TiCl_3(s)$ is negligible.

14-27. At a certain temperature above 300°C, NOCl dissociated into NO and Cl_2, according to the equation $2 \, NOCl(g) \rightleftharpoons 2 \, NO(g) + Cl_2(g)$. The total pressure of the system was 708.1 mm of Hg, the pressure of NOCl was 241.0 mm of Hg, and the pressure of NO was 311.4 mm of Hg. (a) Calculate the equilibrium constant using pressures in the units of atmospheres. (b) What fraction of the total amount of NOCl initially present has dissociated?

14-28. Using the data in Example 14-a, calculate the equilibrium constant employing the partial pressures of the various species present. *Hint:* Calculate the pressures using the ideal gas equation.

14-29. An industrial fuel gas is produced by reducing CO_2 with carbon according to the equation $CO_2(g) + C(s) \rightleftharpoons 2 \, CO(g)$. At a total pressure of 30 atm and 1000°C, the mixture of gases in equilibrium analyzes 17% CO_2 and 83% CO by volume. (a) Determine the equilibrium constant for the reaction. (b) Determine the composition of the gaseous mixture at a total pressure of exactly 10 atm at 1000°C.

14-30. At elevated temperatures, limestone dissociates according to the equation $CaCO_3(s) \rightleftharpoons CaO(s) + CO_2(g)$. At 1000°C, the equilibrium pressure of CO_2 is 3.87 atm. If pure $CaCO_3$ is placed in an evacuated 2.00-liter container and heated to 1000°, how much $CaCO_3$ will decompose to produce the equilibrium pressure? Assume the vapor pressures of the solids are negligible.

14-31. At 2000°K, the formation constant of NO from the elements according to the equation $N_2(g) + O_2(g) \rightleftharpoons 2 \, NO(g)$ is 4×10^{-4}. If the pressure of NO at equilibrium is found to be equal to 0.1 atm, and the pressure of N_2 is equal to the pressure of O_2, what are the equilibrium pressures of N_2 and O_2?

14-32. Derive the relationship between K_c and K_p, that is $K_p = K_c(RT)^{\Delta n}$. *Hint:* Use the general gas law equation $PV = nRT$ where $n/V =$ concentration in moles per liter.

14-33. One liter of an equilibrium mixture of NH_3, N_2, and H_2 at 750°K is composed of 1.60 moles of H_2, 0.10 moles of N_2 and 0.16 moles of NH_3. Consider the equilibrium to be $N_2 + 3 \, H_2 \rightleftharpoons 2 \, NH_3$. (a) Calculate K_c. (b) What is the total pressure of the mixture? (c) Determine the partial pressure of each gas. (d) Calculate K_p from partial pressures and also from the relationship $K_p = K_c(RT)^{\Delta n}$.

14-34. 5.00 moles of PCl_5 is placed in a 100-liter container and heated to 250°C. $K_c = 4.15 \times 10^{-2}$. Calculate the concentrations of each species at equilibrium.

14-35. In the previous problem, if 0.500 mole of Cl_2 is added at constant temperature, what is the concentration of each species when equilibrium has been re-established?

14-36. At $700°K$, $K_p = K_c = 1.83$ for the equilibrium

$$2 HI(g) \rightleftharpoons H_2(g) + I_2(g)$$

25 g of iodine and 0.50 g of hydrogen are heated to $700°K$ in a 5.0-liter flask. What are the equilibrium concentrations in moles/liter?

14-37. K_d for lactic acid between chloroform and water at $25°C$ is 0.0203. If 50 ml of 0.5 M lactic acid in chloroform is shaken with 50 ml of water, what weight of lactic acid (MW = 90) will be extracted?

14-38. Using solubility data given in Example 14-i, determine (a) the weight of iodine extracted from 25 ml of an aqueous solution containing 2 mg I_2 by shaking with 10 ml of CS_2, (b) the amount of I_2 remaining in the water solution after a second extraction with 10 ml CS_2.

14-39. When 100 ml of chloroform ($CHCl_3$) is shaken with 100 ml of 0.001 M aqueous I_2 at $20°C$, 0.025 g of iodine is extracted from the latter solution. Calculate the distribution coefficient for I_2 between $CHCl_3$ and H_2O.

14-40. The distribution coefficient, K_d, for succinic acid between water and ether is 6.0. If 4.0 g of succinic acid is shaken with a mixture of 100 ml water and 100 ml ether, what weight of acid will be dissolved in the ether layer?

CHAPTER 15

Ionic Equilibrium

A. IONIZATION OF WEAK ELECTROLYTES

Weak electrolytes are those that are only partially ionized in solution. Invariably in these solutions there are equilibria between the dissociated and undissociated species. Acetic acid typifies a weak electrolyte and its ionization may be represented by the equation

$$HC_2H_3O_2 \rightleftharpoons H^+ + C_2H_3O_2^-$$

The equilibrium constant for this reaction is written employing the same rules outlined in the previous section on molecular equilibrium, that is, the equilibrium constant is the product of the concentrations of the substances produced in the chemical reaction divided by the product of the concentrations of reacting substances. The concentration of each substance is raised to the power given by the coefficient in the balanced chemical equation. For the acetic acid ionization,

$$K_a = \frac{[H^+][C_2H_3O_2^-]}{[HC_2H_3O_2]}$$

This constant K_a is usually referred to as the dissociation or ionization constant of the acid.*

For the dissociation of a base BOH into B^+ and OH^-, the dissociation or ionization constant of the base is expressed as

$$K_b = \frac{[B^+][OH^-]}{[BOH]}$$

---

* Acids are molecules or ions which donate a proton to the solvent, for example, $HC_2H_3O_2 + H_2O \rightleftharpoons H_3O^+ + C_2H_3O_2^-$ or $HC_2H_3O_2 + 4 H_2O \rightleftharpoons H_9O_4^+ + C_2H_3O_2^-$. For simplicity, it is usual to show the ionization process as giving the simple hydrogen ion, H^+, but it should be remembered that it is aquated in aqueous solution.

A solution of ammonia in water is a typical example of such a weak base. There is a question as to the exact nature of the undissociated species present, and it appears that both NH_3 and hydrogen-bonded species ($HOH—NH_3$) may be present. This means the equilibrium may be complex as shown by the sequence

$$NH_3 + H_2O \rightleftharpoons NH_4OH \rightleftharpoons NH_4^+ + OH^-$$

However, the ionization equilibrium may, for all practical purposes, be adequately represented either as

$$NH_4OH \rightleftharpoons NH_4^+ + OH^- \qquad K_b = \frac{[NH_4^+][OH^-]}{[NH_4OH]} \qquad (1)$$

or as

$$NH_3 + H_2O \rightleftharpoons NH_4^+ + OH^- \qquad K_b = \frac{[NH_4^+][OH^-]}{[NH_3]} \qquad (2)$$

In the second case, the concentration of water has been omitted from the equilibrium constant expression because water is present in large excess, and its concentration does not change measurably during the course of the reaction. Since $[H_2O]$ is a constant, it is included in the value of K_b, so that the values of K_b for reactions 1 and 2 are numerically the same when expressed as above. In these equations, both $[NH_4OH]$ and $[NH_3]$ represent the total concentration of nonionized ammonia. Other nitrogen bases may be treated similarly.

Substances containing more than one acidic hydrogen ion in the molecule (polyprotic acids) undergo successive dissociation reactions. An equilibrium is eventually established involving all the species present in solution. A typical example is phosphoric acid, and the stepwise equilibria are shown below.

$$H_3PO_4 \rightleftharpoons H^+ + H_2PO_4^- \qquad K_1 = \frac{[H^+][H_2PO_4^-]}{[H_3PO_4]}$$

$$H_2PO_4^- \rightleftharpoons H^+ + HPO_4^{--} \qquad K_2 = \frac{[H^+][HPO_4^{--}]}{[H_2PO_4^-]}$$

$$HPO_4^{--} \rightleftharpoons H^+ + PO_4^{---} \qquad K_3 = \frac{[H^+][PO_4^{---}]}{[HPO_4^{--}]}$$

Other common polyprotic acids are sulfuric acid, H_2SO_4; hydrosulfuric acid, H_2S; pyrophosphoric acid, $H_4P_2O_7$; sulfurous acid, H_2SO_3; and carbonic acid, H_2CO_3.

Example 15-a. A 0.1000 M solution of $HC_2H_3O_2$ is found by conductivity measurements to be 1.30% ionized at 25°C. What is the dissociation constant of this acid?

Solution. One and three tenths per cent of the total amount of $HC_2H_3O_2$ present has dissociated into ions, that is, 0.1000 × 0.0130 = 0.00130 mole/liter have dissociated. Therefore, $[H^+]$ = $[C_2H_3O_2^-]$ = 0.00130 mole/liter and the equilibrium concentration of undissociated acetic acid $[HC_2H_3O_2]$ = 0.1000 − 0.0013 = 0.0987 mole/liter. Substituting in the equilibrium constant expression,

$$K_a = \frac{[H^+][C_2H_3O_2^-]}{[HC_2H_3O_2]} = \frac{(1.30 \times 10^{-3})(1.30 \times 10^{-3})}{9.87 \times 10^{-2}}$$

$$= 1.71 \times 10^{-5}$$

Example 15-b. The numerical value of K_b for NH_3 is 1.85×10^{-5}. What percentage of a 0.0200 M solution of NH_3 is ionized?

Solution.

$$NH_3 + H_2O \rightleftharpoons NH_4^+ + OH^-$$

At equilibrium: let X = molar concentration of NH_4^+ produced. Since one OH^- is produced for every NH_4^+,

$$X = [NH_4^+] = [OH^-].$$

The concentration of NH_3 at equilibrium is equal to the total amount present minus the amount ionized:

$$[NH_3] = (0.0200 - X) \text{ mole/liter}$$

Substituting these quantities into the equilibrium constant expression

$$1.85 \times 10^{-5} = \frac{(X)(X)}{0.0200 - X} \tag{1}$$

$$X^2 + (1.85 \times 10^{-5})X - 3.70 \times 10^{-7} = 0$$

On solution,

$$X = 6.0 \times 10^{-4} \text{ mole/liter}$$

Therefore

$$[NH_4^+] = [OH^-] = 6.0 \times 10^{-4} \text{ mole/liter,}$$

and

$$[NH_3] = .00200 - 0.0006 = 0.0194 \text{ mole/liter}$$

$$\text{Percentage dissociation} = \frac{6.0 \times 10^{-4}}{0.0200} \times 100 = 3.0\%$$

Alternate Solution of Equation (1). The solution to this problem can be considerably simplified by making the valid assumption that X is small compared to 0.0200. If we consider that $0.0200 - X \cong 0.0200$, equation (1) transforms to

$$X^2 = 3.7 \times 10^{-7}$$

On solution, $X = \sqrt{3.7 \times 10^{-7}} = 6.1 \times 10^{-4}$ mole/liter which agrees well with the value of X obtained above.

Whether or not this type of approximation procedure is acceptable depends entirely on the accuracy desired in the final answer. This is usually determined by the number of significant figures needed in each individual case. Note that in obtaining the sum or difference of two numbers of different orders of magnitude, the smaller number is neglected. However, in multiplication and division, neglect of the smaller number is *not* permissible.

Example 15-c. The dissociation reactions and their equilibrium constants for H_2S are

$$H_2S \rightleftharpoons H^+ + HS^- \qquad K_1 = 1.10 \times 10^{-7}$$
$$HS^- \rightleftharpoons H^+ + S^{--} \qquad K_2 = 1 \times 10^{-14}$$

(a) What is the concentration of H^+ of a 0.100 M solution of H_2S?
(b) What is the S^{--} concentration of a 0.100 M solution?

Solution. (a) An approximate solution can be obtained as follows.

Since K_1 is very much larger than K_2, it may be assumed that all the H^+ comes from the first reaction.* If this is the case, then the equilibrium concentration of the H^+ is equal to the concentration of HS^-. The equilibrium concentration of H_2S is equal to the initial concentration minus the amount ionized.

Let $[H^+] = X$.

Then $[HS^-] = X$ and $[H_2S] = 0.100 - X$

Substituting these quantities into the equilibrium constant expression

$$K = \frac{[H^+][HS^-]}{[H_2S]} = \frac{(X)(X)}{0.100 - X} \cong \frac{X^2}{0.100}$$

* Note that the hydrogen ion produced by the second ionization is equal to the sulfide ion concentration [which is shown in part (b) to be 1×10^{-14} M]. This value is negligible compared to the hydrogen ion produced from the first ionization, 1.05×10^{-4} M.

if the assumption is made that X is small compared to 0.100.

$$X = \sqrt{K_1 \times 0.100} = \sqrt{1.10 \times 10^{-7} \times 10^{-1}}$$
$$= \sqrt{1.10 \times 10^{-8}} = 1.05 \times 10^{-4} \text{ mole/liter}$$

It is evident that this value of X is small compared to 0.100; hence the assumption that $0.100 - X \cong 0.100$ is justified.

(b) To calculate the S^{--} concentration, the second equilibrium must be employed. From part (a) it is seen that $[HS^-] = 1.05 \times 10^{-4}\ M$ and $[H^+] = 1.05 \times 10^{-4}\ M$. Substituting these values into the equilibrium constant for K_2, $[S^{--}]$ can be obtained.

$$K_2 = \frac{[H^+][S^{--}]}{[HS^-]} = \frac{(1.05 \times 10^{-4})[S^{--}]}{1.05 \times 10^{-4}} = [S^{--}]$$

Substituting the numerical value for K_2, $[S^{--}] = 1 \times 10^{-14}$ mole/liter.

PROBLEMS (Ionization)

15-1. The ionization constant for the weak acid, HA, is 8.0×10^{-6}. Calculate the concentration of H^+ in a 0.020 M solution of the acid.

15-2. Calculate the ionization constant for the weak base, MOH, if a 0.030 M solution is 0.30% ionized.

15-3. K_a for amino-benzoic acid, $NH_2C_6H_4CO_2H$, is 1.2×10^{-5}. Calculate the H^+ concentration in a 0.030 M solution of the acid.

15-4. What is the H^+ concentration in a 0.10 M solution of $HC_2H_3O_2$? The dissociation constant is 1.71×10^{-5}. What percentage of $HC_2H_3O_2$ is dissociated?

15-5. What is the hydroxide ion concentration in a 0.050 M solution of methyl amine, CH_3NH_2? This substance reacts with H_2O as illustrated by the equation $CH_3NH_2 + H_2O \rightleftharpoons CH_3NH_3^+ + OH^-$, for which the equilibrium constant is 5.0×10^{-4}. What percentage of CH_3NH_2 has reacted?

15-6. The aromatic base pyridine, C_5H_5N, is known to have an equilibrium constant $K_b = 1.6 \times 10^{-9}$ for the reaction $C_5H_5N + H_2O \rightleftharpoons C_5H_5NH^+ + OH^-$. (a) What is the OH^- concentration of a 0.50 M solution? (b) What is the OH^- concentration of a 0.020 M solution?

15-7. A 4.0 M solution of piperidine, $C_5H_{10}NH$, is in equilibrium with $C_5H_{10}NH_2^+$ and OH^- as shown by the equation

$$C_5H_{10}NH + H_2O \rightleftharpoons C_5H_{10}NH_2^+ + OH^-$$

At equilibrium, it is found that 1.0% of the $C_5H_{10}NH$ is in the ionized form. What is the equilibrium constant for the reaction?

15-8. The weak acid, HCN, is slightly ionized by the reaction $HCN \rightleftharpoons H^+ + CN^-$. A 0.050 M solution is found to be 1.0×10^{-2} per cent ionized. What is the ionization constant of the acid?

15-9. The ionization constants for oxalic acid, $H_2C_2O_4$, are $K_1 = 5.6 \times 10^{-2}$ and $K_2 = 5.2 \times 10^{-5}$. (a) What is the concentration of H^+ of a 0.20 M solution of $H_2C_2O_4$, and (b) what is the $C_2O_4^{--}$ concentration?

15-10. H_2SO_3 is a diprotic acid and undergoes stepwise dissociation. The numerical value of K_1 is 1×10^{-2} and the value of K_2 is 5×10^{-6}. (a) What is the H^+ concentration of a 0.15 M solution? (b) What is the equilibrium concentration of SO_3^{--}?

B. IONIZATION OF WATER

The solvent water ionizes into H^+ and OH^- as shown by the equation $H_2O \rightleftharpoons H^+ + OH^-$. By well-established procedures, it can be shown that the concentration of each ion in pure H_2O is 1×10^{-7} M at 25°C. From this, the ionization equilibrium constant

$$K_i = \frac{[H^+][OH^-]}{[H_2O]}$$

may be determined. However, in pure water and in dilute aqueous solutions the concentration of water is essentially constant. The number of grams of water per liter is approximately 1000, which is equivalent to 55.5 M. Hence $K_i \times 55.5$ is also a constant, usually called the ion product for water and designated K_w.

$$K_w = [H^+] \times [OH^-]$$

On substitution of the numerical values of the concentrations of H^+ and OH^-, $K_w = 10^{-7} \times 10^{-7} = 10^{-14}$. Under any conditions in dilute aqueous solution, the product of the concentrations of hydrogen and hydroxide ions must equal 10^{-14} at 25°C. Acidic solutions are those in which the hydrogen ion concentration is greater than the hydroxide ion concentration. For example, in 0.1 M HCl, the acid is essentially completely dissociated so that the hydrogen ion concentration is 0.1 M. From the K_w relation

$$[OH^-] = \frac{K_w}{[H^+]} = \frac{10^{-14}}{10^{-1}} = 10^{-13} \text{ mole/liter}$$

Similarly in basic solution, such as 0.01 M NaOH, the $[OH^-] = 10^{-2}$ M and therefore $[H^+] = 10^{-12}$ M. Note that in basic solution, the hydroxide ion concentration is greater than the hydrogen ion concentration.

Example 15-d. The H^+ concentration of the 0.100 M solution of $HC_2H_3O_2$ described in Example 15-a, was 0.0013 mole/liter. What is the OH^- concentration of this solution?

Solution.

$$[OH^-] = \frac{K_w}{[H^+]} = \frac{1.0 \times 10^{-14}}{1.3 \times 10^{-3}} = 7.7 \times 10^{-12} \text{ mole/liter}$$

PROBLEMS (Ionization of Water)

15-11. What are the H^+ and OH^- concentrations of each of the following solutions of strong electrolytes? Assume 100% ionization.

(a) 0.001 M HCl   (f) 5×10^{-7} M NaOH
(b) 0.0001 M HNO_3 (g) 0.02 M $Ba(OH)_2$
(c) 1×10^{-5} M $HClO_4$ (h) 0.01 M KOH
(d) 0.025 M $HClO_4$ (i) 0.1 M KOH
(e) 3×10^{-5} M HCl

C. THE pH VALUES OF AQUEOUS SOLUTIONS

The term pH is used to designate the acidity or basicity of a solution, and is defined as pH $= -\log [H^+]$. This is a convenient scale on which the lower numbers indicate an acidic solution and the higher numbers indicate a basic solution. Similarly, the term pOH may be defined as $-\log [OH^-]$. A direct relationship between pH and pOH exists and is easily derived. It was shown in the preceding section that $[H^+][OH^-] = 10^{-14}$. Taking the logarithms of both sides of the equation, $\log [H^+] + \log [OH^-] = \log 10^{-14} = -14$. Substituting the definitions of pH and pOH, it is found that pH $+$ pOH $= 14$. In the following table are given pH and pOH values for several solutions of strong acids and bases of varying concentrations. Note that the sum of pH and pOH is always 14 for dilute aqueous solutions at 25°C.

Solution	$[H^+]$	pH	$[OH^-]$	pOH	Nature of solution
1.0 M HCl	1.0 M	0	10^{-14} M	14	Acidic
0.1 M HCl	0.1	1	10^{-13}	13	Acidic
0.0001 M HCl	10^{-4}	4	10^{-10}	10	Acidic
Pure water	10^{-7}	7	10^{-7}	7	Neutral
0.001 M NaOH	10^{-11}	11	10^{-3}	3	Basic
0.1 M NaOH	10^{-13}	13	10^{-1}	1	Basic

Example 15-e. (a) What is the pOH of a 0.010 M NH_3 solution? (b) What is the pH? The equilibrium constant is 1.8×10^{-5} for the reaction $NH_3 + H_2O \rightleftharpoons NH_4^+ + OH^-$.

Solution. (a) At equilibrium: Let $X =$ concentration of NH_4^+ produced by reaction of NH_3.

$$X = [NH_4^+] = [OH^-]$$

$$[NH_3] = (0.010 - X)\ \text{mole/liter}$$

Substituting into the equilibrium constant expression

$$1.8 \times 10^{-5} = \frac{X^2}{0.010 - X}$$

Assume $(0.010 - X) \cong 0.010$. Then $X^2 = 1.8 \times 10^{-7}$. $X = 4.2 \times 10^{-4}$ mole/liter, and therefore $[OH^-] = 4.2 \times 10^{-4}$ mole/ liter.

$$\text{pOH} = -\log (4.2 \times 10^{-4}) = 3.38$$

(b) The pH may be obtained in either of two ways. (1) From the K_w relationship,

$$[H^+] = \frac{K_w}{[OH^-]} = \frac{10^{-14}}{4.2 \times 10^{-4}} = 2.4 \times 10^{-11}\ \text{mole/liter}$$

Since pH $= -\log [H^+]$,

$$\text{pH} = -\log (2.4 \times 10^{-11}) = -(\log 2.4 + \log 10^{-11})$$

$$= -(0.38 - 11) = 10.62$$

Or (2) since

$$\text{pH} + \text{pOH} = 14,$$

$$\text{pH} = 14 - \text{pOH} = 14.00 - 3.38 = 10.62$$

Example 15-f. The pH of a 0.500 M solution of formic acid, HCOOH, is 2.05. What is the equilibrium constant?

Solution. Formic acid ionizes as

$$\text{HCOOH} \rightleftharpoons \text{H}^+ + \text{HCOO}^- \qquad K_a = \frac{[\text{H}^+][\text{HCOO}^-]}{[\text{HCOOH}]}$$

At equilibrium, the H^+ concentration may be calculated from the pH

$$\text{pH} = 2.05 = -\log [H^+]; \qquad [H^+] = 8.9 \times 10^{-3}\ \text{mole/liter}$$

Since this is a solution containing only formic acid, then $[H^+] = [HCOO^-]$

$$[HCOOH] = 0.500 - 0.009 = 0.491 \text{ mole/liter}$$

$$K_a = \frac{(8.9 \times 10^{-3})^2}{0.491} = 1.61 \times 10^{-4}$$

PROBLEMS (pH of Aqueous Solution)

15-12. Determine the pH and pOH of each of the following solutions.

(a) 0.020 M NH_3, for which $K_b = 1.85 \times 10^{-5}$.
(b) 0.050 M NH_3, for which $K_b = 1.85 \times 10^{-5}$.
(c) 0.50 M $HC_2H_3O_2$, for which $K_a = 1.7 \times 10^{-5}$.

15-13. The pH of a 0.50 M solution of trichloroacetic acid, $HC_2Cl_3O_2$ is 0.64. What is the equilibrium constant for the acid dissociation reaction $HC_2Cl_3O_2 \rightleftharpoons H^+ + C_2Cl_3O_2^-$?

15-14. The base hydroxyethylamine, $HOCH_2CH_2NH_2$, reacts with water to produce hydroxyethylammonium ion and hydroxide ion as illustrated by the equation

$$HOCH_2CH_2NH_2 + H_2O \rightleftharpoons HOCH_2CH_2NH_3^+ + OH^-$$

The pOH of a 0.20 M solution is 2.48. What is the base equilibrium constant for the reaction?

15-15. Hypochlorous acid dissociates as shown by the equation

$$HClO \rightleftharpoons H^+ + ClO^-$$

If a 0.20 M solution has a pH = 3.85, what is the dissociation constant of the acid?

15-16. A 0.50 M solution of hydrazine, N_2H_4, has a pH of 11.10. If the equilibrium is represented by the equation

$$N_2H_4 + H_2O \rightleftharpoons N_2H_5^+ + OH^-$$

what is the equilibrium constant for the reaction?

D. pK

Another term which is useful in considering equilibrium systems is pK, defined as follows:

$$pK = -\log K$$

where K is the equilibrium constant. In dealing with ionic systems the equilibrium constant may be the ionization constant such as K_a, K_b, K_w, K_{sp}, etc.

For the equilibrium

$$H_2O \rightleftharpoons H^+ + OH^-$$

$$[H^+][OH^-] = K_w = 1 \times 10^{-14}$$

$$\log [H^+] + \log [OH^-] = \log K_w = \log 10^{-14}$$

or

$$-\log [H^+] - \log [OH^-] = -\log K_w = -(-14) = 14 = pK_w$$

Hence

$$pH + pOH = 14 = pK_w$$

Example 15-g. Calculate pK_a for 0.0100 M benzoic acid $(HC_7H_5O_2)$ which is 8% ionized.

Solution.

$$K_a = \frac{[H^+][C_7H_5O_2^-]}{[HC_7H_5O_2]}$$

$$[H^+] = [C_7H_5O_2^-] = (0.08)(0.0100) = 0.0008 \text{ mole/liter}$$

$$[HC_7H_5O_2] = (0.0100 - 0.0008) = 0.0092 \text{ mole/liter}$$

$$K_a = \frac{(8 \times 10^{-4})^2}{9.2 \times 10^{-3}} = 7 \times 10^{-5}$$

$$pK_a = -\log K_a = -\log (7 \times 10^{-5}) = 4.15$$

Example 15-h. What is the pH of a 0.1 molar solution of bromoacetic acid $(HC_2H_2O_2Br)$ if $pK_a = 2.7$

Solution.

$$pK_a = 2.7 = -\log K_a$$

$$K_a = \text{antilog} (-2.7) = 2 \times 10^{-3}$$

$$K_a = \frac{[H^+][C_2H_2O_2Br^-]}{[HC_2H_2O_2Br]} = 2 \times 10^{-3}$$

Since $[H^+] = [C_2H_2O_2Br^-]$

$$\frac{[H^+]^2}{0.1} = 2 \times 10^{-3} \quad \text{and} \quad [H^+] = \sqrt{2 \times 10^{-4}} = 1.4 \times 10^{-2} M$$

$$pH = -\log [H^+] = -\log (1.4 \times 10^{-2}) = 1.9$$

PROBLEMS (pK)

15-17. Calculate the pH of a 0.5 M solution of KOH. Assume complete ionization. $pK_w = 13.8$.

15-18. An aqueous solution of $HC_2H_3O_2$ and $NaC_2H_3O_2$ has a pH of 4.50. What is the ratio of concentrations of $C_2H_3O_2^-$ to $HC_2H_3O_2$? $pK_a = 4.77$

15-19. At 55°C $pK_w = 13.14$. What is the pH of a neutral solution at this temperature?

15-20. pK_a for chloroacetic acid at 25°C $= 2.85$. Calculate the pH of a 0.00100 M solution of the acid.

15-21. A 0.050 M solution of the acid, HX, has a pH of 4.70. Determine pK_a for this acid.

E. THE COMMON ION EFFECT

The addition of a strong electrolyte to a solution of a weak electrolyte has a pronounced effect on the equilibrium involving the weak electrolyte if the two electrolytes have a common ion; for example, the addition of the strong electrolyte NH_4Cl to a solution of NH_3 in H_2O disturbs the equilibrium, $NH_3 + H_2O \rightleftharpoons NH_4^+ + OH^-$.

The introduction of NH_4Cl increases the NH_4^+ concentration. By LeChatelier's principle, the equilibrium shifts in such a way that the stress is minimized, that is, OH^- and NH_4^+ combine to form NH_3 and H_2O. Thus the equilibrium concentration of OH^- decreases and that of free NH_3 increases.

Example 15-i. Show that the OH^- concentration in a solution of NH_3 is greater than that in a solution of NH_3 and NH_4Cl. In Example 15-e, it was demonstrated that the OH^- concentration of a 0.010 M NH_3 solution was 4.2×10^{-4} M. What is the $[OH^-]$ of a solution of 0.020 M NH_4Cl and 0.010 M NH_3?

Solution. Let $X = [OH^-] =$ amount of NH_3 dissociated in moles/liter.

$$[NH_3] = (\text{amount of } NH_3 \text{ added} - \text{amount dissociated})$$
$$= (0.010 - X) \text{ mole/liter}$$
$$[NH_4^+] = (\text{amount of } NH_4Cl \text{ added} + \text{amount formed})$$
$$= (0.020 + X) \text{ mole/liter}$$

Substituting in the equilibrium constant expression for the dissociation of NH_3,

$$K_b = \frac{[NH_4^+][OH^-]}{[NH_3]}$$

$$1.8 \times 10^{-5} = \frac{(0.020 + X)(X)}{0.010 - X}$$

$$0.020X + X^2 = 1.8 \times 10^{-7} - 1.8 \times 10^{-5}X$$

$$X^2 + 0.020X - 1.8 \times 10^{-7} = 0$$

This equation may be solved by use of the quadratic or it may be solved by an approximation procedure as follows.

If X, the concentration of OH^-, is small compared to 0.010, then

$$[NH_3] = 0.010 - X \cong 0.010 \text{ mole/liter}$$

$$[NH_4^+] = 0.020 + X \cong 0.020 \text{ mole/liter}$$

Substituting into the equilibrium constant expression,

$$K_b = \frac{(0.020)(X)}{0.010}$$

$$X = \frac{1.8 \times 10^{-5} \times 10^{-2}}{2.0 \times 10^{-2}} = 9.0 \times 10^{-6} \text{ mole/liter}$$

To show that the assumption made concerning the relative magnitudes of X and 0.010 is valid:

$$[NH_3] = 0.010 - X = 0.010 - 0.000009 = 0.009991 \text{ mole/liter}$$

For all practical purposes, and within the number of significant figures given, 0.009991 is the same as 0.010 and hence the assumption is justified.

In a solution containing only 0.010 M NH_3, $[OH^-] = 4.2 \times 10^{-4}$ M (Example 15-e, Section C) while in a solution of 0.010 M NH_3 + 0.020 M NH_4Cl, $[OH^-] = 9.0 \times 10^{-6}$ M. Thus it is observed that the concentration of the noncommon ion (OH^-) of the weak electrolyte (NH_3) is reduced by the addition of a strong electrolyte (NH_4Cl) having an ion in common (NH_4^+) with the weak electrolyte.

PROBLEMS (Common Ion Effect)

15-22. Show that the H^+ concentration is greater in (a) a solution of 0.20 M formic acid, HCO_2H, than in (b) a solution containing 0.20 M HCO_2H plus

0.10 M sodium formate, $NaHCO_2$. K_a for formic acid, HCO_2H, is 1.6 × 10^{-4}.

15-23. In which solution is the OH^- concentration greater: (a) a solution containing 0.1 M hydrazinium chloride, N_2H_5Cl, and 0.5 M hydrazine, N_2H_4, or (b) in one containing 0.2 M N_2H_5Cl and 0.5 M N_2H_4? K_b for N_2H_4 is 4.2 × 10^{-6}. Establish your answer by calculation of OH^- concentration in each solution.

15-24. In which solution is the $C_2H_3O_2^-$ concentration greater: (a) one containing only 1.0 M $HC_2H_3O_2$ or (b) one containing 1.0 M $HC_2H_3O_2$ plus 0.5 M HCl? K_a for $HC_2H_3O_2$ is 1.7 × 10^{-5}. Establish your answer by calculation of the $C_2H_3O_2^-$ concentration in each solution.

15-25. In which solution is the NH_4^+ concentration greater: (a) one containing 0.10 M NH_3 or (b) one containing 0.10 M NH_3 plus 0.10 M $NaOH$? K_b for NH_3 is 1.85 × 10^{-5}. Establish your answer by calculation of the NH_4^+ concentration in each solution.

F. BUFFER SOLUTIONS

A buffer solution is one composed of substances of such a type that there is little or no change in pH on the addition of relatively small amounts of either acid or base. A solution containing both $NaC_2H_3O_2$ and $HC_2H_3O_2$ is such a solution, and, in general, a buffer solution is one which is composed of either a weak acid (or base) and a salt of that acid (or base).

In the case of an $HC_2H_3O_2$–$NaC_2H_3O_2$ buffer, $NaC_2H_3O_2$ is completely ionized, $NaC_2H_3O_2 \rightarrow Na^+ + C_2H_3O_2^-$; molecular $HC_2H_3O_2$ is in equilibrium with its ions, $HC_2H_3O_2 \rightleftharpoons H^+ + C_2H_3O_2^-$. Such a solution resists a change in pH in the following ways. Addition of H^+ would ordinarily cause a large decrease in pH. As a result of the presence of $C_2H_3O_2^-$ in large amount, the H^+ is converted to nonionized acetic acid by the net reaction $H^+ + C_2H_3O_2^- \rightarrow HC_2H_3O_2$. The decrease in pH is not nearly so large as it would have been in the absence of excess $C_2H_3O_2^-$. Similarly, the addition of OH^- should increase the pH markedly. But because of the presence of excess $HC_2H_3O_2$ molecules, the net reaction

$$HC_2H_3O_2 + OH^- \rightarrow H_2O + C_2H_3O_2^-$$

nullifies most of the effect of the added hydroxide ion.

Example 15-j. (a) A solution contains both 1.00 mole of sodium acetate and 1.00 mole of acetic acid in exactly 1 liter. What is the pH of the solution? The ionization constant of acetic acid is 1.71 × 10^{-5}.

Solution.

$$HC_2H_3O_2 \rightleftharpoons H^+ + C_2H_3O_2^- \qquad K_a = \frac{[H^+][C_2H_3O_2^-]}{[HC_2H_3O_2]}$$

At equilibrium: Let X equal the concentration of H^+ formed by ionization of $HC_2H_3O_2$.

$$[H^+] = X$$

$[C_2H_3O_2^-] = $ (Salt concentration $+ X$) $= (1.00 + X)$ mole/liter
$[HC_2H_3O_2] = $ (Acid concentration $- X$) $= (1.00 - X)$ mole/liter

$$K_a = 1.71 \times 10^{-5} = \frac{(X)(1 + X)}{1 - X}$$

Assume that X is small compared to 1, then $1.00 + X \cong 1.00$. Hence

$$\frac{(X)(1.00)}{1.00} = 1.71 \times 10^{-5}$$

Therefore $[H^+] = 1.71 \times 10^{-5}$ mole/liter and the pH of the solution is 4.77.

(b) Assume that 0.20 mole of NaOH is added to 1 liter of the solution in part (a) in such a fashion that the volume does not change. What is the pH of the solution when equilibrium is reestablished?

Solution. The net reaction on addition of hydroxide ion is $OH^- + HC_2H_3O_2 \rightarrow H_2O + C_2H_3O_2^-$. Hence 0.20 mole of NaOH uses up 0.2 mole of $HC_2H_3O_2$ and produces 0.2 mole of $NaC_2H_3O_2$. When the new equilibrium is established, let X again equal the H^+ concentration.

$[H^+] = X = $ (Amount of $HC_2H_3O_2$ dissociated in moles/liter)

$[C_2H_3O_2^-] = $ (Original salt concentration $+$ Salt concentration formed from added NaOH $+ X$)

$\qquad = 1.00 + 0.20 + X = 1.20 + X \cong 1.20$ moles/liter

$[HC_2H_3O_2] = $ (Original acid concentration $-$ Salt concentration formed by NaOH $- X$)

$\qquad = 1.00 - 0.20 - X = 0.80 - X \cong 0.80$ mole/liter

Substituting into the equilibrium constant expression

$$1.71 \times 10^{-5} = \frac{(X)(1.20)}{0.80}$$

$$X = 1.14 \times 10^{-5} \text{ mole/liter} = [H^+]$$

$$pH = 4.94 \text{ for the buffer solution}$$

If the same amount of NaOH, 0.20 mole, had been added to a liter of water or other unbuffered solution, the pH would be 13.30.

Thus it is observed that there is a tremendous effect of the buffer in resisting the change of pH.

(c) Assume that 0.50 mole of HCl is added to 1.0 liter of the solution in part (a) in such a fashion that the volume is not changed. What is the pH when equilibrium is reestablished?

> **Solution.** The net chemical reaction is the conversion of 0.50 mole of $C_2H_3O_2^-$ to 0.50 mole of $HC_2H_3O_2$. At the new equilibrium, let $X = [H^+]$.
>
> $[H^+] = X =$ (Amount of acetic acid dissociated in moles/liter)
>
> $[C_2H_3O_2^-] =$ (Original salt concentration $-$ Concentration of acid formed from salt $+ X$)
>
> $\qquad = 1.00 - 0.50 + X \simeq 0.50$ mole/liter
>
> $[HC_2H_3O_2] =$ (Original acid concentration $+$ Concentration of acid formed from salt $- X$)
>
> $\qquad = 1.00 + 0.50 - X \simeq 1.50$ moles/liter

Substituting into the equilibrium constant expression and solving for X,

$$X = [H^+] = 1.71 \times 10^{-5} \times \frac{1.50}{0.50} = 5.1 \times 10^{-5} \text{ mole/liter}$$

The pH of the solution is 4.29.

If the same amount of HCl, 0.50 mole, had been added to 1.0 liter of water or other unbuffered solution, the pH of the solution would have been 0.30.

PROBLEMS (Buffer Solutions)

15-26. What is the pH of a buffer solution composed of 0.40 M $HC_2H_3O_2$ and 0.50 M $NaC_2H_3O_2$? $K_a = 1.7 \times 10^{-5}$.

15-27. What are the most abundant and least abundant species (excepting H_2O) in a solution prepared by dissolving 1 mole of NH_4Cl in 1 liter of 1 M NH_4OH?

15-28. What is the pH of a buffer solution composed of 0.050 M NH_3 and 0.40 M NH_4NO_3? $K_b = 1.8 \times 10^{-5}$.

15-29. A buffer solution is made up by dissolving 0.50 mole of $NaC_2H_3O_2$ in 1 liter of 0.10 molar $HC_2H_3O_2$. $K_a = 1.8 \times 10^{-5}$. Calculate the pH of the solution.

15-30. (a) What is the pH of a buffer solution composed of 1.00 M NH$_3$ and 1.00 M NH$_4$Cl? (b) What number of moles of NaOH must one add to a liter of this buffer to increase the pH by 1.00 unit, assuming no volume change? $K_b = 1.85 \times 10^{-5}$.

15-31. A buffer solution is composed of 0.300 M HC$_2$H$_3$O$_2$ and 0.400 M NaC$_2$H$_3$O$_2$. What number of moles of HCl must one add to a liter of this buffer to decrease the pH by one unit? Assume no volume change on addition of the hydrochloric acid. $K_a = 1.71 \times 10^{-5}$.

GENERAL PROBLEMS

15-32. A 0.20 M solution of an acid, HA, is found by experiment to be 2.5% dissociated. What is the ionization constant of the acid?

15-33. A 1.0 molar solution of the weak base XOH is 5% ionized. Determine K_b.

15-34. Acetic acid has a dissociation constant of 1.71×10^{-5}. What fraction of the HC$_2$H$_3$O$_2$ is dissociated in (a) a 1.00 M solution, (b) a 0.100 M solution, and (c) a 0.0100 M solution?

15-35. Calculate the pH of 0.1 molar HXO. $K_a = 2.5 \times 10^{-6}$.

15-36. The base butylamine has an equilibrium constant of 5.1×10^{-4} for the reaction C$_4$H$_9$NH$_2$ + H$_2$O $\rightleftharpoons$ C$_4$H$_9$NH$_3^+$ + OH$^-$. What fraction of the C$_4$H$_9$NH$_2$ has reacted in (a) a 1.0 M solution, (b) a 0.010 M solution, and (c) a 0.0010 M solution?

15-37. A 1.0 M solution of cyanoacetic acid, NCCH$_2$COOH, is approximately 40% dissociated. What is the equilibrium constant for the reaction NCCH$_2$COOH $\rightleftharpoons$ (NCCH$_2$COO)$^-$ + H$^+$?

15-38. A solution of 0.20 M N$_2$H$_4$, and 0.40 M N$_2$H$_5$Cl was found to have a OH$^-$ concentration of 2.0×10^{-6} M. What is the equilibrium constant for the reaction N$_2$H$_4$ + H$_2$O $\rightleftharpoons$ N$_2$H$_5^+$ + OH$^-$?

15-39. Which of the following acids is *strongest*?

(a) Thioacetic acid $K_a = 4.7 \times 10^{-4}$
(b) Valeric acid $K_a = 1.5 \times 10^{-5}$
(c) Glutaric acid $K_a = 3.4 \times 10^{-4}$
(d) Cyanoacetic acid $K_a = 3.7 \times 10^{-3}$
(e) Hypobromous acid $K_a = 2.0 \times 10^{-9}$

15-40. What are the H$^+$ concentrations and pH values of each of the following solutions of strong bases? Assume 100% ionization.

(a) 0.001 M Ca(OH)$_2$
(b) 0.01 M KOH
(c) 1.5×10^{-2} M NaOH
(d) 4.0×10^{-3} M Sr(OH)$_2$
(e) 1×10^{-5} M Ba(OH)$_2$

15-41. What are the OH^- concentrations and pOH values of each of the following solutions of strong acids?

(a) $5 \times 10^{-2} M$ HCl

(b) $1 \times 10^{-3} M$ HClO$_4$

(c) $8 \times 10^{-7} M$ HCl

(d) $1.0 M$ HClO$_4$

(e) $4.0 \times 10^{-4} M$ HNO$_3$

15-42. A certain acid, HA, has an ionization constant of 1.6×10^{-5}. Determine the pH of a $0.010 M$ solution of the acid.

15-43. The ionization constant for $HC_2H_3O_2 \rightleftharpoons H^+ + C_2H_3O_2^-$ is 1.7×10^{-5}. Calculate the quantity of $NaC_2H_3O_2$ in grams to be added to 1.00 liter of $0.100 M$ $HC_2H_3O_2$ to yield a solution with a pH of 4.50.

15-44. A buffer solution is made by dissolving 1 mole of NH_4Cl in 500 ml of $0.1 M$ NH_3. What is the pH of the solution? K_b for $NH_3 = 1.8 \times 10^{-5}$.

15-45. A $0.40 M$ solution of a hypothetical base, BOH, is found to have a pOH of 2.60. What is the dissociation constant of the base?

15-46. A $0.020 M$ solution of a hypothetical acid, HX, is found to have a pOH of 8.50. What is the dissociation constant of the acid?

15-47. A solution of a weak acid (HA) carefully prepared to be $0.050 M$ exhibited a pH of 5.30. What is the ionization constant K_a of this weak acid?

15-48. A diprotic acid, H_2A, dissociates in a stepwise fashion with the equilibrium constants shown.

$$H_2A \rightleftharpoons H^+ + HA^- \qquad K_1 = 3.6 \times 10^{-6}$$
$$HA^- \rightleftharpoons H^+ + A^{--} \qquad K_2 = 2 \times 10^{-11}$$

(a) What is the pH of a $0.10 M$ solution of the acid? (b) What is the concentration of A^{--}?

15-49. A hypothetical base, $E(OH)_2$, dissociates in a stepwise manner with the equilibrium constants shown.

$$E(OH)_2 \rightleftharpoons EOH^+ + OH^- \qquad K_1 = 4 \times 10^{-6}$$
$$EOH^+ \rightleftharpoons E^{++} + OH^- \qquad K_2 = 2 \times 10^{-12}$$

(a) What is the pH of a $2 \times 10^{-2} M$ solution of the base? (b) What is the concentration of EOH^+? (c) What is the concentration of E^{++}?

15-50. The pH of a $2.5 \times 10^{-2} M$ solution of H_2CO_3 is 4.02. What is the value of K_1, the equilibrium constant for the reaction $H_2CO_3 \rightleftharpoons HCO_3^- + H^+$? Assume ionization of HCO_3^- is negligible.

15-51. What is the pH of a solution prepared by mixing 0.50 liter of $0.30 M$ $HC_2H_3O_2$ and 1.0 liter of $0.015 M$ NaOH? $K_a = 1.71 \times 10^{-5}$.

15-52. What is the pH of a solution prepared by mixing 0.5 liter of $0.50 M$ NH_3 and 0.5 liter of $0.10 M$ HCl? $K_b = 1.85 \times 10^{-5}$.

15-53. A solution composed of $0.50 M$ hydrazoic acid, HN_3, and $0.50 M$ sodium azide, NaN_3, has a pH of 4.77. What is the dissociation constant of HN_3?

15-54. What is the dissociation constant for HF if a 0.10 M solution has a pH of 2.02?

15-55. The ionization constant of iodic acid, HIO_3, is 1.9×10^{-1}. What fraction of a 0.40 M solution is dissociated? Determine the fraction dissociated at several concentrations of the acid, and then make a plot of the fraction dissociated versus the concentration of the acid.

15-56. A buffer is prepared containing 0.100 mole of $NaHCO_3$ and 0.100 mole of Na_2CO_3 in 0.500 liter of solution. The second ionization constant of H_2CO_3 is 4.4×10^{-11}, corresponding to the reaction $HCO_3^- \rightleftharpoons H^+ + CO_3^{--}$. (a) What is the pH of the solution? (b) What is the OH^- concentration? (c) How many moles of NaOH must be added to a liter of solution to increase the pH by one unit?

15-57. What is the pH of a solution prepared by mixing 500 ml of 0.44 M NaOH and 1.0 liter of 0.50 M boric acid, H_3BO_3? Boric acid ionizes as shown by the equation $H_3BO_3 \rightleftharpoons H^+ + H_2BO_3^-$, and has an ionization constant equal to 6.4×10^{-10}. The reaction between NaOH and H_3BO_3 is NaOH + $H_3BO_3 \rightleftharpoons NaH_2BO_3 + H_2O$.

CHAPTER 16

Equilibrium in Saturated Solutions

A. SOLUBILITY PRODUCT CONSTANTS

Salts on dissolving in water eventually reach a point of saturation at which an equilibrium is established between the salt and its solution. For many salts the amount dissolved is quite low, and the salts are said to be sparingly soluble or slightly soluble. The equilibrium of a sparingly soluble salt such as $BaSO_4$, with its saturated solution may be represented by an equation of the type

$$BaSO_4(s) \rightleftharpoons Ba^{++} + SO_4^{--}$$

An equilibrium constant may be written for this type of reaction just as for any other equilibrium reaction:

$$K = \frac{[Ba^{++}][SO_4^{--}]}{[BaSO_4(s)]}$$

Since $BaSO_4(s)$ is a pure substance, the "concentration" or "activity" of $BaSO_4$ in the solid is a constant. Therefore the product $K \times [BaSO_4(s)]$ is also a constant, termed the solubility product constant, and is designated as K_{sp}. Hence

$$K_{sp} = [Ba^{++}][SO_4^{--}]$$

For a salt of the general formula M_aX_b, the solubility equilibrium is $M_aX_b(s) \rightleftharpoons a\,M + b\,X$, and the solubility product constant is

$$K_{sp} = [M]^a[X]^b$$

175

The solubility product constant for any salt is thus the product of the concentrations of its ions in a saturated solution, each concentration raised to the power corresponding to its coefficient in the balanced chemical equation.* Concentrations in the equations are always expressed in moles per liter.

Example 16-a. Calculate K_{sp} for TlBr from the data that the experimentally determined solubility at 25°C is 0.0523 g per 0.100 liter of solution. The equation is

$$TlBr(s) \rightleftharpoons Tl^+ + Br^- \quad \text{and} \quad K_{sp} = [Tl^+][Br^-]$$

Solution. The solubility of TlBr (FW = 284) is given in grams per 0.100 liter, and this must be converted to moles per liter, or molarity. The molarity of the thallous bromide solution is

$$\text{Molarity} = \frac{0.0523 \text{ g}}{\dfrac{284 \text{ g}}{\text{mole}} \times 0.100 \text{ liter}} = 1.84 \times 10^{-3} \text{ mole/liter}$$

Since TlBr is completely ionized,

$$[Tl^+] = 1.84 \times 10^{-3} \text{ mole/liter}$$
$$[Br^-] = 1.84 \times 10^{-3} \text{ mole/liter}$$
$$K_{sp} = [Tl^+][Br^-] = 3.38 \times 10^{-6}$$

Example 16-b. The solubility of $Cd_3(PO_4)_2$ in water at 20°C was found to be 1.15×10^{-7} M. What is the K_{sp}?

Solution.

$$Cd_3(PO_4)_3(s) \rightleftharpoons 3 \text{ Cd}^{++} + 2 \text{ PO}_4^{---}$$
$$K_{sp} = [Cd^{++}]^3[PO_4^{---}]^2$$

Since every formula weight of $Cd_3(PO_4)_2$ which dissolves produces three moles of Cd^{++} and two of PO_4^{---},

$$[Cd^{++}] = 3 \times 1.15 \times 10^{-7} = 3.45 \times 10^{-7} \text{ mole/liter}$$
$$[PO_4^{---}] = 2 \times 1.15 \times 10^{-7} = 2.30 \times 10^{-7} \text{ mole/liter}$$
$$K_{sp} = (3.45 \times 10^{-7})^3(2.30 \times 10^{-7})^2$$
$$K_{sp} = (41.1 \times 10^{-21}) \times (5.29 \times 10^{-14}) = 218 \times 10^{-35}$$
$$K_{sp} = 2.18 \times 10^{-33}$$

* The assumption is made in this treatment that the saturated solutions of slightly soluble solids are completely ionized. Whereas this is a good approximation in many instances, it is not so for all electrolytes. In this introductory treatment, however, no attempt will be made to include corrections for the association of electrolytes.

Example 16-c. K_{sp} for AgI in water was found to be 1.2×10^{-17} at 25°C. What is the solubility in (a) moles per liter, and (b) grams per liter?

Solution.

$$AgI(s) \rightleftharpoons Ag^+ + I^-$$

$$K_{sp} = [Ag^+][I^-]$$

Let S = solubility of AgI in moles/liter. The solubility of AgI will be numerically equal to the concentration of Ag^+ and the concentration of I^- since each AgI which dissolves produces one Ag^+ and one I^-, or $S = [Ag^+] = [I^-]$.

$$K_{sp} = S^2$$

$$S = \sqrt{K_{sp}} = \sqrt{1.2 \times 10^{-17}} = \sqrt{12 \times 10^{-18}}$$

$$S = 3.5 \times 10^{-9} \text{ mole/liter}$$

The solubility in grams is obtained by multiplying S by the formula weight 235. Thus the solubility in grams per liter is

$$(3.5 \times 10^{-9} \text{ mole/liter}) \times (235 \text{ g/mole}) = 8.2 \times 10^{-7} \text{ g/liter}.$$

Example 16-d. K_{sp} for CaF_2 is 4.0×10^{-11} at 25°C. What is the solubility of calcium fluoride (FW = 78.1) in moles/liter and in grams/liter?

Solution.

$$CaF_2(s) \rightleftharpoons Ca^{++} + 2 F^-$$

$$K_{sp} = [Ca^{++}][F^-]^2$$

Let S = molar solubility of CaF_2 in H_2O

$$[Ca^{++}] = S$$
$$[F^-] = 2 S \quad \text{(since two } F^- \text{ result from each } CaF_2)$$
$$K_{sp} = (S)(2 S)^2 = 4 S^3$$
$$S = \sqrt[3]{\frac{K_{sp}}{4}} = \sqrt[3]{\frac{4 \times 10^{-11}}{4}} = \sqrt[3]{1 \times 10^{-11}}$$
$$S = \sqrt[3]{10 \times 10^{-12}}$$
$$S = 2.2 \times 10^{-4} \text{ mole/liter}$$

The solubility in grams per liter = 2.2×10^{-4} mole/liter $\times$ 78 g/mole = 1.7×10^{-2} g/liter.

PROBLEMS (Solubility Product)

16-1. Which of the following salts is most insoluble?

- (a) $CaCO_3$ $K_{sp} = 1.7 \times 10^{-8}$
- (b) $BaCO_3$ $K_{sp} = 4.9 \times 10^{-9}$
- (c) $MgCO_3$ $K_{sp} = 2.6 \times 10^{-5}$
- (d) $SrCO_3$ $K_{sp} = 4.6 \times 10^{-9}$
- (e) $PbCO_3$ $K_{sp} = 1.7 \times 10^{-11}$

16-2. The solubility of TlCl is 0.780 g in 0.250 liter of water. What is K_{sp}?

16-3. The solubility of Ag_3PO_4 at 19°C in water is 1.5×10^{-5} M. Assuming the only reaction to be $Ag_3PO_4 \rightleftharpoons 3\,Ag^+ + PO_4^{---}$, what is K_{sp}?

16-4. The solubility of Hg_2I_2 in water is 3.04×10^{-7} g/liter. What is K_{sp}? The equation representing the equilibrium involved is $Hg_2I_2 \rightleftharpoons Hg_2^{++} + 2\,I^-$.

16-5. K_{sp} for TlI has been found to be 4×10^{-8}. (a) What is the solubility of TlI in moles per liter? (b) What is the solubility in grams per liter?

16-6. K_{sp} for PbI_2 is 1.00×10^{-8}. (a) What is the solubility of PbI_2 in moles per liter? (b) What is the solubility in grams per liter?

16-7. K_{sp} for Hg_2Cl_2 is 2×10^{-18}. What is the solubility in water in moles per liter? The equilibrium may be represented by the equation $Hg_2Cl_2 \rightleftharpoons Hg_2^{++} + 2\,Cl^-$.

B. PRECIPITATION

A precipitate can form only when the solubility of a substance is exceeded. In other words, saturation must precede precipitation. In terms of the K_{sp}, a precipitate forms only if the product of ionic concentrations exceeds the K_{sp}.

To predict whether or not precipitation will occur on mixing two solutions, we should first simply determine the product of the concentrations of ions in the resulting mixture *as if no precipitation occurs*. This product, which has the same form as the solubility product constant, is termed the *trial product*. This quantity is then compared to the solubility product constant of the substance that might be expected to precipitate. If the trial product exceeds K_{sp}, precipitation should occur. If the trial product is less than K_{sp}, no precipitation occurs. The following examples clarify this point.

Example 16-e. K_{sp} for silver chloride is 1.1×10^{-10} at 25°C. Should precipitation occur if 1 liter of a solution of 1×10^{-4} M NaCl is mixed with (a) 1 liter of 6×10^{-7} M $AgNO_3$, (b) 2 liters of 9×10^{-3} M $AgNO_3$?

Solution. (a) After mixing, the volume is 2 liters, hence the concentration of each species will be halved:

$$[Cl^-] = \tfrac{1}{2} \times 10^{-4} = 5 \times 10^{-5} \text{ mole/liter}$$
$$[Ag^+] = \tfrac{1}{2} \times 6 \times 10^{-7} = 3 \times 10^{-7} \text{ mole/liter}$$
$$\text{Trial product} = [Ag^+][Cl^-] = (5 \times 10^{-5})(3 \times 10^{-7})$$
$$= 15 \times 10^{-12} = 1.5 \times 10^{-11}$$

Since the trial product is less than the K_{sp} (1.1×10^{-10}), precipitation does *not* occur.

(b) After mixing,

$$[Cl^-] = \tfrac{1}{3} \times 1 \times 10^{-4} = 3.3 \times 10^{-5} \text{ mole/liter}$$
$$[Ag^+] = \tfrac{2}{3} \times 9 \times 10^{-3} = 6 \times 10^{-3} \text{ mole/liter}$$
$$\text{Trial product} = (3.3 \times 10^{-5})(6 \times 10^{-3}) = 20 \times 10^{-8}$$
$$= 2.0 \times 10^{-7}$$

The trial product of the final solution is greater than K_{sp}. Therefore precipitation *should occur* on mixing these solutions.

Example 16-f. What concentration of I^- is just sufficient to cause precipitation of PbI_2 at 20°C from a solution containing 10^{-2} M Pb^{++}, that is, what concentration of iodide ion is just sufficient for the trial product to exceed the K_{sp}? K_{sp} for lead iodide is 1.0×10^{-8} at 20°C.

Solution.

$$PbI_2(s) \rightleftharpoons Pb^{++} + 2\,I^- \qquad K_{sp} = [Pb^{++}][I^-]^2$$
$$[Pb^{++}] = 10^{-2} \text{ mole/liter}$$
$$\text{Let } [I^-] = X$$

Substituting into the K_{sp} expression,

$$1.0 \times 10^{-8} = (10^{-2})(X)^2$$
$$X^2 = 10^{-6}$$
$$X = 10^{-3} \text{ mole/liter}$$

When the concentration of I^- is 10^{-3} M, the solution is just saturated. When the I^- concentration is slightly greater than 10^{-3} M, the trial product is slightly greater than 10^{-8} and, under these conditions, precipitation should begin.

Example 16-g. In the precipitation of PbI_2 by the addition of solid NaI to a solution of lead nitrate, as in Example 16-f, what concentration of Pb^{++}

remains in solution when sufficient NaI has been added so that the final equilibrium concentration of I^- is 2.0×10^{-3} M?

Solution.

$$K_{sp} = [Pb^{++}][I^-]^2 = 1.0 \times 10^{-8}; \qquad [I^-] = 2.0 \times 10^{-3} \text{ mole/liter}$$

$$[Pb^{++}] = \frac{K_{sp}}{[I^-]^2} = \frac{1.0 \times 10^{-8}}{(2.0 \times 10^{-3})^2} = \frac{10 \times 10^{-9}}{4.0 \times 10^{-6}}$$

$[Pb^{++}] = 2.5 \times 10^{-3}$ mole/liter = concentration in final solution

PROBLEMS (Precipitation)

16-8. Exactly $\frac{1}{2}$ liter of 2.5×10^{-4} M $TlNO_3$ solution is mixed with 0.500 liter of 2.5×10^{-4} M KI. K_{sp} for TlI is 4×10^{-8}. Should precipitation occur? Give the reason for your answer.

16-9. 200 ml of 5×10^{-6} M $Hg_2(NO_3)_2$ and 100 ml of 5×10^{-6} M NaCl are mixed. K_{sp} for Hg_2Cl_2 is 2×10^{-18}. Determine whether or not precipitation should occur and justify your answer.

16-10. K_{sp} for $CaCO_3$ is 1.12×10^{-8}. Should precipitation occur when 250 ml of 4.0×10^{-4} M $CaCl_2$ is mixed with 750 ml of 4.0×10^{-4} M Na_2CO_3? Justify your answer.

16-11. K_{sp} for $Be(OH)_2$ is 2.0×10^{-18}. (a) What concentration of OH^- is required to just start precipitation from a 3.0×10^{-4} M $Be(NO_3)_2$ solution? (b) What concentration of Be^{++} remains in solution if the OH^- concentration is raised until the pH of the solution is 11.00?

16-12. K_{sp} for $PbBr_2$ is 9×10^{-6}. (a) What Br^- concentration is required to begin precipitation of $PbBr_2$ from a 0.5 M solution of $Pb(NO_3)_2$? (b) What concentration of Pb^{++} remains after the Br^- concentration has been raised to 5×10^{-2} M?

16-13. K_{sp} for silver bromate, $AgBrO_3$, is 4×10^{-5}. What Ag^+ concentration is required to start precipitation of $AgBrO_3$ from a solution containing 0.500 g of $NaBrO_3$ per 500 ml?

C. SOLUBILITY AND THE COMMON ION EFFECT

As in systems involving dissociation of weak electrolytes, the common ion effect is of importance in solubility equilibria; for example, consider the equilibrium

$$AgCl(s) \rightleftharpoons Ag^+ + Cl^-$$

LeChatelier's principle predicts a decrease in solubility of silver chloride if either a soluble silver salt or a soluble chloride salt is added. Other equilibria behave similarly.

Example 16-h. What is the solubility of AgCl in a 1.0×10^{-2} M NaCl solution? K_{sp} for AgCl is 1.1×10^{-10}.

Solution.

$$AgCl(s) \rightleftharpoons Ag^+ + Cl^-$$

$$K_{sp} = [Ag^+][Cl^-]$$

Let S = Molar solubility of AgCl.

$$[Ag^+] = S \text{ mole/liter}$$

$$[Cl^-] = S + \text{Concentration of NaCl} = (S + 0.010) \text{ mole/liter}$$

Substituting in the solubility product constant expression,

$$1.1 \times 10^{-10} = (S)(S + 0.010)$$

Assume S is small compared to 0.010, that is,

$$(0.010 + S) \cong 0.010$$

$$1.1 \times 10^{-10} = (S)0.010$$

$$S = 1.1 \times 10^{-8} \text{ mole/liter}$$

Note that S is very small compared to 0.010 and hence the approximation that $(0.010 + S) = 0.010$ is justified.*

The solubility of AgCl in pure water is 1.06×10^{-5} M, and thus, the solubility of AgCl is reduced by a factor of about 1000 in 0.010 M NaCl as compared to the solubility in pure water.

PROBLEMS (Solubility and Common Ion Effect)

16-14. K_{sp} for $PbBr_2$ is 9×10^{-6}. (a) What is the molar solubility of $PbBr_2$ in a 0.30 M NaBr solution? (b) What is the solubility in grams per liter?

16-15. K_{sp} for $BaSO_4$ is 1.0×10^{-10}. What is the molar solubility of $BaSO_4$ in a 5.0×10^{-2} M solution of Na_2SO_4?

16-16. K_{sp} for AgI is 2.0×10^{-17}. (a) What is the molar solubility of AgI in a 5.0×10^{-2} M AgNO_3 solution? (b) What is the solubility in grams per

---

* The student is again cautioned to check such assumptions in the solution of problems to be certain that the use of an approximation is justified. See also Example 15-b.

liter? (c) Demonstrate the fact that AgI should be more soluble in water than in the 5×10^{-2} M AgNO$_3$ solution.

16-17. K_{sp} for TlCl is 1.7×10^{-4}. What is the molar solubility of TlCl in 5.0×10^{-2} M NaCl solution?

D. SIMULTANEOUS EQUILIBRIA*

In many instances, the anion of a salt is the anion of a weak acid, and, hence, the solubility of the salt is influenced by the pH of the solution. For example, calcium carbonate is relatively insoluble in water, but in the presence of dilute acid it is readily soluble. Qualitatively, this is illustrated by the following simultaneous equilibria.

$$CaCO_3(s) \rightleftharpoons Ca^{++} + CO_3^{--} \qquad K_{sp} = [Ca^{++}][CO_3^{--}] = 1.12 \times 10^{-8}$$

$$H_2CO_3 \rightleftharpoons H^+ + HCO_3^- \qquad K_1 = \frac{[H^+][HCO_3^-]}{[H_2CO_3]} = 4.0 \times 10^{-7}$$

$$HCO_3^- \rightleftharpoons H^+ + CO_3^{--} \qquad K_2 = \frac{[H^+][CO_3^{--}]}{[HCO_3^-]} = 5.0 \times 10^{-11}$$

LeChatelier's principle predicts that the addition of acid causes the latter two equilibria to be shifted to the left, decreasing the concentration of carbonate ion. This causes the first equilibrium to be shifted to the right, thus increasing the solubility of calcium carbonate.

Example 16-i. (a) What is the solubility of CaCO$_3$ in water, on the assumption that all carbonate dissolving remains as CO$_3^{--}$? K_{sp} for CaCO$_3$ is 1.12×10^{-8}. (b) What is the solubility of calcium carbonate in a buffer at pH $= 7.00$?

Solution. (a) Let $S =$ the molar solubility of CaCO$_3$.

$$S = [Ca^{++}] = [CO_3^{--}]$$

Substituting in the solubility product constant expression

$$K_{sp} = [Ca^{++}][CO_3^{--}] = S \times S = S^2$$

$$S = \sqrt{K_{sp}} = \sqrt{1.12 \times 10^{-8}} = 1.06 \times 10^{-4} \text{ mole/liter}$$

(b) Let S' be the molar solubility of CaCO$_3$ in the buffer. Then

$$S' = [Ca^{++}]$$

* This section may be omitted at the option of the instructor.

S' is also the sum of the concentrations of all carbonate-containing species present:

$$S' = [CO_3^{--}] + [HCO_3^-] + [H_2CO_3]$$

To calculate the solubility from the solubility product expression, the $[CO_3^{--}]$ must be defined in terms of known quantities. From the two dissociation equilibria for carbonic acid, it can be shown that

$$[HCO_3^-] = \frac{[H^+][CO_3^{--}]}{K_2}, \quad \text{and} \quad [H_2CO_3] = \frac{[H^+]^2[CO_3^{--}]}{K_1K_2}$$

By substituting these values into the equation for S',

$$S' = [CO_3^{--}] + \frac{[H^+][CO_3^{--}]}{K_2} + \frac{[H^+]^2[CO_3^{--}]}{K_1K_2}$$

$$S' = [CO_3^{--}]\left\{1 + \frac{[H^+]}{K_2} + \frac{[H^+]^2}{K_1K_2}\right\}$$

Solving for $[CO_3^{--}]$ and substituting into the solubility product constant expression of calcium carbonate, we have

$$K_{sp} = [Ca^{++}][CO_3^{--}] = (S') \times \frac{S'}{1 + \dfrac{[H^+]}{K_2} + \dfrac{[H^+]^2}{K_1K_2}}$$

The values of K_1, K_2, and K_{sp} are known, and the $[H^+]$ may be calculated from the pH of 7.00 to be 1.00×10^{-7}. By substituting the numerical values into this equation, the solubility of $CaCO_3$ is

$$S' = 5.3 \times 10^{-3} \text{ mole/liter}$$

This represents an increase by a factor of 50 over the solubility in water under the conditions outlined in part (a).

PROBLEMS (Simultaneous Equilibria)

16-18. K_{sp} for MgF_2 is 7.1×10^{-9}. If the dissociation constant for HF is 2.2×10^{-4}, what is the solubility of MgF_2 in a solution whose pH is buffered at 2.0?

16-19. (a) What is the solubility of CaC_2O_4 in a solution whose pH is buffered at 4.0? K_{sp} for CaC_2O_4 is 3.0×10^{-9} and the acid dissociation constants for $H_2C_2O_4$ are $K_1 = 5.6 \times 10^{-2}$ and $K_2 = 5.2 \times 10^{-5}$. (b) What is the solubility of CaC_2O_4 in a solution whose pH is buffered at 1.0?

16-20. (a) What is the solubility of ZnS in a solution whose pH is buffered at 3.0? (b) What is the solubility in a solution whose pH is buffered at 1.0? K_{sp} for ZnS is 1.2×10^{-23}, and the ionization constants for H_2S are $K_1 = 1.1 \times 10^{-7}$ and $K_2 = 1 \times 10^{-14}$.

E. SOLUBILITY AND SEPARATION PROCEDURES

A knowledge of solubility product constants allows the prediction of separation procedures based on the differences in the solubility of the compounds in question. In many instances, considerable wasted time and effort can be avoided by use of relatively simple and straightforward calculations. It is not always possible to determine that a process will work in practice, even though it should be successful in principle. However, an indication that a particular process is not possible in principle is almost always demonstrated by experiment to be a correct prediction. Thus, in this case, negative information is extremely valuable.

1. Separation of the Silver Group Cations as Insoluble Chlorides

In a list of solubility product constants of metal chlorides, we find only a few relatively small values, that is, only a few chlorides are relatively insoluble. The more insoluble ones are listed in Table 16-1. In working

Table 16-1   Relatively Insoluble Chlorides

Salt	K_{sp}	Theoretical minimum [Cl$^-$] necessary to start precipitation from a 10^{-3} M solution of metal ion
AgCl	1.1×10^{-10}	1.1×10^{-7} mole/liter
PbCl$_2$	2×10^{-5}	1.4×10^{-1} mole/liter
Hg$_2$Cl$_2$	2.0×10^{-18}	4.5×10^{-8} mole/liter
HgCl$_2$	9.5×10^{-3}	3.1 moles/liter

out a separation procedure, we ask several questions:

1. What concentration of Cl$^-$ is necessary to begin precipitation of each of the ions listed in Table 16-1 from a solution of particular concentration, for example, a 10^{-3} M solution?

This chloride concentration is easily calculated from the solubility product principles outlined earlier. For AgCl the minimum concentration of Cl⁻ required is given by the relationship

$$[Cl^-] = \frac{K_{sp}}{[Ag^+]} = \frac{1.1 \times 10^{-10}}{10^{-3}} = 1.1 \times 10^{-7} \text{ mole/liter}$$

For PbCl₂ the concentration of chloride required is

$$[Cl^-] = \sqrt{\frac{K_{sp}}{[Pb^{++}]}} = \sqrt{\frac{2 \times 10^{-5}}{10^{-3}}} = \sqrt{2 \times 10^{-2}}$$

$$= 1.4 \times 10^{-1} \text{ mole/liter}$$

Similarly, these quantities may be calculated for Hg_2^{++}, Hg^{++}, and other metal ions if the solubility product constants are known. The commonly encountered chlorides of other metals are all more soluble than mercuric chloride, and hence require even greater concentrations of chloride ion to cause precipitation.

2. The second question we ask is, What concentration of Cl⁻ is appropriate for a practical separation of Ag^+, Pb^{++}, and Hg_2^{++} from Hg^{++} and other soluble ions? A convenient procedure is to select a Cl⁻ concentration large enough such that the trial products of those to be precipitated exceed the solubility product constants, but also one which is small enough that the solubility product constant of $HgCl_2$ is not exceeded. In this instance a Cl⁻ concentration of 1.0 M would suffice since this exceeds the PbCl₂ requirement but is less than that required for $HgCl_2$ precipitation. See Table 16-1.

3. Our third question is, What concentration of metal ion is left in solution after the [Cl⁻] is made 1.0 M?

This too can be obtained by use of the solubility product constant expression.

$$[Pb^{++}] = \frac{K_{sp}}{[Cl^-]^2} = \frac{(2 \times 10^{-5})}{(1)^2} = 2 \times 10^{-5} \text{ mole/liter}$$

$$[Ag^+] = \frac{K_{sp}}{[Cl^-]} = \frac{1.1 \times 10^{-10}}{1} = 1.1 \times 10^{-10} \text{ mole/liter}$$

$$[Hg_2^{++}] = \frac{K_{sp}}{[Cl^-]^2} = \frac{2.0 \times 10^{-18}}{(1)^2} = 2.0 \times 10^{-18} \text{ mole/liter}$$

Thus it is observed that Ag^+ and Hg_2^{++} are quite effectively removed from solution, whereas the concentration of Pb^{++} is reduced by a factor of 50 from its original concentration of 10^{-3} M.

2. Separation of the Copper Group as Insoluble Sulfides

Many metals form insoluble sulfides of quite varying degrees of solubility. In attempting to work out a scheme for the separation of certain sulfides into two groups, the same sort of questions are raised as for the chlorides.

1. What minimum concentrations of S^{--} are required to begin precipitation of each of the metal ions in Table 16-2 if the concentration of each of the metal ions is 10^{-3} M? (These have been calculated as indicated below and included in Table 16-2.)

In solutions of divalent metal sulfides, the equilibria may be represented as $MS(s) \rightleftharpoons M^{++} + S^{--}$, and K_{sp} may be written as $K_{sp} = [M^{++}][S^{--}]$. The theoretical minimum S^{--} concentration to just start precipitation from a 10^{-3} M Hg^{++} solution is

$$[S^{--}] = \frac{K_{sp}}{[Hg^{++}]} = \frac{10^{-54}}{10^{-3}} = 10^{-51} \text{ mole/liter}$$

For bismuth ion, $K_{sp} = [Bi^{+++}]^2[S^{--}]^3$, and hence the theoretical minimum sulfide ion concentration necessary to cause precipitation from a 10^{-3} M solution of bismuth ion is

$$[S^{--}] = \sqrt[3]{\frac{K_{sp}}{[Bi^{+++}]^2}} = \sqrt[3]{\frac{10^{-96}}{10^{-6}}} = 10^{-30} \text{ mole/liter.}$$

2. What is a convenient point for separation into two groups, and what $[S^{--}]$ is required to make this separation?

Table 16-2   Some Insoluble Sulfides

Metal sulfide	K_{sp}	Theoretical minimum $[S^{--}]$ necessary to start precipitation from a 10^{-3} M solution of metal ion
HgS	10^{-54}	10^{-51} mole/liter
CuS	10^{-40}	10^{-37} mole/liter
Bi_2S_3	10^{-96}	10^{-30} mole/liter
CdS	10^{-28}	10^{-25} mole/liter
ZnS	10^{-24}	10^{-21} mole/liter
FeS	10^{-22}	10^{-19} mole/liter
CoS	10^{-21}	10^{-18} mole/liter
MnS	10^{-16}	10^{-13} mole/liter

The choice of separation of the ions into two groups depends on subsequent alternative procedures. From a chemical standpoint it is convenient to precipitate Cd^{++} and the less soluble ones (ions above Cd^{++} in Table 16-2) and leave Zn^{++} and the more soluble ones (ions below Zn^{++}) in solution. Therefore, a S^{--} concentration just slightly less than $10^{-21}\,M$ would be ideal. Specifically, the concentration $9 \times 10^{-22}\,M$ would work very nicely since this concentration would cause the solubility product constant of CdS to be exceeded, and yet it would be just short of that required to precipitate ZnS.

3. How is this concentration of S^{--} obtained?

Earlier, H_2S in aqueous solution was described as a weak acid which dissociates in the stepwise fashion below.

$$H_2S \rightleftharpoons H^+ + HS^- \qquad K_1 = \frac{[H^+][HS^-]}{[H_2S]} = 1.1 \times 10^{-7}$$

$$HS^- \rightleftharpoons H^+ + S^{--} \qquad K_2 = \frac{[H^+][S^{--}]}{[HS^-]} = 1 \times 10^{-14}$$

In this instance, an overall reaction combining the two stepwise reactions, which is somewhat more convenient, may be written

$$H_2S \rightleftharpoons 2\,H^+ + S^{--} \qquad K_a = \frac{[H^+]^2[S^{--}]}{[H_2S]}$$

and

$$K_a = K_1 \times K_2 = 1.1 \times 10^{-21}$$

From LeChatelier's principle and from the common ion effect, we can predict that addition of acid would repress the ionization, that is, form more H_2S and decrease the concentration of S^{--} in solution. The addition of OH^- would tend to increase the S^{--} concentration. It was shown in Example 15-c that the concentration of S^{--} in H_2S saturated water was $1 \times 10^{-14}\,M$. Therefore, acid must be added to decrease the S^{--} concentration to $9 \times 10^{-22}\,M$. What concentration of H^+ produces this S^{--} concentration?

From the preceding equation and its equilibrium constant

$$[H^+]^2 = \frac{[H_2S]\,K_a}{[S^{--}]}$$

The concentration of dissolved H_2S in a saturated aqueous solution at 1 atm pressure and room temperature is approximately $10^{-1}\,M$. The desired concentration of S^{--} is $9 \times 10^{-22} M$ and $K_a = 1.1 \times 10^{-21}$. Making the appropriate substitutions,

$$[H^+]^2 = \frac{(10^{-1})(1.1 \times 10^{-21})}{9 \times 10^{-22}} = 0.12$$

$$[H^+] = 0.35 \text{ mole/liter}$$

$$pH = 0.45$$

Therefore, to precipitate the ions Hg^{++}, Bi^{+++}, Cu^{++}, Cd^{++} as insoluble sulfides and to leave Zn^{++}, Fe^{++}, Co^{++}, Mn^{++} in solution, a pH = 0.45 is required to produce the necessary S^{--} concentration. This pH is easily obtained by preparing a 0.35 M solution of HCl.

PROBLEMS (Separations)

16-21. A solution contains both I^- and Cl^- at a concentration of $5.0 \times 10^{-2} M$. What concentration of Pb^{++} is required to precipitate the maximum amount of I^- without precipitating any Cl^- from solution? K_{sp} for PbI_2 is 1.0×10^{-8} and that for $PbCl_2$ is 2×10^{-5}.

16-22. A solution contains both Tl^+ and Ag^+ at a concentration of $3.0 \times 10^{-3} M$. What concentration of Cl^- is needed to precipitate the maximum amount of one of these ions without precipitating any of the other? Which ion precipitates first? K_{sp} for TlCl is 1.7×10^{-4} and that for AgCl is 1.1×10^{-10}.

16-23. A solution contains both Ba^{++} and Ca^{++} each at a concentration of $4 \times 10^{-3} M$. (a) What concentration of $C_2O_4^{--}$ is required to precipitate the maximum amount of Ca^{++} and little or no Ba^{++}? (b) What pH must be maintained in the solution to achieve this $C_2O_4^{--}$ concentration if the amount of undissociated $H_2C_2O_4$ is $1 \times 10^{-3} M$? K_{sp} for BaC_2O_4 is 2×10^{-7} and that for CaC_2O_4 is 3×10^{-9}. The value of K_1 for $H_2C_2O_4$ is 5.6×10^{-2} and K_2 is 5.2×10^{-5}.

16-24. A solution contains both Co^{++} and Zn^{++} at a concentration of $2 \times 10^{-1} M$. What concentration of S^{--} would be required to precipitate the maximum amount of Zn^{++} while leaving as much Co^{++} in solution as possible? K_{sp} for CoS is 1×10^{-21} and that for ZnS is 1×10^{-23}. What pH is required in a saturated H_2S solution ($[H_2S] = 0.1 M$) in order to obtain the needed S^{--} concentration?

GENERAL PROBLEMS

16-25. K_{sp} for TlCl is 1.7×10^{-4}. What is the solubility in moles per liter?
16-26. What is the solubility of TlCl in a 0.25 M solution of NaCl? K_{sp} for TlCl is 1.7×10^{-4}.

16-27. K_{sp} for AgBr in water is 5×10^{-13}, and K_{sp} for AgBr in methyl alcohol is 6.3×10^{-16} at the same temperature. Calculate the molar solubility of AgBr in water and in methyl alcohol.

16-28. What is the solubility for AgBr in 0.1 M KBr? K_{sp} for AgBr is 5×10^{-13}.

16-29. Calculate K_{sp} for $PbCl_2$ at 25°C. The experimentally determined solubility at this temperature is 0.475 g per 100 ml of solution.

16-30. The solubility of Ag_2CrO_4 is 3.32×10^{-2} *grams* per liter. Calculate K_{sp}.

16-31. The solubility of PbI_2 in water was found to be 0.0626 g per 0.100 liter. What is K_{sp}?

16-32. K_{sp} for BiI_3 is 8×10^{-19}. What is the solubility of BiI_3 in 0.05 M sodium iodide?

16-33. K_{sp} for TlI is 8×10^{-9}. (a) What concentration of I^- is required to just cause precipitation of Tl^+ from a 0.04 M solution of $TlNO_3$? (b) What concentration of I^- is it necessary to have at equilibrium so that the Tl^+ concentration has been reduced to 10^{-5} M?

16-34. K_{sp} for $Pb(IO_3)_2$ is 1.2×10^{-13}. (a) What concentration of Pb^{++} is necessary to cause precipitation from a 0.10 M $NaIO_3$ solution? (b) What concentration of Pb^{++} is necessary to reduce the equilibrium concentration of IO_3^- to 2×10^{-4} M?

16-35. Should precipitation occur on mixing 200 ml of 5×10^{-4} M $Pb(NO_3)_2$ and 300 ml of 1×10^{-4} M $NaIO_3$? K_{sp} for $Pb(IO_3)_2$ is 1.2×10^{-13}.

16-36. Should precipitation occur on mixing 0.5 liter of a solution containing 1 g of $NaIO_3$ and 0.5 liter of a solution containing 1 g of $AgNO_3$? K_{sp} for $AgIO_3$ is 1×10^{-8}.

16-37. What is the solubility of $AgBrO_3$ in a solution of 0.4 M $AgNO_3$? K_{sp} for $AgBrO_3$ is 4×10^{-5}.

16-38. What minimum concentration of Tl^+ is required to start precipitation of $TlBrO_3$ from a 2.5×10^{-3} M solution of $NaBrO_3$? K_{sp} for $TlBrO_3$ is 8.5×10^{-5}.

16-39. K_{sp} for PbS is 4×10^{-28}. What is the solubility of PbS in a solution of 0.3 M HCl which is saturated with H_2S? The concentration of H_2S in a saturated solution is approximately 0.1 M. K_1 for H_2S is 1.1×10^{-7}; K_2 is 1×10^{-14}.

16-40. If a solution containing 0.5 M $MnCl_2$ and 0.001 M HCl is saturated with H_2S (0.1 M), does precipitation occur? K_{sp} for MnS is 1.4×10^{-15}.

16-41. The solubility of $TlIO_3$ in water was found to be 6.3×10^{-4} M. What is K_{sp} for $TlIO_3$?

16-42. K_{sp} for $AgBrO_3$ is 4×10^{-5}. What is the molar solubility of $AgBrO_3$ in 0.10 M $NaBrO_3$?

16-43. What concentration of PO_4^{---} is required to just begin precipitation of Ag^+ from a 5.0×10^{-2} M solution of $AgNO_3$? K_{sp} for Ag_3PO_4 is 1.5×10^{-21}.

16-44. A solution contains the following metal ions at a concentration of $2 \times 10^{-3}\ M$. If the concentration of S^{--} is slowly increased, in what order should the metal ions precipitate?

	K_{sp}			K_{sp}
Ag_2S	5.6×10^{-51}		ZnS	1×10^{-24}
Tl_2S	6.3×10^{-23}		Sb_2S_3	1.7×10^{-93}
CdS	1.6×10^{-28}		Bi_2S_3	3.1×10^{-90}
HgS	1.0×10^{-54}			

16-45. Sodium sulfate was added to a $0.01\ M$ $BaCl_2$ solution until the solution concentration of SO_4^{--} was $0.005\ M$. What concentration of Ba^{++} (in moles per liter) remained in solution? K_{sp} $(BaSO_4) = 1.2 \times 10^{-10}$.

16-46. One liter of a solution originally contained $5.00 \times 10^{-3}\ M$ $Pb(ClO_4)_2$ and $5.00 \times 10^{-3}\ M$ $Sr(ClO_4)_2$. Sodium sulfate was added until the equilibrium SO_4^{--} concentration was $5.00 \times 10^{-3}\ M$. (a) What are the concentrations of Pb^{++} and Sr^{++} in this equilibrium mixture? (b) What weight of $PbSO_4$ and what weight of $SrSO_4$ have been precipitated? K_{sp} for $PbSO_4 = 2.00 \times 10^{-8}$; K_{sp} for $SrSO_4 = 4.00 \times 10^{-7}$.

16-47. A solution is prepared by mixing $TlNO_3$, $AgNO_3$, and NaI. This solution is saturated with respect to both TlI and AgI. (a) If the concentration of Tl^+ is found to be $4 \times 10^{-4}\ M$, what is the concentration of Ag^+? K_{sp} for TlI is 4×10^{-8} and for AgI, K_{sp} is 1.2×10^{-17}. (b) What is the concentration of I^-?

16-48. The pH of a saturated $Mg(OH)_2$ solution in pure H_2O is 10.38. Assuming the principal equilibrium to be $Mg(OH)_2(s) \rightleftharpoons Mg^{++} + 2\ OH^-$, what is K_{sp} for $Mg(OH)_2$?

16-49. A $5.0 \times 10^{-2}\ M$ solution of Al^{+++} begins to show precipitation of $Al(OH)_3$ at a pH of 3.76. What is K_{sp}? Assume you can observe precipitation at the instant the solubility product is exceeded.

16-50. K_{sp} for $Fe(OH)_3$ is 1.0×10^{-36}. For a solution of $1.25 \times 10^{-4}\ M$ $Fe(ClO_4)_3$, at what pH does precipitation begin? Assume the reaction to be $Fe(OH)_3 \rightleftharpoons Fe^{+++} + 3\ OH^-$.

16-51. K_{sp} for $AgCl$ is 1.1×10^{-10}. Calculate the concentrations of Ag^+ which can be in equilibrium with solid $AgCl$ at each of the chloride ion concentrations given below. Make a plot of $[Ag^+]$ versus $[Cl^-]$; plot points for each of these concentrations of Cl^-: $(0.1, 0.5, 0.7, 0.8, 0.9, 1.0, 2.0, 3.0, 5, 10) \times 10^{-5}\ M Cl^-$. Is it possible to obtain complete precipitation?

CHAPTER 17

Hydrolysis and Complex Ion Formation

A. HYDROLYSIS

1. Equilibria and pH

Hydrolysis is the reaction of a substance with water to produce H^+ or OH^- plus some other solute species. This reaction obviously affects the equilibrium between water and its ions. The production of H^+ by a hydrolysis reaction brings about a decrease in OH^- concentration, and similarly the production of OH^- decreases the H^+ concentration.

For example, an aqueous solution of $NaC_2H_3O_2$ has a pH greater than 7, indicating an excess of OH^- over H^+. $NaC_2H_3O_2$ is completely ionized as Na^+ and $C_2H_3O_2^-$, and because $HC_2H_3O_2$ is a weak acid, the reaction

$$C_2H_3O_2^- + H_2O \rightleftharpoons HC_2H_3O_2 + OH^-$$

takes place to some extent with the consequent production of OH^-. There is no similar tendency for Na^+ to react with H_2O to form undissociated $NaOH$ and H^+; hence there is a net increase of both the concentration of OH^- and the pH of the $NaC_2H_3O_2$ solution above those of pure water.

Just as other equilibria can be treated quantitatively, hydrolysis reactions lend themselves to mathematical description. For the above hydrolysis of acetate ion, the equilibrium constant may be written as

$$K_h' = \frac{[HC_2H_3O_2][OH^-]}{[C_2H_3O_2^-][H_2O]}$$

As in other cases of dilute solutions, $[H_2O]$ is essentially constant and $K'_h \times [H_2O]$ is a constant denoted as K_h and called an hydrolysis constant. Hence

$$K_h = \frac{[HC_2H_3O_2][OH^-]}{[C_2H_3O_2^-]}$$

Similarly, certain cations can be considered to undergo hydrolysis. The fact that a solution of $FeCl_3$ is acidic can be explained on the basis of the hydrolysis of Fe^{+++}. Because HCl is a strong acid, there is virtually no tendency of Cl^- to react with water to form HCl molecules and OH^-. The hydrolysis of Fe^{+++} may be represented by the equation* $Fe^{+++} + H_2O \rightleftharpoons FeOH^{++} + H^+$.

2. Relationship between Hydrolysis Constants and Dissociation Constants of Weak Acids and Bases

In the hydrolysis of $C_2H_3O_2^-$, the reaction can be considered to be the sum of two reactions: first, the dissociation of water,

$$H_2O \rightleftharpoons H^+ + OH^-; \qquad K_w = [H^+][OH^-] = 1.00 \times 10^{-14} \qquad (1)$$

and second, the combination of $C_2H_3O_2^-$ with H^+ to produce undissociated $HC_2H_3O_2$,

$$C_2H_3O_2^- + H^+ \rightleftharpoons HC_2H_3O_2; \qquad \frac{1}{K_a} = \frac{[HC_2H_3O_2]}{[H^+][C_2H_3O_2^-]}$$

$$= \frac{1}{1.71 \times 10^{-5}} \qquad (2)$$

Adding the chemical equations (1) and (2) we obtain

$$C_2H_3O_2^- + H_2O \rightleftharpoons HC_2H_3O_2 + OH^-; \qquad K_h = \frac{[HC_2H_3O_2][OH^-]}{[C_2H_3O_2^-]} \qquad (3)$$

* An alternative and probably more precise method of explaining the acidity of aquo cations is to consider that molecules of water coordinated to metal ions lose one or more protons to the solvent. The equations for successive ionizations by the aquated ferric ion may be written as

$$[Fe(H_2O)_6]^{+++} + H_2O \rightleftharpoons [Fe(H_2O)_5OH]^{++} + H_3O^+$$
$$[Fe(H_2O)_5OH]^{++} + H_2O \rightleftharpoons [Fe(H_2O)_4(OH)_2]^+ + H_3O^+$$
$$[Fe(H_2O)_4(OH)_2]^+ + H_2O \rightleftharpoons [Fe(H_2O)_3(OH)_3] + H_3O^+$$

It should be apparent that these equilibria may be treated similarly to those of polyprotic acids such as sulfuric and phosphoric acids. For the sake of simplicity, the coordinated water molecules of the metal ions and that of the hydronium ion have been omitted in the treatment presented here.

Now if we solve algebraic equation (1) for $[OH^-]$

$$[OH^-] = \frac{K_w}{[H^+]}$$

and substitute in algebraic equation (3), we obtain

$$K_h = \frac{[HC_2H_3O_2] K_w}{[H^+][C_2H_3O_2^-]}$$

But since

$$\frac{[HC_2H_3O_2]}{[H^+][C_2H_3O_2^-]} = \frac{1}{K_a}$$

$$K_h = \frac{K_w}{K_a}$$

For the hydrolysis of acetate ion, the equilibrium constant is

$$K_h = \frac{1.00 \times 10^{-14}}{1.71 \times 10^{-5}} = 5.8 \times 10^{-10}$$

For the hydrolysis of the cations of weak bases, it is possible to show that

$$K_h = \frac{K_w}{K_b}$$

The student should derive this in a manner similar to the above.

Example 17-a. What is the hydrolysis constant for the $HCOO^-$ ion? K_a for $HCOOH$ is 2×10^{-4}.

   Solution.

$$HCOO^- + H_2O \rightleftharpoons HCOOH + OH^-$$

$$K_h = \frac{[HCOOH][OH^-]}{[HCOO^-]} = \frac{K_w}{K_a} = \frac{1 \times 10^{-14}}{2 \times 10^{-4}} = 5 \times 10^{-11}$$

Example 17-b. What is the pH of a $1.0\ M$ solution of $NaC_2H_3O_2$? The hydrolysis constant is 5.8×10^{-10}.

   Solution.

$$C_2H_3O_2^- + H_2O \rightleftharpoons HC_2H_3O_2 + OH^-$$

$$K_h = \frac{[HC_2H_3O_2][OH^-]}{[C_2H_3O_2^-]}$$

Let X = the number of moles per liter of $C_2H_3O_2^-$ which has undergone hydrolysis. Then, at equilibrium,

$$[OH^-] = [HC_2H_3O_2] = X$$

$$[C_2H_3O_2^-] = \text{total concentration of } NaC_2H_3O_2$$
$$- \text{ concentration of } C_2H_3O_2^- \text{ hydrolyzed}$$
$$= (1.0 - X)$$

Substituting into the equilibrium constant expression,

$$5.8 \times 10^{-10} = \frac{(X)(X)}{1.0 - X}$$

On the assumption that X is small compared to 1, then $(1.0 - X) \cong$ 1.0, and $X^2 = 5.8 \times 10^{-10}$.

$$X = [OH^-] = 2.4 \times 10^{-5} \text{ mole/liter}$$

$$[H^+] = \frac{K_w}{[OH^-]} = \frac{1.0 \times 10^{-14}}{2.4 \times 10^{-5}} = 4.2 \times 10^{-10} \text{ mole/liter}$$

$$pH = -\log (4.2 \times 10^{-10}) = 9.38$$

Example 17-c. The pH of a 0.10 M solution of $FeCl_3$ is 2.00. On the assumption that the only hydrolysis reaction of importance is

$$Fe^{+++} + H_2O \rightleftharpoons FeOH^{++} + H^+$$

what is the hydrolysis constant?

Solution.

$$K_h = \frac{[FeOH^{++}][H^+]}{[Fe^{+++}]}$$

From the pH,

$$[H^+] = \text{antilog} (-2.00)$$
$$= 1.00 \times 10^{-2} \text{ mole/liter}$$
$$[FeOH^{++}] = [H^+]$$
$$[Fe^{+++}] = (0.10) - [FeOH^{++}] = (0.10 - 0.01) = 0.09 \text{ mole/liter}$$

Substituting in the equilibrium constant expression,

$$K_h = \frac{(1.00 \times 10^{-2})(1.00 \times 10^{-2})}{0.09} = 1.1 \times 10^{-3}$$

PROBLEMS (Hydrolysis)

17-1. (a) Calculate the hydrolysis constant for the hydrolysis reaction

$$NO_2^- + H_2O \rightleftharpoons HNO_2 + OH^- \text{ if } K_a = 4.5 \times 10^{-4}$$

(b) What is the pH of a 0.50 M solution of $NaNO_2$?

17-2. What is the pH of a solution of 0.30 M $Zn(ClO_4)_2$ if K_h is 1.0×10^{-9} for the reaction $Zn^{++} + H_2O \rightleftharpoons ZnOH^+ + H^+$?

17-3. Assume the hydrolysis of NH_4^+ to proceed according to the equation $NH_4^+ + H_2O \rightleftharpoons NH_3 + H_3O^+$ and that K_b for NH_3 is 1.82×10^{-5}. (a) What is the equilibrium constant for the hydrolysis reaction? (b) For a 0.90 M solution of NH_4NO_3 what is the H^+ concentration? (c) What is the OH^- concentration of the solution? (d) What is the pH of the solution? (e) What is the equilibrium concentration of NH_3?

17-4. A 0.030 M solution of NaCN is prepared. What is the pH of the solution if K_a for HCN is 4.5×10^{-10}?

17-5. The hydrolysis constant for the reaction

$$Be^{++} + H_2O \rightleftharpoons BeOH^+ + H^+$$

is 2.0×10^{-7}. (a) What is the pH of a solution of 2.0×10^{-2} M $BeCl_2$? (b) What is the concentration of $BeOH^+$?

17-6. The pH of a 0.160 M solution of $Th(ClO_4)_4$ is 2.54. (a) What is the hydrolysis constant for the reaction

$$Th^{++++} + H_2O \rightleftharpoons ThOH^{+++} + H^+?$$

(b) What is the concentration of $ThOH^{+++}$ at equilibrium?

17-7. What is the pH at the end point of a titration of 0.20 M HNO_2 and 0.20 M KOH? K_a for HNO_2 is 4.5×10^{-4}.

17-8. What is the pH at the end point of a titration of 0.4 M NH_3 with 0.4 M HNO_3? K_b for ammonia is 1.82×10^{-5}.

B. COMPLEX ION FORMATION

Many positive ions associate in aqueous solution with negative ions or neutral molecules to form complex ions. An equilibrium exists between the various species present and may be written in the form of a dissociation reaction analogous to that for a weak electrolyte.

In a solution of $AgNO_3$ and NH_3, the diamminesilver complex ion, $Ag(NH_3)_2^+$, is formed and is in equilibrium with Ag^+ and NH_3 molecules. This is represented by the equation

$$Ag(NH_3)_2^+ \rightleftharpoons Ag^+ + 2\,NH_3$$

and the equilibrium or dissociation constant for the reaction is

$$K_d = \frac{[Ag^+][NH_3]^2}{[Ag(NH_3)_2^+]} = 6.2 \times 10^{-8}$$

Similarly, Ag^+ reacts with CN^- to form the complex $Ag(CN)_2^-$, whose equilibrium can be represented $Ag(CN)_2^- \rightleftharpoons Ag^+ + 2\ CN^-$. The dissociation constant is given by the equation

$$K_d = \frac{[Ag^+][CN^-]^2}{[Ag(CN)_2^-]} = 1.40 \times 10^{-20}$$

Complex ion formation is especially important in understanding the chemistry of ions in solution. The study of solubilities of normally insoluble substances in the presence of complexing agents is one of the more important areas of solution chemistry.

For example, $AgCl(s)$ is soluble in excess NH_3 owing to the formation of the $Ag(NH_3)_2^+$ complex ion. This problem may be treated as an example of simultaneous equilibria analogous to those described previously. The equilibria involved are

$$AgCl(s) \rightleftharpoons Ag^+ + Cl^- \qquad K_{sp} = [Ag^+][Cl^-] \qquad (1)$$

$$Ag(NH_3)_2^+ \rightleftharpoons Ag^+ + 2\ NH_3 \qquad K_d = \frac{[Ag^+][NH_3]^2}{[Ag(NH_3)_2^+]} \qquad (2)$$

Addition of ammonia should, according to LeChatelier's principle, cause the conversion of Ag^+ to $Ag(NH_3)_2^+$. Therefore more $AgCl(s)$ should dissolve to replace part of the Ag^+ used up. The solubility equilibrium under these conditions is represented by the equation $AgCl(s) + 2\ NH_3 \rightleftharpoons Ag(NH_3)_2^+ + Cl^-$. The equilibrium constant for this reaction has the form

$$K_{eq} = \frac{[Ag(NH_3)_2^+][Cl^-]}{[NH_3]^2}$$

and can be related to the K_{sp} of AgCl and the K_d of the complex. Solving the K_{sp} and K_d equations for $[Ag^+]$ and equating them because the $[Ag^+]$ is the same for both equilibria, we have

$$[Ag^+] = \frac{K_{sp}}{[Cl^-]} = \frac{K_d[Ag(NH_3)_2^+]}{[NH_3]^2}$$

Solving for K_{sp}/K_d, we obtain

$$\frac{K_{sp}}{K_d} = \frac{[Ag^+][Cl^-]}{\dfrac{[Ag^+][NH_3]^2}{[Ag(NH_3)_2^+]}} = \frac{[Ag(NH_3)_2^+][Cl^-]}{[NH_3]^2}$$

This equation is in the same form as that of the equilibrium constant for the net chemical reaction, and hence the ratio K_{sp}/K_d is numerically equal to K_{eq}.

Other systems may be treated similarly.

Example 17-d. A solution of 0.050 M $KAg(CN)_2$ was prepared in which the equilibrium concentration of Ag^+ was found to be 5.6×10^{-8} M. What is the value of K_d for the dissociation of $Ag(CN)_2^-$ ion?

Solution.

$$Ag(CN)_2^- \rightleftharpoons Ag^+ + 2\ CN^-$$

At equilibrium $[Ag^+] = 5.6 \times 10^{-8}$ mole/liter

$$[CN^-] = 2\ [Ag^+] = 2 \times 5.6 \times 10^{-8} = 11.2 \times 10^{-8} \text{ mole/liter}$$
$$[Ag(CN)_2^-] = [\text{total } Ag(CN)_2^-] - [\text{dissociated } Ag(CN)_2^-]$$
$$= 0.050 - 5.6 \times 10^{-8} \cong 0.050 \text{ mole/liter}$$

$$K_d = \frac{5.6 \times 10^{-8}(1.12 \times 10^{-7})^2}{5.0 \times 10^{-2}} = 1.40 \times 10^{-20}$$

Example 17-e. A 3.0×10^{-2} M solution of $[Ag(NH_3)_2]NO_3$ is prepared by dissolving the solid in water. If the dissociation constant of the complex ion is 6.2×10^{-8}, what is the concentration of Ag^+ and NH_3 in the equilibrium solution (assuming NH_3 does not dissociate as a base)? What fraction of the complex ion has dissociated?

Solution.

$$Ag(NH_3)_2^+ \rightleftharpoons Ag^+ + 2\ NH_3$$

At equilibrium let

$$[Ag^+] = X$$

then

$$[NH_3] = 2\ X$$
$$[Ag(NH_3)_2^+] = 3.0 \times 10^{-2} - X \cong 3.0 \times 10^{-2} \text{ mole/liter}$$

Substituting into the equilibrium constant expression

$$K_d = \frac{X(2X)^2}{3.0 \times 10^{-2}} = \frac{4X^3}{3.0 \times 10^{-2}} = 6.2 \times 10^{-8}$$

Solving for X, $X = 7.8 \times 10^{-4}$ mole/liter

The fraction of complex dissociated is

$$\frac{[Ag^+]}{[\text{total Ag}]} = \frac{7.8 \times 10^{-4}}{3.0 \times 10^{-2}} = 2.6 \times 10^{-2}$$

Example 17-f. Excess AgCl(s) is treated with a 0.100 M solution of NH_3. The equation for the reaction is

$$AgCl(s) + 2\,NH_3 \rightleftharpoons Ag(NH_3)_2^+ + Cl^-$$

What are the equilibrium concentrations of NH_3, of $Ag(NH_3)_2^+$, of Cl^-, and of Ag^+?

Solution. K_{eq} for the equilibrium above was shown to be K_{sp}/K_d. Substitution of the appropriate numerical values yields

$$K_{eq} = \frac{1.71 \times 10^{-10}}{6.2 \times 10^{-8}} = 2.76 \times 10^{-3}$$

At equilibrium let $X = [NH_3]$

$$[\,Ag(NH_3)_2^+] = [Cl^-] = \tfrac{1}{2}([NH_3 \text{ total}] - [NH_3 \text{ unreacted}])$$

$$= \frac{(0.100 - X)}{2}\text{ mole/liter}$$

Substituting into the equilibrium constant expression,

$$K_{eq} = \frac{[Ag(NH_3)_2^+][Cl^-]}{[NH_3]^2} = \frac{\left(\dfrac{0.100 - X}{2}\right)^2}{X^2}$$

Taking the square root of both sides,

$$\sqrt{K_{eq}} = \frac{\dfrac{0.100 - X}{2}}{X}$$

Solving for X, $X = 0.090$ mole/liter
Therefore

$$[NH_3] = 9.0 \times 10^{-2}\text{ mole/liter}$$

$$[Ag(NH_3)_2^+] = \frac{0.100 - 0.090}{2} = 0.005\text{ mole/liter}$$

$$[Cl^-] = 0.005\text{ mole/liter}$$

$$[\,Ag^+] = \frac{K_{sp}}{[Cl^-]} = \frac{1.71 \times 10^{-10}}{5 \times 10^{-3}} = 3.4 \times 10^{-8}\text{ mole/liter}$$

PROBLEMS (Complex Ions)

17-9. When a solution is prepared containing $0.100\ M\ Cd^{++}$ and $0.400\ M\ NH_3$, it is found that at equilibrium the concentration of free Cd^{++} is $1.00 \times 10^{-2}\ M$. What is the equilibrium constant for the dissociation reaction $Cd(NH_3)_4{}^{++} \rightleftharpoons Cd^{++} + 4\ NH_3$ if this is the only reaction of significance occurring in solution?

17-10. The complex ion $CdBr^+$ forms on mixing solutions of Cd^{++} and Br^-. In a particular solution containing $0.50\ M$ total Cd^{++}, it is found that only 20% of the total exists as free Cd^{++} when the free Br^- concentration is $5.0 \times 10^{-2}\ M$. What is the dissociation constant for the reaction $CdBr^+ \rightleftharpoons Cd^{++} + Br^-$?

17-11. A solution of $0.41\ M\ [Cu(NH_3)_4](NO_3)_2$ was prepared. On analysis it was found that the equilibrium concentration of NH_3 was $2.4 \times 10^{-3}\ M$. Assuming the only equilibrium to be

$$Cu(NH_3)_4{}^{++} \rightleftharpoons Cu^{++} + 4\ NH_3$$

what is the value of K_d?

17-12. A solution of $K[Au(CN)_2]$ was prepared by dissolving 52.5 g of the complex salt in 0.50 liter of water. At equilibrium the concentration of CN^- was found to be $2.0 \times 10^{-16}\ M$. Assuming that the only equilibrium involved is the dissociation of the complex ion according to the equation $Au(CN)_2{}^- \rightleftharpoons Au^+ + 2\ CN^-$, what is the value of the dissociation constant?

17-13. The dissociation constant for $HgCl_4{}^{--}$ in aqueous solution is 8.50×10^{-16} for the reaction $HgCl_4{}^{--} \rightleftharpoons Hg^{++} + 4\ Cl^-$. Assuming this to be the only equilibrium involved, what is the concentration of each of the above species in the equilibrium mixture prepared by dissolving 1.00 mole of $K_2[HgCl_4]$ in 1.00 liter of solution?

17-14. The reaction for the dissociation of $Hg(SCN)_4{}^{--}$ may be represented by the equation $Hg(SCN)_4{}^{--} \rightleftharpoons Hg^{++} + 4\ SCN^-$, which has an equilibrium constant $K_d = 1.15 \times 10^{-21}$. (a) What percentage of a $1.00\ M$ solution of $K_2[Hg(SCN)_4]$ dissociates, assuming that this is the only reaction occurring? (b) What are the concentrations of the various species involved when the equilibrium is established?

17-15. $Pb(OH)_2$ dissolves to a certain extent in a $1.0\ M$ solution of NaOH. Assuming an excess of solid $Pb(OH)_2$, (a) what are the equilibrium concentrations of Pb^{++}, $Pb(OH)_3{}^-$, and OH^-? (b) What is the total solubility of $Pb(OH)_2$ in the $1.0\ M$ solution of NaOH? Given the following equilibria,

$$Pb(OH)_2(s) \rightleftharpoons Pb^{++} + 2\ OH^-, \qquad K_{sp} = 1.0 \times 10^{-16}$$

and

$$Pb(OH)_3{}^- \rightleftharpoons Pb^{++} + 3\ OH^-, \qquad K_d = 1.26 \times 10^{-14}.$$

17-16. (a) What is the equilibrium constant for the reaction

$$Co(NH_3)_6^{+++} + 6\,CN^- \rightleftharpoons Co(CN)_6^{---} + 6\,NH_3?$$

K_d for $Co(NH_3)_6^{+++} = 2.2 \times 10^{-34}$ and K_d for $Co(CN)_6^{---}$ is 1.0×10^{-64}.
(b) What must the concentration of NH_3 be in an equilibrium solution when the same amount of each cobalt complex exists and the concentration of uncomplexed CN^- is 1×10^{-5} mole/liter?

GENERAL PROBLEMS

17-17. What is the pH of a solution of $0.25\,M\,NaHCO_2$? K_a for $HCOOH$ is 1.7×10^{-4}.

17-18. What is the pH of a $0.40\,M$ solution of NH_4NO_3? K_b for NH_3 is 1.8×10^{-5}.

17-19. The pH of a $0.80\,M$ solution of the salt NaX is found to be 8.60. Assuming that HX is a weak acid, what is the ionization constant of the acid?

17-20. The equilibrium constant for the dissociation of CdI^+, by the reaction $CdI^+ \rightleftharpoons Cd^{++} + I^-$, is 5.2×10^{-3}. In a solution containing $0.100\,M\,Cd^{++}$ total and $0.100\,M\,I^-$ total, what fraction of the cadmium present exists in the form of the complex ion?

17-21. A $0.0100\,M$ solution of $HgCl_2$ is found to be 0.029% dissociated in aqueous solution. Calculate K_d for the reaction

$$HgCl_2(aq) \rightleftharpoons Hg^{++} + 2\,Cl^-$$

assuming that this is the only important equilibrium established in the solution.

17-22. The dissociation constant of $Ag(NH_3)_2^+$ has been established as 6.2×10^{-8}. If NH_3 is added to a solution containing $0.20\,M\,AgNO_3$, what is the concentration of free uncomplexed NH_3 when 90% of the silver has been converted to $Ag(NH_3)_2^+$?

17-23. A solution is prepared by mixing 500 ml of $0.0200\,M\,Fe(ClO_4)_3$ and 500 ml of $0.0200\,M\,NaF$. A reaction occurs which involves formation of FeF^{++}. When equilibrium is established, the concentration of FeF^{++} is found to be $9.8 \times 10^{-3}\,M$. (a) What are the equilibrium concentrations of Fe^{+++} and F^-? (b) What is K_d for FeF^{++}?

17-24. The dissociation constant for $Cu(NH_3)_4^{++}$ is 4.9×10^{-14}. If a solution is to be prepared in which 50% of the complex has dissociated as in the reaction $Cu(NH_3)_4^{++} \rightleftharpoons Cu^{++} + 4\,NH_3$, what concentration of uncomplexed NH_3 must there be in the solution if the above equilibrium is the only one to be considered?

17-25. Excess solid $AgIO_3$ ($K_{sp} = 3.1 \times 10^{-8}$) is shaken with a solution of NH_3 of initial concentration $0.20\,M$. (a) When equilibrium has been achieved, what amount of $AgIO_3$ has been dissolved? K_d for $Ag(NH_3)_2^+$ is 6.2×10^{-8}. (b) What is the equilibrium concentration of NH_3?

17-26. The hydroxide $Cr(OH)_3$ is slightly soluble in dilute NaOH. An excess of solid $Cr(OH)_3$ is equilibrated with a solution containing initially 0.60 M NaOH, and at equilibrium the concentration of OH^- is found to be 0.43 M. What is the equilibrium constant for the reaction $Cr(OH)_3(s) + OH^- \rightleftharpoons Cr(OH)_4^-$? If the solubility product constant of $Cr(OH)_3$ is 2×10^{-30}, what is the dissociation constant for the reaction $Cr(OH)_4^- \rightleftharpoons Cr^{+++} + 4\,OH^-$?

17-27. Hydrolysis of a cation in aqueous solution is an alternative way of describing the formation of hydroxide ion complexes of metals. (a) Show that the hydrolysis constant of Be^{++} is directly related to the equilibrium constant for the reaction

$$Be(OH)^+ \rightleftharpoons Be^{++} + OH^-$$

The hydrolysis constant for Be^{++} to form $BeOH^+$ is 1×10^{-6}. (b) What is the dissociation constant for $BeOH^+$?

17-28. A solution contains 0.200 M NH_3, 0.200 M NaCN, and 0.020 M $AgNO_3$. K_d for $Ag(NH_3)_2^+ = 6.2 \times 10^{-8}$, and K_d for $Ag(CN)_2^- = 1.4 \times 10^{-20}$. (a) What is the equilibrium constant for the reaction $Ag(NH_3)_2^+ + 2\,CN^- \rightleftharpoons Ag(CN)_2^- + 2\,NH_3$? (b) What is the ratio of the concentrations of NH_3 and CN^- when equilibrium is established?

CHAPTER 18

Nuclear Changes

A. RADIOACTIVITY

Radioactivity is the spontaneous emission of high energy radiation due to decay or disintegration of the nuclei of atoms. Atoms of atomic number greater than 82 and a few of lower atomic number are naturally radioactive (for example, U, Th, Pa, etc.), and emit alpha, beta, and/or gamma radiation. Many isotopes of the lighter elements have been produced by nuclear reactions, and some of these also exhibit the property of radioactivity. In addition to the decay processes mentioned, lighter isotopes may also decay by positron emission or K-electron capture.

Radioactivity is a property of the nucleus and the rate of disintegration is independent of external effects such as temperature, pressure, etc. Each radioactive isotope has a characteristic rate of decay which is expressed as a *half-life*. The half-life is the time required for one-half of any given amount of the element to decompose. For example, the half-life of $_{88}Ra^{226}$ is 1620 years. Starting with 1 g of Ra today, after 1620 years, $\frac{1}{2}$ g would remain; in two half-lives (3240 years) only $\frac{1}{4}$ g would remain undecomposed; in three half-lives (4860 years) only $\frac{1}{8}$ of the original amount remains, etc. This process is analogous to dividing the original amount by 2 a number of times equal to the number of half-lives elapsed. Therefore, the amount of isotope left after three half-lives is

$$\frac{\text{Original amount}}{2 \times 2 \times 2} = \frac{\text{Original amount}}{2^3} = \frac{\text{Original amount}}{8}$$

In this case $\frac{1}{8}$ g would be left after 4860 years. In a more general statement, the amount remaining after n half-lives = original amount/$(2)^n$.

Example 18-a The half-life of $_{86}Rn^{222}$ gas is 4 days. Starting with 10 mg of the isotope, what amount would decay in 16 days and what amount would remain?

Solution. Sixteen days represent four ($\frac{16}{4}$) half-lives. Starting with 10 mg, 5 mg would decay in one half-life period (4 days).

$$5 \text{ mg} + 2\tfrac{1}{2} \text{ mg} = 7\tfrac{1}{2} \text{ mg} \text{ would decay in two half-life periods (8 days)}$$

$$5 \text{ mg} + 2\tfrac{1}{2} \text{ mg} + 1\tfrac{1}{4} \text{ mg} = 8\tfrac{3}{4} \text{ mg} \text{ would decay in three half-life periods (12 days)}$$

$$5 \text{ mg} + 2\tfrac{1}{2} \text{ mg} + 1\tfrac{1}{4} \text{ mg} + \tfrac{5}{8} \text{ mg} = 9\tfrac{3}{8} \text{ mg} \text{ would decay in four half-life periods (16 days)}$$

The amount remaining after 16 days would be $10 - 9\tfrac{3}{8} = \tfrac{5}{8}$ mg; or, using the formula above,

$$\text{Amount remaining} = \frac{10 \text{ mg}}{(2)^4} = \frac{10 \text{ mg}}{16} = \frac{5}{8} \text{ mg}$$

Example 18-b. A certain isotope has a half-life of 15 min. In what period of time would the activity be reduced to approximately 10% of the original?

Solution.

In one half-life, or 15 min, activity will be reduced to 50%.
In two half-lives, or 30 min, activity will be reduced to 25%.
In three half-lives, or 45 min, activity will be reduced to $12\tfrac{1}{2}$%.
In four half-lives, or 60 min, activity will be reduced to $6\tfrac{1}{4}$%.

Thus the activity would be reduced to 10% in approximately 50 min. A more precise estimation of the time can be obtained by plotting the data as shown in Fig. 18-1. Reading from the graph, the time is about 49.5 min.

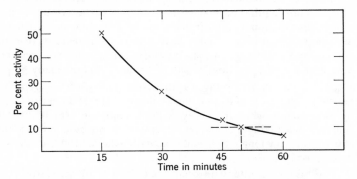

Fig. 18-1. Decay curve for a substance with $t_{1/2} = 15$ min.

PROBLEMS

18-1. The half-life of $_{82}Pb^{214}$ is 27 min. Starting with 64 mg of this isotope, what quantity would be left at the end of 2 hr and 15 min?

18-2. The $_{92}U^{238}$ isotope, which occurs naturally, has a half-life of 4.5×10^9 yr. In what period of time would the activity be reduced to $\frac{1}{8}$ of the original?

18-3. The half-life of $_6C^{14}$ is 5730 yr. Plot a decay curve for this isotope similar to the one shown in Fig. 18-1. Estimate the time to reduce the original activity to 20%.

18-4. A piece of charcoal found in a cave from a wood fire of long ago, shows an activity of 15.0% of freshly prepared charcoal. Using the graph from Problem 18-3, estimate the age of the charcoal sample.

18-5. Four grams of $_{83}Bi^{214}$ decays to $\frac{1}{2}$ g in 1 hr. What is the half-life of this isotope?

18-6. Explain what changes in mass and charge take place in the nucleus on the emission of (a) an alpha particle, (b) an electron, (c) gamma radiation, and (d) a positron.

18-7. A radioactive isotope has a half-life of 2.0 hr. Starting with 3.2×10^{21} atoms, how many atoms will be left at the end of 10 hr?

18-8. Determine the half-life of an element, 87.5% of which decays in 2 yr.

18-9. The half-life of $_{16}S^{35}$ is 87 days. Starting with 8×10^5 atoms of S^{35}, how many atoms would *decompose* in 261 days?

18-10. What is the half-life of a radioactive element $\frac{3}{4}$ of which decomposes in 10 hr?

18-11. The half life of francium is 21 min. Starting with 4×10^{18} atoms of francium, how many atoms would *decompose* in 1 hr and 3 min?

18-12. The half life of a radioactive element is 3 hr. In what period of time would the activity of the sample be reduced to $\frac{1}{16}$ of the original activity?

18-13. In one of the radioactive disintegration series, $_{92}U^{238}$ ultimately forms $_{82}Pb^{206}$. Knowing that the emission of an alpha particle results in a decrease of two units in atomic number and a loss of four units in mass number and that the emission of a β^- particle results in an increase of one in atomic number and no change in mass number, (a) how many α particles must have been emitted per atom of $_{82}Pb^{206}$ formed, (b) how many β^- particles?

18-14. $_{88}Ra^{226}$ decays according to the equation $_{88}Ra^{226} \rightarrow {}_{86}Rn^{222} + {}_2He^4$. Starting with 1 mole of Ra, (a) what volume of He (at STP) will be produced by this reaction when the decay is 100% complete? (b) If this He gas is collected in a 1.00-liter container, what will the pressure in the container be at 0°C? (c) How long a time will it actually take for 100% of the sample to decompose?

18-15. $_{94}Pu^{239}$ is an α emitter with a half-life of 2.70×10^4 yr. Starting with 5 mg of $_{94}Pu^{239}$, how many Pu atoms will have decayed after 8.10×10^4 yr?

18-16. $_{83}Bi^{212}$ is an α emitter with a half-life of 60 min. Starting with 21.2 g of $_{83}Bi^{212}$, what period of time will be required to form 2.17 liters of He gas measured at STP?

B. RADIOACTIVE DECAY AS A FIRST-ORDER RATE PROCESS

Since the decay of a radioactive element is a first-order process, precise calculations may be made from the rate equation for first-order reactions

$$2.303 \log \frac{A}{A_0} = -kt \qquad (1)$$

where A_0 is the amount of substance at the start, A is the amount present at time t, and k is the rate constant. At the end of one half-life period, $t_{1/2}$, the amount remaining is one-half the original amount. That is, when $t = t_{1/2}$, $A = A_0/2$. By substitution of the quantities into equation (1), it can then be shown that

$$t_{1/2} = \frac{0.693}{k} \qquad (2)$$

Example 18-c. The half-life of $_{38}Sr^{90}$ is 28 yr. In what period of time would the activity of a given sample be reduced to $\frac{1}{10}$ of the original activity?

Solution. This is analogous to asking the question, "When will $A/A_0 = \frac{1}{10}$?" First evaluate k from equation (2)

$$k = \frac{0.693}{28 \text{ yr}} = 0.025/\text{yr}$$

Substitute for A/A_0 and k in equation (1) above and solve for t.

$$2.303 \log \frac{1}{10} = -\frac{0.025 \, t}{\text{yr}}$$

$$t = \frac{2.303 \, (-1)}{-0.025/\text{yr}} = 92 \text{ yr}$$

Example 18-d. 5.0% of a certain radioactive element decays in 21.0 min. Determine its half-life.

Solution. The fraction remaining after 21.0 min is 0.95 or $A/A_0 = 0.95$. Substituting into (1)

$$2.303 \log 0.950 = -k(21.0 \text{ min})$$

$$k = \frac{(2.303)(-0.0220)}{-21.0 \text{ min}} = 0.00241/\text{min}$$

$$t_{1/2} = \frac{0.693}{k} = \frac{0.693}{0.00241/\text{min}} = 288 \text{ min}$$

Example 18-e. Amounts of radioactive substances may be measured in terms of a unit of radiation called the *Curie*. One curie of a radioactive material is that weight of substance which produces 3.70×10^{10} disintegrations per sec. If the half-life of radium is 1620 yr, determine the amount of radium which will furnish one curie of radiation.

Solution. For first-order reactions, the rate of decay is proportional to the amount of substance or number of atoms present at any time. Hence

$$\text{Rate} = \text{Number of disintegrations/sec} = k \times N$$

where k is the rate constant expressed in reciprocal seconds and N is the number of atoms present.

$$k = \frac{0.693}{t_{1/2}} = \frac{0.693}{1620 \text{ yr} \times 365 \text{ d/yr} \times 24 \text{ hr/d} \times 3600 \text{ sec/hr}}$$
$$= 1.35 \times 10^{-11}/\text{sec}$$

$$\text{Rate} = 3.70 \times 10^{10} \text{ atoms/sec} = (1.35 \times 10^{-11}/\text{sec}) \times N$$

$$\text{Number of atoms present} = N = \frac{3.70 \times 10^{10} \text{ atoms/sec}}{1.35 \times 10^{-11}/\text{sec}}$$
$$= 2.74 \times 10^{21} \text{ atoms}$$

$$\text{Number of g atoms of Ra} = \frac{2.74 \times 10^{21} \text{ atoms}}{6.02 \times 10^{23} \text{ atoms/g atom}}$$
$$= 4.55 \times 10^{-3} \text{ g atoms}$$

$$\text{Number of grams of Ra} = (4.55 \times 10^{-3} \text{ g atoms})(226 \text{ g/g atom})$$
$$= 1.028 \text{ g}$$

PROBLEMS   (Using First-Order Rate Equations)

18-17. If 80.0% of a radioactive isotope decays in 30.0 min, what is its half-life?

18-18. A certain isotope has a half-life of 15.0 min. In what period of time would the activity be reduced to 10.0% of the original? Compare this value with the result obtained by graphical means in Example 18-b.

18-19. Work exercise 18-3 using the first-order rate expression. Then compare the answer with that obtained graphically. The half-life of $_6C^{14}$ is 5730 yr.

18-20. $_{53}I^{131}$ has a half-life of 8.0 days. Starting with 5.00×10^{21} atoms, how many would be left at the end of 85 days?

18-21. A certain isotope has a half-life of 1.0×10^5 yr. What fraction of the original sample would be left at the end of 1.0×10^6 yr?

18-22. A certain radioactive element X has a half-life of 1.00 yr. (a) Starting with 1.00 g of X (atomic weight 233), how many disintegrations will occur in 10 sec? (b) How many curies of radiation are produced by 1.00 g of X?

18-23. One g atom of a certain radioactive element emits 2.06×10^{10} β particles per second. Assuming each beta particle corresponds to the decay of one atom of the radioactive element, what is the half-life of the latter?

18-24. A sample of a radioactive substance registers 1000 counts per sec, but 6.0 hr later the activity is 100 counts per sec. What is the half-life of this substance?

18-25. A particular mineral of uranium was found to contain 0.604 g of $_{82}Pb^{206}$ for every 1.340 g of $_{92}U^{238}$. If all of the $_{82}Pb^{206}$ present was formed by radioactive decay of $_{92}U^{238}$, what is the age of the mineral? Note that numbers derived from such a source are usually taken as indicative of the age of the earth. The half-life of $_{92}U^{238}$ is 4.51×10^9 yr; the half-lives of the elements between U and Pb are short in comparison with $_{92}U^{238}$.

18-26. The procedure for determining the age of carbon-containing objects was worked out by Professor Willard F. Libby while he was at the University of Chicago. This method is based on the fact that $_6C^{14}$, a β^- emitter with a half-life of 5730 yr, is produced by the interaction of cosmic rays with nitrogen of the atmosphere. It is presumed to have achieved a steady-state concentration so that it decomposes at the same rate as it is produced. Hence a constant ratio of $_6C^{14}$ to $_6C^{12}$ exists in the CO_2 of the atmosphere. Living organisms are composed of carbon with essentially the same ratio of C^{14}/C^{12} as atmospheric CO_2. On death, no further carbon is taken into the organism, and from this time on, the ratio of C^{14}/C^{12} decreases because of the decay of C^{14}. A sample of wood charcoal obtained from an Indian fishing site at the confluence of the Snake and Palouse Rivers in Washington State was found to have a C^{14}/C^{12} ratio of 27.5% of present-day charcoal. Assuming the C^{14}/C^{12} ratio in the atmosphere has been constant over the intervening years, how long ago was the site occupied?

18-27. A dating procedure for water has also been proposed by Professor Libby based on the β^- activity of $_1H^3$. Ordinary water contains tritium or hydrogen of mass 3 which is formed in the earth's atmosphere from cosmic rays. It decomposes by beta emission at a rate equal to that at which it is formed, and hence water exposed to the atmosphere will contain a constant mole per cent of tritium. Wine which has been bottled will not have its tritium replaced after it decays, and hence, the older the wine, the lower the tritium content. A certain sample was found to have 33.7% of the tritium content of present-day water. Assuming tritium is lost only by β^- decay, what is the age of the wine? $t_{1/2}$ for $_1H^3 = 12.26$ yr.

18-28. A sample of uranium ore is found to contain 4.75 g of $_{92}U^{238}$. On heating, helium gas is driven off. If all of the helium originally came from the decay of $_{92}U^{238}$ to $_{82}Pb^{206}$ during the 2.50×10^9 yr of existence of the rock, what volume of He was collected at 25°C and 1 atm pressure? The half-life of $_{92}U^{238}$ is 4.51×10^9 yr; the half-lives of the elements between U and Pb are short in comparison with $_{92}U^{238}$.

18-29. $_{38}Sr^{90}$ is one of the fission products of the atomic bomb which is present in radioactive fall-out. If a given sample of soil has 1.15×10^{-11} g of $_{38}Sr^{90}$, (a) What is its activity in curies? (b) What is the activity in picocuries? (c) How long will it take for the activity to decrease to $\frac{1}{10}$ of its present amount? The half-life of $_{38}Sr^{90}$ is 28.0 yr.

C. NUCLEAR EQUATIONS

In writing nuclear equations it is customary to show the mass number and the atomic number or charge of the atoms or particles involved. Usually the mass number is shown as a superscript to the right of the symbol and the charge or atomic number as a subscript to the left, for example, $_{92}U^{238}$. For bombarding and ejected particles, the following representations may be used:

alpha particle	$_2He^4$	or (α)
proton	$_1H^1$	or (p)
deuteron	$_1D^2$	or (d)
tritium	$_1T^3$	or (t)
beta particle		
(negative electron)	$_{-1}e^0$	or (β^- or e^-)
positron	$_{+1}e^0$	or (β^+ or e^+)
neutron	$_0n^1$	or (n)
gamma ray	γ	

In a balanced nuclear equation, the sum of the superscripts (mass number) must be the same on both sides of the equation; likewise the sum of the subscripts (atomic number or charge) must balance on the two sides of the equations. Examples are

1. $_7N^{14} + _2He^4 \rightarrow _8O^{17} + _1H^1$
2. $_7N^{14} + _0n^1 \rightarrow _6C^{14} + _1H^1$
3. $_{30}Zn^{68} + _0n^1 \rightarrow _{28}Ni^{65} + _2He^4$
4. $_{25}Mn^{55} + _1D^2 \rightarrow _{26}Fe^{55} + 2\,_0n^1$
5. $_{17}Cl^{35} + _0n^1 \rightarrow _{16}S^{35} + _1H^1$
6. $_{11}Na^{23} + _1D^2 \rightarrow _{11}Na^{24} + _1H^1$
7. $_3Li^6 + _0n^1 \rightarrow _2He^4 + _1T^3$

Note that in all of these equations both the subscripts and the superscripts balance on the two sides of the equations.

A somewhat shorter notation may be employed to show nuclear processes in which the bombarding particles (the *in* particle) and the ejected particles (the *out* particle) are shown in parenthesis between the target atom and the product atom, thus reactions 1–4 above may be shown as:

1. $_7N^{14}(\alpha, p)_8O^{17}$
2. $_7N^{14}(n, p)_6C^{14}$

3. $_{30}Zn^{68}(n, \alpha)_{28}Ni^{65}$
4. $_{25}Mn^{55}(d, 2n)_{26}Fe^{55}$

etc.

PROBLEMS (Nuclear Equations)

18-30. Complete the following:

(a) $_5B^{11} + ? \rightarrow _7N^{14} + _0n^1$
(b) $_4Be^9 + _2He^4 \rightarrow _6C^{12} + ?$
(c) $_{15}P^{30} \rightarrow _{14}Si^{30} + ?$
(d) $_3Li^7 + _1H^1 \rightarrow 2?$
(e) $_{92}U^{235} + _0n^1 \rightarrow _{38}Sr^{94} + _{54}Xe^{139} + ?$

18-31. Write balanced equations for the following nuclear processes:

(a) $_{19}K^{39}(p, n)$
(b) $_{15}P^{31}(n, \gamma)$
(c) $_{77}Ir^{191}(\alpha, n)$
(d) $_{12}Mg^{24}(\alpha, n)$
(e) $_{96}Cm^{242}(\alpha, 2n)$

18-32. From your text book select nuclear reactions which represent the following: (a) fission, (b) fusion, (c) artificial radioactivity, (d) natural radioactivity, (e) proton bombardment, and (f) K-electron capture.

D. UNITS OF RADIOACTIVITY

The *curie* is the standard unit of radioactivity and is defined as the amount of a radioactive substance which produces 3.70×10^{10} disintegrations per second. Smaller units are the *millicurie* (3.70×10^7 dis/sec) and the *microcurie* (3.70×10^4 dis/sec).

Example 18-f. Calculate the weight of radium which has a radioactivity of one microcurie. $t_{1/2}$ for Ra $= 1622$ yr.

> **Solution.** Since the decay of a radioactive element is a first order process,
>
> $$\frac{-d[A]}{dt} = k[A]$$ where $[A]$ is the number of atoms of the substance disintegrating (see Section 13A).
>
> In this case $\dfrac{-d[A]}{dt} = 1$ microcurie $= 3.70 \times 10^4$ atoms/sec
>
> $$t_{1/2} = \frac{0.693}{k} \quad \text{hence} \quad k = \frac{0.693}{t_{1/2}}$$
>
> $$k = \frac{0.693}{5.12 \times 10^{10} \text{ sec}} = 1.35 \times 10^{-11}/\text{sec}$$
>
> Substituting in the first order rate equation,
>
> $$3.70 \times 10^4 \text{ atoms/sec} = (1.35 \times 10^{-11}/\text{sec})[A]$$
>
> or
>
> $$[A] = \frac{3.70 \times 10^4 \text{ atoms/sec}}{1.35 \times 10^{-11}/\text{sec}} = 2.74 \times 10^{15} \text{ atoms}$$

This is the number of Ra atoms present. To convert this to weight of Ra:

$$\text{Mass Ra} = 2.74 \times 10^{15} \text{ atoms} \times \frac{1 \text{ g atom}}{6.02 \times 10^{23} \text{ atoms}}$$

$$\times 226 \text{ g Ra/g atom} = 1.03 \times 10^{-6} \text{ g Ra}$$

PROBLEMS (Units of Radioactivity)

18-33. The half life of $_{83}\text{Bi}^{214}$ is 19.7 min. Calculate the weight of this radioisotope which will give an activity of 1 millicurie.

18-34. What weight of $_{92}U^{238}$ has an activity of 5 microcuries? $t_{1/2} = 4.5 \times 10^9$ yr.

18-35. The half-life of $_{94}Pu^{239}$ is 2.4×10^4 yr. (a) Determine the amount of Pu which has an activity of one curie. (b) Determine the rate of decay in disintegrations per second per milligram of Pu.

E. MASS-ENERGY RELATIONSHIPS

The very large amount of energy liberated to the surroundings in a nuclear change is a result of a loss in mass during the process. Mass is converted into *its energy equivalent* as given by the Einstein relation

$$E = Mc^2$$

Energy liberated (ergs)

= Mass decrease (grams) [Velocity of light (cm/sec)]²

Since a number of units of both mass and energy are employed in describing these transformations, the student must know the definition of these units and the relationship between them.

An *atomic mass unit, amu,* is defined as $\frac{1}{12}$ the mass of the $_6C^{12}$ atom, whose isotopic mass is exactly 12 *amu*. As shown on page 17, 1 amu is equivalent to 1.660×10^{-24} g.

Energies may be expressed in *electron volts*. An electron volt is defined as the energy required to raise one electron through a potential difference of one volt. This is a very small unit; hence, one million electron volts, Mev, is commonly used as a unit of energy for nuclear processes. Other energy units used are calories and ergs,

$$1 \text{ Mev} = 1.602 \times 10^{-6} \text{ ergs}; \quad or \quad 1 \text{ erg} = 6.24 \times 10^5 \text{ Mev}$$

Since 4.184×10^7 ergs = 1 cal,

$$1 \text{ Mev} = 3.829 \times 10^{-14} \text{ cal}; \quad or \quad 1 \text{ cal} = 2.612 \times 10^{13} \text{ Mev}$$

Now let us return to the Einstein equation and determine the relationship between mass and energy in ergs, Mev, and calories.

$$\text{Energy (ergs)} = 1 \text{ g} \times (2.998 \times 10^{10} \text{ cm/sec})^2 = 8.988 \times 10^{20} \text{ ergs}$$

Thus *the energy equivalent of mass* is

$$8.988 \times 10^{20} \text{ ergs/g}$$

or

$$\frac{8.988 \times 10^{20} \text{ ergs}}{g} \times \frac{1 \text{ cal}}{4.184 \times 10^7 \text{ ergs}} = 2.148 \times 10^{13} \text{ cal/g}$$

or

$$\frac{8.988 \times 10^{20} \text{ ergs}}{\text{g}} \times \frac{6.24 \times 10^5 \text{ Mev}}{1 \text{ erg}} = 5.61 \times 10^{26} \text{ Mev/g}$$

Since 1.660×10^{-24} g $= 1$ amu, the *energy equivalent of mass* is also

$$5.61 \times 10^{26} \frac{\text{Mev}}{\text{g}} \times 1.660 \times 10^{-24} \frac{\text{g}}{\text{amu}} = 931 \text{ Mev/amu}$$

Example 18-g. Consider the following nuclear fusion reaction:

$$_3\text{Li}^{7.01601*} + {}_1\text{H}^{1.00782} \rightarrow 2 \; _2\text{He}^{4.00265}$$

Determine (a) the mass lost in amu per 2 atoms of He formed, (b) the corresponding energy release in Mev and calories, and (c) the energy release per 2 g atoms of He formed.

Solution.

(a) Mass of reactants: $7.01601 + 1.00782 = 8.02383$ amu
   Mass of product: $2 \times 4.00265 = 8.00530$ amu
                        mass loss $= \overline{0.01853 \text{ amu}}$

(b) Since 1 amu $= 931$ Mev, the energy release per 2 atoms of He formed

$$= \frac{0.01853 \text{ amu}}{(2 \text{ atoms He})}\left(931 \frac{\text{Mev}}{\text{amu}}\right) = 17.25 \text{ Mev/(2 atoms He)}$$

or, expressed in calories, the energy release

$$= \frac{17.25 \text{ Mev/(2 atoms He)}}{\left(6.24 \times 10^5 \frac{\text{Mev}}{\text{erg}}\right)\left(4.184 \times 10^7 \frac{\text{erg}}{\text{cal}}\right)}$$

$$= 6.61 \times 10^{-13} \text{ cal/(2 atoms He)}$$

(c) For 1 g atom of Li reacting with 1 g atom of H to form 2 g atoms of He,

$$\text{Energy release} = \frac{17.25 \text{ Mev}}{(2 \text{ atoms He})}\left(6.02 \times 10^{23} \frac{\text{atoms}}{\text{g atom He}}\right)$$

$$= 1.04 \times 10^{25} \text{ Mev/(2 g atoms He)}$$

---

* To conserve space and as a matter of convenience, the exact isotopic mass has been used in place of the mass number in stating the problems in this chapter. For example, the $_3\text{Li}^7$ isotope is written $_3\text{Li}^{7.01601}$ to indicate that the isotope of mass number 7 has an exact isotopic mass of 7.01601.

or

$$\text{Energy release} = \frac{6.61 \times 10^{-13} \text{ cal}}{(2 \text{ atoms He})}\left(6.02 \times 10^{23} \frac{\text{atoms}}{\text{g atom}}\right)$$

$$= 3.98 \times 10^{11} \text{ cal}/(2 \text{ g atoms He})$$

F. BINDING ENERGY

The mass of an atom is always less than the sum of the masses of its constituent particles. Thus the helium atom may be regarded as being derived from 2 hydrogen atoms and 2 neutrons:

$$\text{Mass of 2 } _1\text{H}^{1.007825} = 2.01565 \text{ amu}$$
$$\text{Mass of 2 } _0\text{n}^{1.008675} = 2.01735 \text{ amu}$$
$$\text{Mass of constituent particles} = \overline{4.03300 \text{ amu}}$$
$$\text{Mass of } _2\text{He}^{4.00265} = 4.00265 \text{ amu}$$
$$\text{Mass difference} = \overline{0.03035 \text{ amu}}$$

$$\text{Energy equivalent} = (0.03035 \text{ amu})\left(931 \frac{\text{Mev}}{\text{amu}}\right) = 28.3 \text{ Mev}$$

This energy is termed the *binding energy* of the nucleus, that is, the energy which holds it together. Nuclear particles, protons and neutrons, are termed *nucleons*. The binding energy per nucleon for He, which has 4 nucleons, is $28.3/4 = 7.1$ Mev. The binding energy is related to stability and the most stable of naturally occurring elements are those with high binding energies.

Example 18-h. Compute the binding energy per nucleon for the isotope

$$_{26}\text{Fe}^{55.9349}$$

Solution. The iron atom may be regarded as composed of 26 H atoms and 30 neutrons.

$$26 \ _1\text{H}^{1.00782} \text{ weigh } 26 \times 1.00782 = 26.2033 \text{ amu}$$
$$30 \ _0\text{n}^{1.00868} \text{ weigh } 30 \times 1.00868 = 30.2604 \text{ amu}$$
$$\text{Mass of constituent particles} = \overline{56.4637 \text{ amu}}$$
$$\text{Mass of } _{26}\text{Fe}^{56} \text{ isotope} = 55.9349 \text{ amu}$$
$$\text{Mass difference} = \overline{0.5288 \text{ amu}}$$

$$\text{Binding energy} = (0.529 \text{ amu})\left(931 \frac{\text{Mev}}{\text{amu}}\right) = 492 \text{ Mev}$$

$$\text{Binding energy per nucleon} = 492 \text{ Mev}/56 \text{ nucleons}$$
$$= 8.8 \text{ Mev/nucleon}$$

PROBLEMS

18-36. The conversion of 5 g of matter to energy by the Einstein equation yields how many *ergs*?

18-37. Calculate the energy equivalent in Mev of 10 g of matter.

18-38. For the nuclear reaction $_1D^{2.01410} + _1T^{3.01605} \rightarrow _2He^{4.00265} + _0n^{1.00868}$, determine (a) the mass loss in amu per He atom formed, (b) the energy released in MeV and calories, and (c) the energy released per g atom of He formed.

18-39. For each of the following nuclear reactions calculate (a) the mass loss in amu per atom of bombarded substance, (b) the energy released per atom, and (c) the energy released per mole of reaction.

(1) $_5B^{11.00931} + _2He^{4.00265} \rightarrow _7N^{14.00310} + _0n^{1.00868}$

(2) $_4Be^{9.01220} + _2He^{4.00265} \rightarrow _6C^{12.00000} + _0n^{1.00868}$

(3) $_3Li^{6.01515} + _0n^{1.00868} \rightarrow _2He^{4.00265} + _1T^{3.01605}$

18-40. Calculate the binding energy per nucleon for each of the following isotopes. (Follow the procedure in Example 18-h.)

(a) $_5B^{11.00931}$ (c) $_{56}Ba^{136.9056}$ (e) $_{38}Sr^{87.9056}$

(b) $_{17}Cl^{34.96885}$ (d) $_{18}Ar^{39.96238}$ (f) $_{92}U^{235.044}$

CHAPTER 19

Electrochemistry

A. FARADAY'S LAWS

Electrochemistry is concerned with the interconversion of electrical and chemical energy. During electrolysis, electrical energy is expended in bringing about chemical change. In a voltaic cell or battery, a chemical change produces electrical energy. The relationships between electrical current and chemical change are given by *Faraday's laws*.

1. The weight of a given substance undergoing chemical change is directly proportional to the quantity of electricity passed.

2. The weights of different substances undergoing chemical change by passage of a given quantity of electricity are proportional to their equivalent weights.

The quantity of electricity required to deposit 1 gram equivalent weight of a substance is 96,500 coulombs. This quantity of electricity consists of 6.02×10^{23} electrons and is often referred to as an *equivalent* or 1 *Faraday* of electrons. A coulomb is an ampere-second, that is, a current of 1 ampere flowing for 1 second.

Example 19-a. What weight of Cu and Cl_2 will be produced at inert electrodes on passage of a current of 5.00 amp through a solution of $CuCl_2$ for 30.0 min? At the cathode, the reaction $Cu^{++} + 2\,e^- \rightarrow Cu$ occurs, and at the anode, the reaction $2\,Cl^- \rightarrow Cl_2 + 2\,e^-$ occurs.

Solution.

$$\text{Quantity of electricity} = 5.00 \text{ amp} \times \left(30.0 \text{ min} \times \frac{60 \text{ sec}}{\text{min}}\right)$$

$$= 9000 \text{ amp-sec or } 9000 \text{ coulombs}$$

Equivalent weights:

$$Cu = 63.6/2 = 31.8 \qquad Cl_2 = 71/2 = 35.5$$

Weight of Cu produced

$$= \frac{9000 \text{ coul}}{96,500 \text{ coul/equiv}} \times 31.8 \text{ g/equiv} = 2.97 \text{ g}$$

Weight of Cl_2 produced

$$= \frac{9000 \text{ coul}}{96,500 \text{ coul/equiv}} \times 35.5 \text{ g/equiv} = 3.31 \text{ g}$$

Example 19-b. What quantity of electricity is required to deposit 2.158 g of Ag from a solution of $AgNO_3$? $Ag^+ + e^- \rightarrow Ag$

Solution. Equivalent weight of Ag = 107.9/1 = 107.9

$$\text{Number of equivalents of Ag} = \frac{2.158 \text{ g}}{107.9 \text{ g/equiv}}$$

$$= 0.02000 \text{ equivalent}$$

$$\text{Quantity of electricity} = 0.02000 \text{ equiv} \times 96,500 \text{ coul/equiv}$$

$$= 1.930 \times 10^3 \text{ coulombs}$$

Example 19-c. A quantity of electricity is passed through a series of solutions of $AgNO_3$, $CrCl_3$, $ZnSO_4$, and $CuSO_4$. (a) If 1.000 g of Ag is deposited from the first solution, what weights of the metallic ions in the other three solutions are simultaneously deposited? (b) What quantity of electricity in coulombs is used?

Solution. (a) The equations for the half-reactions at the cathodes are:

$$Ag^+ + e^- \rightarrow Ag \qquad Zn^{++} + 2\,e^- \rightarrow Zn$$
$$Cr^{+++} + 3\,e^- \rightarrow Cr \qquad Cu^{++} + 2\,e^- \rightarrow Cu$$

Equivalent weights:

$$Ag = 107.9/1 = 107.9 \qquad Zn = 65.4/2 = 32.7$$
$$Cr = 52.0/3 = 17.3 \qquad Cu = 63.6/2 = 31.8$$

According to Faraday's laws the weights deposited will be proportional to their equivalent weights and to the number of equivalents of electricity passed.

$$\text{Number of equivalents of electricity} = \frac{1.000 \text{ g Ag}}{107.9 \text{ g/equiv}}$$

$$= 0.00927 \text{ equiv}$$

$$\text{Wt Cr} = 0.00927 \text{ equiv} \times \frac{17.3 \text{ g}}{\text{equiv}} = 0.160 \text{ g}$$

$$\text{Wt Zn} = 0.00927 \text{ equiv} \times \frac{32.7 \text{ g}}{\text{equiv}} = 0.303 \text{ g}$$

$$\text{Wt Cu} = 0.00927 \text{ equiv} \times \frac{31.8 \text{ g}}{\text{equiv}} = 0.295 \text{ g}$$

(b) Quantity of electricity

$$= \frac{1.000 \text{ g}}{107.9 \text{ g/equiv}} \times 96,500 \text{ coul/equiv} = 894 \text{ coulombs}$$

PROBLEMS (Faraday's Laws)

19-1. A steady current was passed through a solution of $CuSO_4$ until 1.59 g of metallic Cu had been produced. How many coulombs of electricity were used?

19-2. A current of 3.00 amp is passed for exactly 10 hr through a solution of $CoSO_4$ between Pt electrodes. What weight of cobalt will be deposited on the cathode?

19-3. A steady current is passed for 3.000 hr through the following series of solutions: $AgNO_3$; $FeCl_2$; $NiSO_4$; $CrCl_3$. 2.158 g of Ag are deposited from the first solution. (a) What weights of metals are simultaneously produced in the remaining solutions? (b) What strength of current in amperes was used?

19-4. How long must a current of 5.00 amp be passed through a solution of $ZnCl_2$ to deposit 3.27 g of Zn?

19-5. In the electrolysis of an aqueous KCl solution the following reaction occurs:

$$2 \text{ KCl} + 2 \text{ H}_2\text{O} \rightarrow 2 \text{ KOH} + \text{H}_2 + \text{Cl}_2$$

(a) What volume of H_2 gas at 25°C and 1 atm would be obtained if 7.10 g of Cl_2 is produced? (b) If a current of 5.0 amp is used, how long must the electrolysis proceed?

19-6. Calculate the charge in coulombs of a single electron.

19-7. An aqueous solution containing both $ZnSO_4$ and $CdSO_4$ was electrolyzed until all of the Zn and Cd were electrodeposited. If the weight of the mixture of metals was 95.0 g and 2.00 Faradays of electricity were used, what weight of Zn was in the mixture?

B. OXIDATION POTENTIALS

1. Voltaic Cells

A voltaic cell or battery converts chemical energy to electrical energy. To accomplish this an oxidation half-reaction takes place at the anode of the cell and a reduction half-reaction occurs at the cathode. Simultaneously, a transfer of electrons from the anode to the cathode is effected by means of a metallic conductor connecting the electrodes outside the cell. The arrangement shown in Fig. 19-1 constitutes a voltaic cell in which Zn

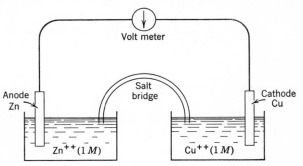

Fig. 19-1.   The Cu–Zn voltaic cell (Daniell cell).

metal is in contact with 1 M Zn^{++} in the anode compartment and Cu metal is in contact with 1 M Cu^{++} in the cathode compartment. This cell generates an electromotive force (emf) of about 1.10 volts.

A system in which a metal electrode is in contact with its ions is called a *half-cell*; two half-cells connected by a salt bridge are required to complete the voltaic cell or battery. The salt bridge, composed of a solution of an electrolyte whose ions do not take part in the chemical reaction, serves to complete the circuit and allow movement of ions between the half-cells. During operation of the battery in Fig. 19-1, the following chemical changes occur:

$$\text{At the anode:} \quad Zn \rightarrow Zn^{++} + 2\,e^- \quad \text{(oxidation)}$$

$$\text{At the cathode:} \quad Cu^{++} + 2\,e^- \rightarrow Cu \quad \text{(reduction)}$$

Adding the two half-reactions, the overall *cell reaction* is obtained:

$$Zn + Cu^{++} \rightarrow Zn^{++} + Cu$$

Any combination of two half-cells gives, in principle, a battery in which a potential difference exists between the two electrodes. By convention, such a cell is often represented by the notation

$$M \text{ (electrode)} \mid M^+ \text{ (solution)} \parallel N^+ \text{ (solution)} \mid N \text{ (electrode)}*$$

when the cell reaction is $M + N^+ \rightarrow M^+ + N$. Thus the cell in Fig. 19-1 would be represented as

$$Zn \mid Zn^{++} \text{ (1 } M) \parallel Cu^{++} \text{ (1 } M) \mid Cu$$

The cell reaction is $Zn + Cu^{++} \rightarrow Zn^{++} + Cu$ and electrons flow from Zn to Cu in the metallic circuit outside the cell.

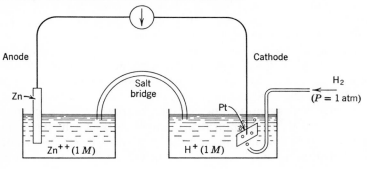

Fig. 19-2.   Using the hydrogen electrode as a standard.

The Hydrogen Electrode. Another voltaic cell is shown schematically in Fig. 19-2, in which the $Zn \mid Zn^{++}$ half-cell is similar to the one in Fig. 19-1. The second half-cell is made up from a piece of platinum foil dipping into a solution of 1 M HCl or some other strong acid ($[H^+] = 1 \ M$) with hydrogen gas (1 atm pressure) bubbling over the Pt electrode. The equilibrium, $H_2 \rightleftharpoons 2 \ H^+ + 2 \ e^-$, is established. This second electrode assembly is called a standard hydrogen electrode.

The emf generated by this cell (Fig. 19-2) is 0.762 volt and the half-cell reactions are

$$\text{At the anode:} \quad Zn \rightarrow Zn^{++} + 2 \ e^- \quad \text{(oxidation)}$$

$$\text{At the cathode:} \quad 2 \ H^+ + 2 \ e^- \rightarrow H_2 \quad \text{(reduction)}$$

* ‖ is the notation for a salt bridge connecting the two half-cells.

The overall cell reaction:

$$Zn + 2\,H^+ \rightarrow Zn^{++} + H_2$$

2. Standard Oxidation Potentials

The observed emf of a battery is simply the difference in potential between the two electrodes and is a measure of the relative tendency of substances to undergo oxidation. In the cell shown in Fig. 19-2 the emf 0.762 measures the relative tendencies of the following two half-reactions to occur:

$$Zn(s) \rightleftharpoons Zn^{++}\,(1\,M) + 2\,e^-$$
$$H_2\,(1\,atm) \rightleftharpoons 2\,H^+\,(1\,M) + 2\,e^-$$

By convention, the value 0.0000 volt has been assigned to the hydrogen electrode as its standard oxidation potential (designated $\mathscr{E}^\circ$) when the pressure of H_2 gas is 1 atm and the hydrogen ion concentration is 1 M. Reactions which have a greater tendency to occur are assigned positive values of $\mathscr{E}^\circ$ while those with a lesser tendency to occur are assigned negative values for $\mathscr{E}^\circ$. Since it is known that when a piece of zinc is placed in a 1 M HCl solution, hydrogen gas and zinc chloride form, then zinc has a greater tendency to be oxidized than hydrogen gas. Hence the value of the standard oxidation potential of Zn is +0.762 volt. By setting up other half-cells in conjunction with the standard hydrogen electrode and determining the emf of the resulting voltaic cells, other standard oxidation potentials may be assigned. Table 19-1 includes $\mathscr{E}^\circ$ values for a few half-reactions. All standard oxidation potentials are for 1 M solutions of the ions. Note that in Table 19-1 the half-reactions are written as oxidation processes, hence the potentials are termed *oxidation* potentials. By reversing the reaction and *changing the sign* of the potential, the corresponding *reduction* potentials are obtained. There are also some reactions listed where an inert metal such as platinum must be used to remove electrons from the reducing agent or to supply electrons to the oxidizing agent. This is a situation analogous to that of the hydrogen electrode.

3. The Qualitative Use of Standard Oxidation Potentials

By combining an oxidation half-reaction with a reduction half-reaction and adding the voltages of the two half-reactions, we may determine whether or not the reaction as written will proceed. If the net emf is positive, then the reaction as written is spontaneous and will proceed to a greater or lesser extent depending upon the magnitude of the voltage. If

Table 19-1   Oxidation Potentials ($\mathscr{E}^\circ$)

Half-Reaction	$\mathscr{E}^\circ$ (Volts)
$K \rightleftharpoons K^+ + e^-$	2.92
$Al \rightleftharpoons Al^{+++} + 3\,e^-$	1.66
$Zn \rightleftharpoons Zn^{++} + 2\,e^-$	0.762
$AsH_3 \rightleftharpoons As + 3\,H^+ + 3\,e^-$	0.60
$H_2C_2O_4 \rightleftharpoons 2\,CO_2 + 2\,H^+ + 2\,e^-$	0.49
$Cr^{++} \rightleftharpoons Cr^{+++} + e^-$	0.41
$Cd \rightleftharpoons Cd^{++} + 2\,e^-$	0.40
$Pb + SO_4^{--} \rightleftharpoons PbSO_4 + 2\,e^-$	0.36
$In \rightleftharpoons In^{+++} + 3\,e^-$	0.342
$Cr \rightleftharpoons Cr^{++} + 2\,e^-$	0.28
$Ni \rightleftharpoons Ni^{++} + 2\,e^-$	0.25
$Sn \rightleftharpoons Sn^{++} + 2\,e^-$	0.14
$H_2 \rightleftharpoons 2\,H^+ + 2\,e^-$	0.00
$H_2S \rightleftharpoons S + 2\,H^+ + 2\,e^-$	-0.14
$Sn^{++} \rightleftharpoons Sn^{++++} + 2\,e^-$	-0.15
$H_2SO_3 + H_2O \rightleftharpoons SO_4^{--} + 4\,H^+ + 2\,e^-$	-0.20
$Cu \rightleftharpoons Cu^{++} + 2\,e^-$	-0.337
$S + 3\,H_2O \rightleftharpoons H_2SO_3 + 4\,H^+ + 4\,e^-$	-0.45
$2\,I^- \rightleftharpoons I_2 + 2\,e^-$	-0.53
$H_3AsO_3 + H_2O \rightleftharpoons H_3AsO_4 + 2\,H^+ + 2\,e^-$	-0.56
$H_2O_2 \rightleftharpoons O_2 + 2\,H^+ + 2\,e^-$	-0.68
$Fe^{++} \rightleftharpoons Fe^{+++} + e^-$	-0.77
$Ag \rightleftharpoons Ag^+ + e^-$	-0.799
$NO_2 + H_2O \rightleftharpoons NO_3^- + 2\,H^+ + e^-$	-0.81
$NO + 2\,H_2O \rightleftharpoons NO_3^- + 4\,H^+ + 3\,e^-$	-0.96
$2\,Br^- \rightleftharpoons Br_2 + 2\,e^-$	-1.06
$Mn^{++} + 2\,H_2O \rightleftharpoons MnO_2 + 4\,H^+ + 2\,e^-$	-1.28
$2\,Cr^{+++} + 7\,H_2O \rightleftharpoons Cr_2O_7^{--} + 14\,H^+ + 6\,e^-$	-1.33
$2\,Cl^- \rightleftharpoons Cl_2 + 2\,e^-$	-1.36
$Pb^{++} + 2\,H_2O \rightleftharpoons PbO_2 + 4\,H^+ + 2\,e^-$	-1.46
$Mn^{++} + 4\,H_2O \rightleftharpoons MnO_4^- + 8\,H^+ + 5\,e^-$	-1.52
$Ce^{+++} \rightleftharpoons Ce^{++++} + e^-$	-1.61
$2\,H_2O \rightleftharpoons H_2O_2 + 2\,H^+ + 2\,e^-$	-1.77
$2\,F^- \rightleftharpoons F_2 + 2\,e^-$	-2.85

the voltage is negative, the reaction as written occurs to only a very slight extent, and the reverse of the reaction written will take place to a greater extent.

Example 19-d. Will Ag metal reduce Sn^{++}? Consider the two half-reactions

$$Ag \rightleftharpoons Ag^+ + e^- \quad \text{(oxidation)} \quad \mathscr{E}^\circ = -0.80 \text{ volt}$$

$$Sn^{++} + 2\,e^- \rightleftharpoons Sn \quad \text{(reduction)} \quad \mathscr{E}^\circ = -0.14 \text{ volt}$$

The sign for the second half-reaction is the reverse of that in Table 19-1 since this is reduction.

> **Solution.** To cancel electrons, the first equation is multiplied by 2 and the half-reactions are added. The voltage, however, is not changed.

$$2\,Ag \rightleftharpoons 2\,Ag^+ + 2\,e^- \qquad\qquad \mathscr{E}^\circ = -0.80 \text{ volt}$$

$$\underline{Sn^{++} + 2\,e^- \rightleftharpoons Sn \qquad\qquad\qquad \mathscr{E}^\circ = -0.14 \text{ volt}}$$

Cell reaction

$$2\,Ag + Sn^{++} \rightleftharpoons 2\,Ag^+ + Sn \quad \mathscr{E}^\circ_{cell} = -0.94 \text{ volt}$$

Since the voltage is negative, the reaction as written will proceed to only a very slight extent; that is, Ag will not reduce Sn^{++}. The reverse reaction will go to a much greater extent, however, that is, Sn will reduce Ag^+.

Example 19-e. Determine whether or not Fe^{++} will reduce MnO_4^- in acid solution.

> **Solution.**

$$5\,Fe^{++} \rightleftharpoons 5\,Fe^{+++} + 5\,e^- \qquad\qquad\qquad \mathscr{E}^\circ = -0.77 \text{ volt}$$

$$\underline{MnO_4^- + 8\,H^+ + 5\,e^- \rightleftharpoons Mn^{++} + 4\,H_2O \qquad \mathscr{E}^\circ = +1.52 \text{ volt}}$$

$$5\,Fe^{++} + MnO_4^- + 8\,H^+$$

$$\rightleftharpoons Mn^{++} + 4\,H_2O + 5\,Fe^{+++} \quad \mathscr{E}^\circ_{cell} = +0.75 \text{ volt}$$

> The emf is positive and therefore the reaction as written will proceed, that is, Fe^{++} in acid solution will reduce MnO_4^-.

It may be noted that in the table of oxidation potentials just given, the oxidized form will oxidize the reduced form of any substance which lies *above* it in the table; for example, Cu^{++} will oxidize Zn but not Ag. Similarly, the reduced form will reduce the oxidized form of any substance which lies *below* it; for example, Ni will reduce Ag^+ but not Al^{+++}.

4. Oxidation-Reduction Potentials and Determination of Equilibrium Constants

It is possible to treat oxidation-reduction equilibria in exactly the same fashion as any other in that equilibrium constants for the reactions may be written. For the reaction

$$5\ Fe^{++} + MnO_4^- + 8\ H^+ \rightleftharpoons Mn^{++} + 4\ H_2O + 5\ Fe^{+++}$$

the equilibrium constant is written

$$K = \frac{[Mn^{++}][Fe^{+++}]^5}{[Fe^{++}]^5[MnO_4^-][H^+]^8}$$

The equilibrium constant for an oxidation-reduction reaction is easily obtained at 25°C from the relation

$$\mathscr{E}^\circ_{cell} = 2.303\ \frac{RT}{nF}\ \log K$$

where n is the number of electrons transferred in the chemical equation for the cell reaction; $\mathscr{E}^\circ_{cell}$ is the difference in the standard oxidation potentials of the two half-reactions making up the cell; F is the Faraday constant expressed as 23,060 cal/volt-equivalent. At 25°C this relation becomes:

$$\mathscr{E}^\circ_{cell} = \frac{0.059}{n}\ \log K \tag{1}$$

For the permanganate-ferrous ion reaction just cited, the cell emf was 0.75 volt and n was 5. On substitution into the log K expression,

$$0.75 = \frac{0.059}{5}\ \log K$$

$$\log K = \frac{5 \times 0.75}{0.059} = 63.5$$

$$K = 3 \times 10^{63}$$

From the magnitude of this equilibrium constant, it is obvious that the reaction should proceed virtually to completion.

Similarly, for the reaction $2\ Ag^+ + Sn \rightleftharpoons 2\ Ag + Sn^{++}$, $\mathscr{E}^\circ_{cell}$ is 0.94 volt, n is 2, and the equilibrium constant for the reaction is

$$K = \frac{[Sn^{++}]}{[Ag^+]^2}$$

Substituting into the expression (1) relating $\mathscr{E}^{\circ}_{cell}$ and $\log K$ above,

$$0.94 = \frac{0.059}{2} \log K$$

$$\log K = \frac{2 \times 0.94}{0.059} = 31.8$$

$$K = 6 \times 10^{31}$$

It should be noted that if the value of $\mathscr{E}^{\circ}_{cell}$ is negative, the value of the equilibrium constant is less than 1. For example, the cell voltage in Example 19-d was -0.94 volt. This corresponds to a value of $K = 1.7 \times 10^{-32}$ for the reaction $2\,Ag + Sn^{++} \rightleftharpoons 2\,Ag^{+} + Sn$.

5. Oxidation Potential Changes with Concentration

The standard oxidation potentials in Table 19-1 are for standard conditions, that is, all solids as pure substances, all gases at 1 atm pressure, and all solutions at 1 M. Often we are dealing with other concentrations and need to correct the standard potential for change in concentration from 1 M. The relationship between concentration and oxidation potential, at 25°C, is given by the Nernst equation

$$\mathscr{E} = \mathscr{E}^{\circ} - \frac{0.059}{n} \log \frac{[oxid]}{[red]} \tag{2}$$

where $[oxid]$ = concentration of oxidized form and $[red]$ = concentration of the reduced form.

Example 19-h. Determine the oxidation potential for the $Zn \rightleftharpoons Zn^{++} + 2\,e^{-}$ half-reaction at a concentration of $Zn^{++} = 0.100\ M$.

Solution.

$$\mathscr{E} = \mathscr{E}^{\circ} - \frac{0.059}{n} \log \frac{[Zn^{++}]}{[Zn]}$$

$$\mathscr{E} = 0.762 - \frac{0.059}{2} \log \frac{0.100}{1}$$

where the concentration or "activity" of pure metallic Zn is taken as unity.

$$\mathscr{E} = 0.762 - \frac{0.059}{2} \times (-1.00) = 0.762 + 0.030 = 0.790 \text{ volt}$$

PROBLEMS (Oxidation Potentials)

19-8. For the cell Ni $|$ Ni^{++} (1 M) $\|$ Cu^{++} (1 M) $|$ Cu, (a) determine the emf of the cell. (b) Write the equation for the cell reaction which occurs. (c) In which direction do electrons flow outside the cell? (d) Calculate the equilibrium constant for the cell reaction.

19-9. From the table of standard oxidation potentials, (a) determine whether or not the reaction 2 Al^{+++} + 3 Zn $\rightleftharpoons$ 2 Al + 3 Zn^{++} will occur to an appreciable extent. (b) Illustrate your answer by determining the equilibrium constant for the reaction.

19-10. (a) What is the value of the $\mathscr{E}^{\circ}_{cell}$ for the cell Sn $|$ Sn^{++} $\|$ Br^{-}, Br$_2$ $|$ Pt? (b) Write the cell reaction. (c) In what direction do electrons flow outside the cell? (d) What is the equilibrium constant for the cell reaction?

19-11. What is the oxidation potential of an Ag electrode in a 0.100 M AgNO$_3$ solution? (b) Of Cu in a 0.100 M Cu(NO$_3$)$_2$ solution? (c) Of In in a 0.100 M In(NO$_3$)$_3$ solution? (d) Of In in a 0.100 M In$_2$ (SO$_4$)$_3$ solution?

19-12. What is the electromotive force of a cell which is composed of a Ag electrode in a 0.20 M AgNO$_3$ solution and a Cd electrode in a 2.0 M Cd(NO$_3$)$_2$ solution? The cell may be represented

$$Cd \mid Cd^{++} (2.0\ M) \parallel Ag^{+}(0.20\ M) \mid Ag$$

19-13. Calculate the equilibrium constants for the following reactions:

(a) NO$_3^-$ + H$^+$ + Br$^-$ $\rightleftharpoons$ Br$_2$ + NO$_2$ + H$_2$O
(b) H$_2$O$_2$ + NO$_3^-$ + H$^+$ $\rightleftharpoons$ O$_2$ + NO + H$_2$O
(c) Zn + As + H$^+$ $\rightleftharpoons$ Zn^{++} + AsH$_3$
(d) H$_2$S + H$_2$SO$_3$ $\rightleftharpoons$ S + H$_2$O
(e) Cl$_2$ + H$_2$SO$_3$ + H$_2$O $\rightleftharpoons$ Cl$^-$ + SO$_4^{--}$ + H$^+$

GENERAL PROBLEMS

19-14. A current of 3.50 amp is passed through a solution of ZnSO$_4$ between Pt electrodes for exactly 1 hr. What weight of Zn will be deposited on the cathode?

19-15. What steady current in amp will be needed to deposit 0.100 equiv of Cu from a CuSO$_4$ solution in 4.00 hr?

19-16. A quantity of electricity is passed through a series of electrolytic cells containing AgNO$_3$, CuCl$_2$, Ni(NO$_3$)$_2$, Sc$_2$(SO$_4$)$_3$, Au(CN)$_2^-$ solutions. If 0.050 equiv of Ag is deposited in the first cell, determine the weight in grams of each of the metals deposited from solution. Determine the quantity of electricity employed.

19-17. What is the equivalent weight of X if exactly 5.0 g of X is deposited using a current of 5.00 amp for 40.0 min?

19-18. Water is electrolyzed at a current of 10.0 amp for exactly 2 hr; H_2 and O_2 gases are liberated at the cathode and anode, respectively. What would the dry volume of each be at STP?

19-19. Using the table of standard oxidation potentials, determine the values of $\mathscr{E}^{\circ}_{cell}$ for the following cells (concentrations all $1 \cdot M$):

(a) $Zn \mid Zn^{++} \parallel Ag^+ \mid Ag$ (d) $Al \mid Al^{+++} \parallel Sn^{++} \mid Sn$

(b) $Zn \mid Zn^{++} \parallel Ni^{++} \mid Ni$ (e) $Cu \mid Cu^{++} \parallel Ag^+ \mid Ag$

(c) $Cu \mid Cu^{++} \parallel Fe^{++}, Fe^{+++} \mid Pt$

19-20. Write the cell reaction for each of the voltaic cells in problem 19-19 as current is being drawn from them.

19-21. (a) Which of the following will be oxidized by Cu^{++}?

(1) Ni,   (2) Ag,   (3) Al,   (4) Cr^{+++},   (5) F^-

(b) Which of the following will be reduced by Sn?

(1) Ni^{++},   (2) Ag^+,   (3) Br_2,   (4) Al^{+++},   (5) Cu^{++}

19-22. Balance and then determine which of the following reactions will proceed spontaneously to an appreciable extent.

(a) $H_2O_2 + MnO_4^- + H^+ \rightleftharpoons O_2 + Mn^{++} + H_2O$

(b) $Fe^{+++} + Cr^{+++} + H_2O \rightleftharpoons Fe^{++} + Cr_2O_7^{--} + H^+$

(c) $Ni + H^+ \rightleftharpoons Ni^{++} + H_2$

(d) $PbO_2 + H^+ + Sn \rightleftharpoons Sn^{++} + Pb^{++} + H_2O$

(e) $Mn^{++} + H_2O + AsO_4^{---} \rightleftharpoons MnO_4^- + AsO_3^{---} + H^+$

19-23. Determine the oxidation potential of copper at 25°C in a $5.00 \times 10^{-4}\ M$ solution of Cu^{++}.

19-24. Calculate the equilibrium constant for each of the following reactions:

(a) $Cu^{++} + Ni \rightleftharpoons Cu + Ni^{++}$

(b) $Cu + NO_3^- + H^+ \rightleftharpoons Cu^{++} + NO_2 + H_2O$

(c) $Ag + Ni^{++} \rightleftharpoons Ag^+ + Ni$

(d) $I_2 + Br^- \rightleftharpoons Br_2 + I^-$

(e) $Br_2 + I^- \rightleftharpoons I_2 + Br^-$

(f) $F_2 + Br^- \rightleftharpoons Br_2 + F^-$

19-25. Determine the equilibrium constant for each of the following reactions:

(a) $Ce^{++++} + Cl^- \rightleftharpoons Cl_2 + Ce^{+++}$

(b) $Cl_2 + I^- \rightleftharpoons I_2 + Cl^-$

(c) $H_2O_2 + H_2SO_3 \rightleftharpoons SO_4^{--} + H_2O + H^+$

(d) $H_2O_2 + Br^- + H^+ \rightleftharpoons H_2O + Br_2$

(e) $H_2O_2 + Cr_2O_7^{--} + H^+ \rightleftharpoons Cr^{+++} + O_2 + H_2O$

19-26. The equilibrium constant for the reaction,

$$Sn^{++} + 2\ Hg^{++} \rightleftharpoons Hg_2^{++} + Sn^{++++}$$

is 5.0×10^{25}. (a) Calculate the emf of the cell if each of the species in solution is present at a concentration of $1\ M$. (b) From the table of oxidation potentials, obtain the value of $\mathscr{E}^{\circ}$ for the half-reaction $Sn^{++} \rightleftharpoons Sn^{++++} + 2\ e^-$, and calculate the value of $\mathscr{E}^{\circ}$ for the reaction $Hg_2^{++} \rightleftharpoons 2\ Hg^{++} + 2\ e^-$.

CHAPTER 20

Chemical Thermodynamics

Energy changes accompany all chemical and physical changes. Chemical thermodynamics is a study of the relationships between all forms of energy as they pertain to chemical processes.

SOME THERMODYNAMIC TERMS AND EQUATIONS

It is assumed that the student's text will be a source of an elaboration and discussion of these terms in more detail. Therefore the definitions and relationships between terms will be described briefly here.

$E = Internal\ Energy$. All of the energy in a system* including potential, chemical, bond energy, lattice energy, etc. The internal energy is a function of the temperature, pressure, and the chemical state of the system.

$H = Enthalpy$. This term is sometimes called the *Heat Content*. H and E are related by the equation: $H = E + PV$ where $P = $ pressure and $V = $ volume. The heat content or enthalpy is equal to the internal energy plus the pressure–volume product.

$S = Entropy$. A measure of degree of disorder or randomness in a system. The entropy of a gas is greater than that of a liquid which in turn is greater than that of a solid, since a solid is the most ordered of the three physical states and a gas is the least ordered.

$G = Free\ Energy$. $G = H - TS$ where T is absolute temperature. The free energy of a system is a measure of the energy which is available for useful work.

* A system is an isolated portion of the universe having physical boundaries separating it from other portions of the universe.

Thermodynamics is concerned with the *change* in these functions in passing from one state* to another. The mechanism by which the system gets from one state to a second is *not* a consideration. Therefore we may set up the following:

$$\Delta E = E_2 - E_1 \quad \text{where } E_2 \text{ is energy of state 2 and } E_1 \text{ is energy of state 1}$$
$$\Delta H = H_2 - H_1$$
$$\Delta S = S_2 - S_1 \quad \text{etc.}$$

The change in enthalpy, ΔH, can be related to the *change* in internal energy, ΔE, by the equation: $\Delta H = \Delta E + \Delta(PV)$. For a process at constant pressure, $\Delta H = \Delta E + P\,\Delta V$, where $P\,\Delta V$ represents the work done in increasing the volume of a system by ΔV against a constant pressure P.

A. THE FIRST LAW OF THERMODYNAMICS

This is essentially a statement of the Law of Conservation of Energy, that is, "Energy may be converted from one form to another but can neither be created nor destroyed." The first law is concerned only with the conservation of energy during changes in a system and not with the conditions under which changes might or might not occur.

When a system changes from State 1 to State 2, the change in internal energy (ΔE) will be the quantity of heat absorbed (q) in the process less the amount of work (w) done by the system on the surroundings, that is, $\Delta E = q - w$. If the work done is pressure-volume work, that is, expansion of gas at constant pressure, $w = P\,\Delta V$ and thus $\Delta E = q_p - P\,\Delta V$ where q_p is used to denote the heat absorbed at constant pressure. Since $\Delta E = \Delta H - P\,\Delta V$ (constant P) then $\Delta H = q_p$. Also, if no work is done, then the heat absorbed at constant volume, denoted as q_v, is equal to ΔE.

In considering $P\,\Delta V$ work the volume change in liquids and solids is relatively small and may be neglected—that is, where only liquids and solids are involved, it may be assumed that $P\,\Delta V = 0$ and therefore $q_v = q_p$. When gases are present in the reaction, there may be considerable volume change and the work may be calculated from the expression, work $= P\,\Delta V = (\Delta n)RT$, where Δn = number of moles of gaseous products minus the number of moles of gaseous reactants as shown by the balanced equation. R is the molar gas constant of 1.987 cal/(mole-°K) and T is the absolute temperature.

---

* The state of a system is defined by its T, P, V, and chemical composition.

Example 20-a. Given the value of ΔE, calculate ΔH for the reaction at 25°C and one atmosphere pressure.

$$H_2(g) + \tfrac{1}{2} O_2(g) \rightarrow H_2O(l) \qquad \Delta E = -67.43 \text{ kcal}$$

Solution.

$$\Delta H = \Delta E + P \,\Delta V$$

$$P \,\Delta V = (\Delta n)RT$$

The volume of the liquid phase is assumed to be negligible compared to the volume of the gaseous phase, therefore

$$\Delta n = 0 - (1 + 0.500) = -1.500 \text{ moles}$$

and

$$(\Delta n)RT = (-1.5 \text{ moles})(1.987 \text{ cal/mole-°K})(298°\text{K})$$

$$= -888 \text{ cal} = -0.89 \text{ kcal}$$

$$\Delta H = -67.43 - 0.89 = -68.32 \text{ kcal}$$

For reactions at one atmosphere and 25°C, ΔH and ΔE usually differ only slightly.

1. Isothermal and Adiabatic Processes*

An *isothermal* process is one conducted at constant temperature in a thermostat; heat flows in or out of the system to maintain the constant temperature. If a gas expands isothermally, work is done on the surroundings and energy is absorbed by the system equivalent to the work done. For the expansion of an ideal gas at constant temperature, the internal energy is constant, that is, $\Delta E = 0$. Since $\Delta E = q - w = 0$, then $q = w$. The work done is given by the equation, $w = nRT(2.303) \log \dfrac{V_2}{V_1}$ where V_2 is the final volume and V_1 is the initial volume. If a process is carried out in an insulated system such that no heat flows in or out of the system, the process is said to be *adiabatic*. In these cases, $q = 0$ and $\Delta E = -w$. If a gas expands adiabatically, work is done on the surroundings and the temperature of the gas falls.

Example 20-b. Calculate the energy required to expand isothermally one mole of gas at 273°K from 1 atm to 0.5 atm.

* For ideal gases.

Solution.

$$\text{Volume of 1 mole at } 273°\text{K and 1 atm} = 22.4 \text{ liters}$$

$$\text{Volume of 1 mole at } 273°\text{K and 0.5 atm} = 44.8 \text{ liters}$$

$$\Delta E = 0 = q - w \quad \text{so} \quad q = w$$

$$w = \text{work} = nRT(2.303) \log \frac{44.8}{22.4}$$

$$= (1 \text{ mole})(1.987 \text{ cal/mole-}°\text{K})(273°\text{K})(2.303) \log 2$$

$$q = w = 376 \text{ cal.}$$

2. Standard States and Changes

Of particular interest to chemists are changes of ΔH and ΔE for chemical reactions. Because ΔE and ΔH are both functions of temperature and pressure, it is necessary to establish certain conventions:

Standard State. The standard state of an element or compound is its most stable state at one atmosphere pressure and at some specified temperature, usually 25°C.

Standard Enthalpy of Formation. The standard enthalpy of formation, ΔH_f°, of any *compound* in its standard state is the enthalpy change when the compound in its standard state is formed from the elements in their standard states. The standard enthalpy of formation for any *element* in its standard state is *zero.*

Standard Enthalpy Changes. The standard enthalpy change, $\Delta H°$, for a reaction where the reactants and products are in their standard states, is the difference between the sum of the standard enthalpies of formation of the products and the sum of the standard enthalpies of formation of the reactants.

Example 20-c. For the reaction shown by the equation:

$$S(s) + O_2(g, 1 \text{ atm}) \rightarrow SO_2(g, 1 \text{ atm})$$

$\Delta H°$ was found by experiment to be -70.76 kcal/mole. Determine the standard enthalpy of formation, $\Delta H_{SO_2}°$, and the standard internal energy of formation, $\Delta E_{SO_2}°$.

Solution.

$$\Delta H° = \Delta H°_{SO_2} - (\Delta H°_S + \Delta H°_{O_2})$$

$$-70.76 = \Delta H°_{SO_2} - (0 + 0)$$

$$\Delta H°_{SO_2} = -70.76 \text{ kcal/mole}$$

$$\Delta H°_{SO_2} = \Delta E°_{SO_2} + (\Delta n)\, RT$$

$\Delta n = 0$ assuming only the number of moles of gaseous products and reactants are important.

$$\Delta H°_{SO_2} = \Delta E°_{SO_2} = -70.76 \text{ kcal/mole}$$

Note that $\Delta H°_{SO_2}$ corresponds to a rather unusual set of conditions, that is, the reactants in their standard states form the product in its standard state. If the conditions are different from standard, then ΔH will be different from $\Delta H°$.

Example 20-d. Using the standard enthalpies of formation at 25°C in Table 20-1, determine the standard enthalpy change and also $\Delta E°$ for the reaction:

$$4\, NH_3(g, 1 \text{ atm}) + 5\, O_2(g, 1 \text{ atm}) \rightarrow 4\, NO(g, 1 \text{ atm}) + 6\, H_2O(g, 1 \text{ atm})$$

Solution.

$$\Delta H° = 4\Delta H°_{NO} + 6\Delta H°_{H_2O} - (4\Delta H°_{NH_3} + 5\Delta H°_{O_2})$$

$$= 4(21.60) + 6(-57.80) - [4(-11.04) + 5(0)]$$

$$= 86.40 - 346.80 - (-44.16 + 0) = -216.24 \text{ kcal}$$

$$\Delta H° = \Delta E° + (\Delta n)\, RT$$

$$\Delta n = 10 - 9 = 1$$

$$\Delta E° = -216.24 - (1 \text{ mole} \times 1.987 \text{ cal/mole-°K} \times 298°K)$$

$$\times (10^{-3} \text{ kcal/cal})$$

$$= -216.24 - 0.59 = -216.83 \text{ kcal}$$

Example 20-e. Calculate $\Delta H°$ and $\Delta E°$ for the following reaction at 25°C and 1 atm.

$$NH_4NO_2(s) \rightarrow N_2(g) + 2\, H_2O(l)$$

Refer to Table 20-1 for enthalpies of formation at 25°C and 1 atm.

Table 20-1   Thermodynamic Properties

Values are for substances at 25°C and 1 atm pressure. Enthalpies of formation (ΔH_f°) and free energies of formation (ΔG_f°) are expressed in kilocalories per mole—entropy (S°) values in calories per mole per degree absolute temperature.

Substance	ΔH_f°	ΔG_f°	S°
$NH_3(g)$	−11.04	−3.98	46.0
$HBr(g)$	−8.66	−12.7	47.4
$BrCl(g)$	3.51	−0.21	57.3
$CO_2(g)$	−94.1	−94.3	51.1
$CO(g)$	−26.4	−32.8	47.3
$HCl(g)$	−22.1	−22.8	44.6
$HF(g)$	−64.2	−64.7	41.5
$H_2O(g)$	−57.80	−54.64	45.1
$H_2O(l)$	−68.32	−56.69	16.7
$HI(g)$	6.20	0.31	49.3
$I_2(g)$	14.9	4.63	62.3
$Hg(l)$	14.5	7.60	41.8
$PBr_3(g)$	−35.9	−41.2	83.1
$PCl_3(g)$	−73.2	−68.4	74.5
$PCl_5(g)$	−95.4	−77.6	84.3
$NaCl(s)$	−98.3	−91.8	17.3
$CH_4(g)$	−17.9	−12.1	44.5
$CH_3OH(l)$	−57.0	−39.7	30.3
$COCl_2(g)$	−53.3	−50.3	69.1
$CH_3Cl(g)$	−19.6	−14.0	56.0
$C_2H_4(g)$	12.5	16.3	52.5
$C_2H_5OH(l)$	−66.4	−41.8	38.4
$C_2H_6(g)$	−20.2	−7.86	54.9
$NH_4NO_2(s)$	−63.1	—	—
$NO(g)$	21.60	20.72	50.3
$NOCl(g)$	12.57	15.86	63.0

Solution.

$$\Delta H^\circ = \Delta H_f^\circ \text{ (products)} - \Delta H_f^\circ \text{ (reactants)}$$
$$= [(0) + 2(-68.3)] - (-63.1)$$
$$= -136.6 + 63.1 = -73.5 \text{ kcal}$$
$$\Delta E^\circ = \Delta H^\circ - (\Delta n)RT \qquad \Delta n = (1) - (0) = 1$$

assuming volumes of solid NH_4NO_2 and liquid H_2O are relative small and may be neglected.

$$\Delta E^\circ = -73,500 \text{ cal} - (1)(1.987 \text{ cal/mole-}^\circ K)(298^\circ K)$$
$$= -73,500 \text{ cal} - 592 \text{ cal} = -74,092 \text{ cal} = -74.1 \text{ kcal}$$

B. THE SECOND LAW

1. Entropy Change

The second law specifies that only certain processes will occur spontaneously; for example, heat flows only from a hot object to a cooler one. A simple statement of the law is as follows: A spontaneous change takes place only with a net increase in total entropy of the system and its surroundings, that is, $\Delta S_{total} > 0$. When a process occurs reversibly* at constant temperature and pressure, the change in entropy of the system, ΔS, is equal to the heat absorbed, q_p, divided by the absolute temperature at which the change occurs, that is, $\Delta S = \dfrac{q_p}{T}$.

Example 20-f. Calculate ΔS for the melting of one mole of ice to liquid H_2O at 0°C. Heat of fusion $= 79.7$ cal/g.

Solution. Molar heat of fusion $= 79.7$ cal/g $\times$ 18.01 g/mole $=$ 1435 cal/mole.

$$\Delta S = \frac{q_p}{T} = \frac{1435}{273} = 5.26 \text{ cal/mole-°K}$$

Since ΔS is positive, the entropy is increasing. Note that the disorder or randomness (entropy) is greater in the liquid state than in the solid state.

2. Free Energy Change

Usually it is more convenient to think only in terms of the system rather than the system and its surroundings. The change in free energy of the system, rather than the change in entropy, provides a useful criterion of spontaneity *at constant temperature and pressure*. The free energy of a system, G, may be thought of as the energy possessed by the system which may be converted to useful work—that is, it is the maximum *available* energy. The *change* in free energy of a system ΔG, is related to ΔH, ΔS, and T by the equation: $\Delta G = \Delta H - T \Delta S$.

---

* A reversible change is one which is carried out essentially under equilibrium conditions. For example, the freezing of water at its melting point may be carried out reversibly by the slow removal of heat at constant temperature and pressure; the process may be reversed by the addition of heat from the surroundings.

ΔG is a combination of terms depending on the changes in enthalpy and entropy in the system, and serves to determine whether a chemical reaction will occur under these conditions. If ΔG is negative for the equation as written, then that process will occur spontaneously; if ΔG is positive the reverse reaction occurs spontaneously. If ΔG is 0, the system is at equilibrium and no net reaction occurs. Basically, the same standard state conventions have been adopted for ΔG as for ΔH. The free energy of formation, ΔG_f°, of the free element in its most stable state at one atmosphere is taken as zero. Free energies of formation for a number of compounds are given in Table 20-1.

Example 20-g. Calculate $\Delta G°$ for the reaction:

$$NOCl(g) \rightarrow NO(g) + \tfrac{1}{2} Cl_2(g)$$

Refer to Table 20-1 for standard free energies of formation.

> **Solution.**
>
> $$\Delta G° = \Delta G_{NO}^\circ + \tfrac{1}{2}\Delta G_{Cl_2}^\circ - \Delta G_{NOCl}^\circ$$
> $$= (20.72 + 0) - 15.86 = 4.86 \text{ kcal.}$$

Example 20-h. Calculate $\Delta G_{NH_3}^\circ$ from data in Table 20-1 and the experimental fact that $\Delta G° = -229.04$ kcal for the reaction:

$$4 NH_3(g, 1 \text{ atm}) + 5 O_2(g, 1 \text{ atm}) \rightarrow 4 NO(g, 1 \text{ atm}) + 6 H_2O(g, 1 \text{ atm})$$

$$\Delta G° = (4\Delta G_{NO}^\circ + 6\Delta G_{H_2O}^\circ) - (4\Delta G_{NH_3}^\circ + 5\Delta G_{O_2}^\circ)$$
$$-229.04 = 4(20.72) + 6(-54.64) - 4\Delta G_{NH_3}^\circ$$
$$-229.04 = 82.88 - 327.84 - 4\Delta G_{NH_3}^\circ$$

$$\Delta G_{NH_3}^\circ = -\frac{15.92}{4} = -3.98 \text{ kcal/mole}$$

Once $\Delta G°$ and $\Delta H°$ are known for a particular reaction, calculation of $\Delta S°$ is a relatively simple matter. At constant temperature, $\Delta S°$ may be obtained from the relationship:

$$\Delta G° = \Delta H° - T \Delta S° \quad \text{or} \quad \Delta S° = \frac{\Delta H° - \Delta G°}{T}$$

Example 20-i. For the reaction

$$4 NH_3(g, 1 \text{ atm}) + 5 O_2(g, 1 \text{ atm}) \rightarrow 4 NO(g, 1 \text{ atm}) + 6 H_2O(g, 1 \text{ atm})$$

Calculate $\Delta S°$. Use values of $\Delta H°$ and $\Delta G°$ from Examples 20-d and 20-h. Note that $\Delta S°$ is expressed in calories, hence $\Delta H°$ and $\Delta G°$ must be converted to calories.

Solution.

$$\Delta H^\circ = -216.24 \text{ kcal} \qquad \Delta G^\circ = -229.04 \text{ kcal}$$

$$\Delta S^\circ = \frac{[-216,240 - (-229,040)]}{298} = 43.0 \text{ cal/(mole-}^\circ\text{K)}$$

C. THE THIRD LAW

This law states that the entropy of a perfect crystalline substance at absolute zero (0°K) is zero. The units in a crystal are arranged in a very definite way. This orderly arrangement possesses no disorder or randomness, hence may be regarded as having a minimum value for entropy, that is, 0. So-called standard entropies may be calculated in their standard states at 25°C and these may be used to calculate ΔS for reactions by subtracting the sum of the entropy values of reactants from products. A few such entropies are recorded in Table 20-2.

Table 20-2 Standard Entropies (S°) at 25°C in cal/(mole-deg)

$H_2(g)$	31.2	$HCl(g)$	44.62	CO_2	51.06
$O_2(g)$	49.00	$H_2O(l)$	16.72	CH_4	44.50
$Cl_2(g)$	53.29	$H_2O(g)$	45.11	$ZnSO_4(aq, 1M)$	−21.3
$F_2(g)$	48.6	$Zn(s)$	38.45	CO	47.3
$Na(s)$	12.2	$H_2SO_4(aq, 1M)$	4.1	NO	50.34
		$C(s, \text{graphite})$	1.36	NH_3	46.01

Example 20-j. Determine ΔS° for the reaction

$$H_2(g) + \tfrac{1}{2} O_2(g) \rightarrow H_2O(l)$$

from the standard entropy values in Table 20-2.

Solution.

$$\Delta S^\circ = S^\circ_{H_2O} - [S^\circ_{H_2} + \tfrac{1}{2}S^\circ_{O_2}]$$

$$= 16.72 - [31.2 + \tfrac{1}{2}(49.0)]$$

$$= -39.0 \text{ cal/(mole-}^\circ\text{K)}$$

Example 20-k. From the values of S° in Table 20-2, determine ΔS° for the reaction:

$$4 \text{ NH}_3(g, 1 \text{ atm}) + 5 \text{ O}_2(g, 1 \text{ atm}) \rightarrow 4 \text{ NO}(g, 1 \text{ atm}) + 6 \text{ H}_2\text{O}(g, 1 \text{ atm})$$

Solution.

$$\Delta S^\circ = \sum S^\circ(\text{products}) - \sum S^\circ(\text{reactants})$$
$$= 4S^\circ_{NO} + 6S^\circ_{H_2O} - (4S^\circ_{NH_3} + 5S^\circ_{O_2})$$
$$= 4(50.34) + 6(45.11) - [4(46.01) + 5(49.00)]$$
$$= 42.98 \text{ cal}/(\text{mole-}^\circ K)$$

(This is essentially equivalent to the 43.0 value in Example 20-i.)

D.   CHEMICAL EQUILIBRIUM AND FREE ENERGY

The standard free energy changes ΔG° calculated above are limited to reactions with both reactants and products in their standard states. Only rarely are these circumstances encountered in actual situations. Under conditions different from those of the standard states, the sign of ΔG is the controlling factor in predicting spontaneity of reactions at constant temperature and pressure. The equation

$$\Delta G = \Delta G^\circ + 2.303 RT \log Q_p$$

allows values of ΔG to be evaluated for any set of conditions. The second term on the right hand side corrects for the differences between the actual initial conditions and the standard state conditions of the reactants and products. Q_p has the same form as the equilibrium constant, and for gaseous reactions, the partial pressures (in atmospheres) of the constituents are used as a measure of the initial amounts of the reacting species. For the general reaction

$$mA + nB \rightleftharpoons rC + sD$$

$$Q_p = \left[\frac{(p_C)^r (p_D)^s}{(p_A)^m (p_B)^n} \right]_{\text{initial}}$$

If any of the reactants or products are pure liquids or solids, their vapor pressures at constant temperature are constant and do not change during the course of the reaction. Such terms are not included in expressions for K_p or Q_p (see Examples 14-f, g).

Example 20-l. Calculate ΔG at 298°K for the reaction for the conditions listed in the equation:

$$4 \text{ NH}_3(g, 0.1 \text{ atm}) + 5 \text{ O}_2(g, 10 \text{ atm})$$
$$\rightleftharpoons 4 \text{ NO}(g, 2.0 \text{ atm}) + 6 \text{ H}_2\text{O}(g, 1 \text{ atm})$$

where ΔG° is -229.04 kcal

Solution.

$$Q_p = \frac{(p_{NO})^4(p_{H_2O})^6}{(p_{NH_3})^4(p_{O_2})^5}$$

$$\Delta G = \Delta G^\circ + (2.303)(1.987 \text{ cal/mole-}^\circ K)(10^{-3} \text{ kcal/cal})$$

$$\times (298^\circ K) \log \frac{(2.0)^4(1)^6}{(0.1)^4(10)^5}$$

$$\Delta G = -229.04 \text{ kcal} + [1.364 \log 1.6] \text{ kcal}$$
$$= -228.76 \text{ kcal}$$

Because ΔG is *negative* 228.76 kcal, NH_3 at a pressure of 0.1 atm and O_2 at a pressure of 10 atm *will* react to produce NO at a pressure of 2 atm and H_2O at a pressure of 1 atm.

For a system in chemical equilibrium, there is no tendency for any further net chemical change. Under these conditions, $\Delta G = 0$ and Q_p becomes numerically equal to K_p. ΔG° then is given by the equation: $\Delta G^\circ = -2.303RT \log K_p$. Values of the equilibrium constant may be calculated from ΔG° and vice versa.

Example 20-m. Calculate K_p for the equilibrium:

$$COCl_2(g) \rightleftharpoons CO(g) + Cl_2(g)$$

See Table 20-1 for free energies of formation.

Solution.

$$K_p = \frac{p_{CO}p_{Cl_2}}{p_{COCl_2}}$$

$$\Delta G^\circ = -(2.303)(RT \log K_p)$$

ΔG° may be calculated from free energies of formation.

$$\Delta G^\circ = \Delta G_f^\circ(\text{products}) - \Delta G_f^\circ(\text{reactants})$$

$$\Delta G^\circ = (-32.8) + (0) - (-50.3) = 17.5 \text{ kcal}$$

$$\Delta G^\circ = -(2.303)(1.987 \text{ cal/mole-}^\circ K)(298^\circ) \log K = 17,500 \text{ cal}$$

$$\log K_p = \frac{17,500 \text{ cal}}{-1364 \text{ cal}} = -12.9$$

$$K = 1.1 \times 10^{-13}$$

Example 20-n. Calculate K_p for the equilibrium in Example 20-m at 1000°K from the value at 25°C (1.1×10^{-13}) assuming $\Delta H°$ and $\Delta S°$ are the same at the two temperatures.

$$COCl_2(g) \rightleftharpoons CO(g) + Cl_2(g)$$

Solution.

$$\Delta G° = \Delta H° - T \Delta S° = -(2.303)(RT) \log K_p$$
$$\Delta H° = \Delta H_f°(\text{products}) - \Delta H_f°(\text{reactants})$$
$$= (-26.4) + (0) - (-53.3) = 26.9 \text{ kcal}$$
$$\Delta S° = S°(\text{products}) - S°(\text{reactants})$$
$$\Delta S° = (47.4) + (53.3) - (69.1) = 31.5 \text{ cal/mole-°K}$$

or from Example 20-m $\Delta G° = 17.5$ kcal

Solve for ΔS in the equation $\Delta G° = \Delta H° - T \Delta S°$

$$\Delta S° = \frac{\Delta H° - \Delta G°}{T} = \frac{26,900 - 17,500}{298} = 31.5 \text{ cal/mole-°K}$$

Since $-2.303 RT \log K = \Delta G° = \Delta H° - T \Delta S°$

$$\log K = \frac{\Delta H° - T \Delta S°}{-2.303 RT} = \frac{-\Delta H°}{2.303 RT} + \frac{\Delta S°}{2.303 R}$$

Substituting,

$$\log K = \frac{-(26,900 \text{ cal})}{(2.303)(1.987 \text{ cal/mole-°K})(1000°K)}$$
$$+ \frac{31.5 \text{ cal/°K}}{(2.303)(1.987 \text{ cal/mole-°K})}$$
$$= \frac{-26,900 \text{ cal}}{4576} + 6.88 = -5.88 + 6.88 = 1.00$$

$$K = 10.0$$

E. ELECTROCHEMICAL CELLS AND FREE ENERGY

Just as equilibrium constants for gaseous reactions can be related to standard free energy changes, equilibrium constants for reactions in solution can also be related to standard free energy changes, that is

$$\Delta G° = -2.303 RT \log K \tag{1}$$

In Chapter 19, Section 4, it was pointed out that the standard electromotive force of an electrochemical cell was related to the equilibrium constant for the cell reaction by the equation:

$$\mathscr{E}^\circ_{cell} = 2.303 \frac{RT}{nF} \log K \tag{2}$$

where n = number of equivalents of electricity transferred in the balanced chemical equation; $\mathscr{E}^\circ_{cell}$ is the difference in the standard oxidation potentials of the two half-reactions making up the cell, and F is the Faraday constant expressed as 23,060 cal/volt-equiv. By combining equations (1) and (2)

$$\Delta G^\circ = -nF\mathscr{E}^\circ_{cell}$$

For reactions in solution, K is usually expressed in terms of concentrations (molarity) of reactants and products. As in the case of standard oxidation potentials, standard state conditions are: all solids are as pure substances in their most stable form; all gases are at 1 atm pressure, and all solutes are at a concentration of 1 molar. The solvent in dilute solutions is considered the same as a pure liquid, hence is in its standard state.

For conditions other than standard conditions,

$$\Delta G = -nF\mathscr{E}_{cell} \tag{3}$$

where $\mathscr{E}_{cell}$ is given by the equation:

$$\mathscr{E}_{cell} = \mathscr{E}^\circ_{cell} - \frac{RT}{nF}(2.303) \log Q \tag{4}$$

where Q is as defined in Section 20-D. Combining (3) and (4) above, we may obtain

$$\Delta G = -nF\mathscr{E}^\circ_{cell} + (2.303)RT \log Q$$

Example 20-o. Determine ΔG for the reaction in aqueous solution:

$$2\,Ag^+(0.1\,M) + Sn(s) \rightleftharpoons 2\,Ag(s) + Sn^{++}(0.5\,M) \qquad \mathscr{E}^\circ_{cell} = 0.94 \text{ volts}$$

Solution.

$$\Delta G = -nF\mathscr{E}^\circ_{cell} + 2.303\,RT \log \frac{[Sn^{++}]}{[Ag^+]^2}$$

$$\Delta G = -(2)\text{ equiv }(23,060 \text{ cal/v-equiv})(0.940 \text{ v})$$
$$+ (2.303)(1.987 \text{ cal/mole-}°K)298°K\left(\log \frac{0.5}{(0.1)^2}\right)$$
$$= -43,350 \text{ cal} + 1364\,(\log 50)$$
$$= -43,350 \text{ cal} + 2320 \text{ cal}$$
$$= -41,000 \text{ cal} = -41.0 \text{ kcal}$$

PROBLEMS

20-1. Calculate ΔH for the reaction at 25°C

$$CH_4(g) + 2 O_2(g) \rightarrow CO_2(g) + 2 H_2O(l)$$

20-2. The heat of combustion of C_2H_2 at constant volume (ΔE) and 20°C is −311.1 kcal/mole. Calculate heat of combustion at constant pressure (ΔH) of 1 atm and 20°C.

20-3. Calculate the work done in isothermally expanding one mole of gas from 5 liters to 10 liters at 27°C.

20-4. Using standard entropy values in Table 20-2, calculate $\Delta S°$ for the following reactions:
(a) $\frac{1}{2} H_2(g, 1\ atm) + \frac{1}{2} Cl_2(g, 1\ atm) \rightarrow HCl(g, 1\ atm)$
(b) $C(s, 1\ atm) + O_2(g, 1\ atm) \rightarrow CO_2(g, 1\ atm)$
(c) $Zn(s) + H_2SO_4(aq, 1\ M) \rightarrow ZnSO_4(aq, 1\ M) + H_2(g, 1\ atm)$
(d) $Na(s) + \frac{1}{2} Cl_2(g, 1\ atm) \rightarrow NaCl(s)$

20-5. Calculate $\Delta H°$ and $\Delta E°$ for the following reaction at 25°C.

$$C_2H_6(g, 1\ atm) + 3\frac{1}{2} O_2(g, 1\ atm) \rightarrow 2 CO_2(g, 1\ atm) + 3 H_2O(l)$$

20-6. Calculate $\Delta E°$ and $\Delta H°$ for the following reactions at 25°C and 1 atm pressure. $\Delta H_f°$ for $(NH_4)_2C_2O_4$ and NH_4HCO_3 are −267.15 and −203.1 kcal/mole respectively.
(a) $(NH_4)_2C_2O_4(s) \rightarrow 2 NH_3(g) + CO_2(g) + CO(g) + H_2O(l)$
(b) $NH_4HCO_3(s) \rightarrow NH_3(g) + CO_2(g) + H_2O(l)$

20-7. Determine the boiling point of Hg from the following data: heat of vaporization = 70.6 cal/g. At the boiling point and standard pressure the entropy change is 22.5 cal/mole-°K.

20-8. Use data from Example 20-n in this exercise. $COCl_2$ is pumped into a cylinder at 1000°K and allowed to come to equilibrium with its dissociation products. At equilibrium the partial pressure of $COCl_2$ is 5 atm. Calculate the partial pressures of CO and Cl_2.

20-9. Calculate $\Delta G°$ and K_p at 25°C for the equilibrium

$$PCl_5(g) \rightleftharpoons PCl_3(g) + Cl_2(g)$$

20-10. Calculate K_p at 600°K for the equilibrium in exercise 20-9. Heats of formation $PCl_3 = -73.2$ kcal/mole, $PCl_5 = -95.4$ kcal/mole. (*Hint:* calculate $\Delta S°$ from $\Delta G°$ above.)

20-11. The heat of vaporization of Hg at its boiling point 356.6°C is 70.6 cal/g. Calculate ΔH, ΔE, ΔS, ΔG for the vaporization of 1 mole of Hg at the B.P. and 1 atm pressure.

20-12. Heats of formation of HCl, $CO_2(g)$, NaCl(s) are respectively −22.1, −94.1 and −98.2 kcal/mole. Calculate $\Delta G°$ at 298°K for parts (a), (b), and (d) in exercise 20-4.

20-13. Determine $\Delta G°$ for the following reactions:
(a) $CH_4(g) + 2 O_2(g) \rightarrow CO_2(g) + 2 H_2O(l)$
(b) $2 P(s) + 3 Br_2(l) \rightarrow 2 PBr_3(g)$
(c) $C(s) + O_2(g) \rightarrow CO_2(g)$
(d) $C_2H_6(g) + 3\frac{1}{2} O_2(g) \rightarrow 2 CO_2(g) + 3 H_2O(l)$

20-14. The heat of fusion of ethyl alcohol, C_2H_5OH, at its melting point of $-114°C$ is 24.9 cal/g. If one mole of liquid ethyl alcohol is converted to solid at $-114°C$ calculate $\Delta S°$. Is the latter positive or negative and how do you determine this? Discuss the value of ΔG for the change.

20-15. Using data in Table 20-1 and Table 20-2, calculate for the following reaction at 25°C.

$$CO(g) + \tfrac{1}{2} O_2(g) \rightarrow CO_2(g)$$

$\Delta H°$, $\Delta S°$, $\Delta G°$. What do you conclude regarding the spontaneity of the reaction under standard conditions?

20-16. Calculate ΔG at 298°K for the reactions under the conditions indicated in the following equations:
(a) $CO(g, 0.5\text{ atm}) + Cl_2(g, 0.2\text{ atm}) \rightarrow COCl_2(g, 3\text{ atm})$
(b) $CH_4(g, 0.2\text{ atm}) + Cl_2(g, 0.3\text{ atm}) \rightarrow CH_3Cl(g, 1\text{ atm}) + HCl(g, 1\text{ atm})$
(c) $CH_3Cl(g, 0.1\text{ atm}) + H_2O(g, 0.01\text{ atm}) \rightarrow CH_3OH(l) + HCl(g, 0.3\text{ atm})$
(d) $N_2(g, 40\text{ atm}) + 3 H_2(g, 50\text{ atm}) \rightarrow 2 NH_3(g, 1\text{ atm})$
(e) $H_2(g, 10\text{ atm}) + I_2(g, 0.05\text{ atm}) \rightarrow 2 HI(g, 0.3\text{ atm})$
(f) $2 NOCl(g, 0.1\text{ atm}) + CO(g, 0.3\text{ atm}) \rightarrow 2 NO(g, 0.5\text{ atm})$
$$+ COCl_2(g, 1\text{ atm})$$
(g) $CO(g, 50\text{ atm}) + H_2O(g, 20\text{ atm}) \rightarrow CO_2(g, 3\text{ atm}) + H_2(g, 2\text{ atm})$

20-17. Calculate K_p at 298°K for the equilibria represented by the following equations:
(a) $Br_2(l) + Cl_2(g) \rightarrow 2 BrCl(g)$
(b) $2 CO(g) + O_2(g) \rightarrow 2 CO_2(g)$
(c) $2 HI(g) + O_2(g) \rightarrow H_2O(g) + I_2(g)$
(d) $C_2H_4(g) + H_2(g) \rightarrow C_2H_6(g)$
(e) $CH_4(g) + Cl_2(g) \rightarrow CH_3Cl(g) + HCl(g)$
(f) $2 Na(s) + Cl_2(g) \rightarrow 2 NaCl(s)$
(g) $2 NOCl(g) + CO(g) \rightarrow 2 NO(g) + COCl_2(g)$
(h) $H_2(g) + I_2(g) \rightarrow 2 HI(g)$
(i) $H_2(g) + F_2(g) \rightarrow 2 HF(g)$

20-18. Using data in Table 19-1, determine $\Delta G°$ for each of the following reactions in aqueous solution as shown by the equations. *Note:* the standard state for a solvent is taken as the pure liquid; however, in dilute solutions the solvent is assumed to be in its standard state. This assumption is made here and a term for the "concentration of water" does not appear in the expression for either Q or K.
(a) $Zn(s) + 2 Cr^{+++} \rightarrow Zn^{++} + 2 Cr^{++}$
(b) $Sn^{++} + 2Fe^{+++} \rightarrow Sn^{++++} + 2Fe^{++}$
(c) $Cr_2O_7^{--} + 14 H^+ + 6 Fe^{++} \rightarrow 6 Fe^{+++} + 2 Cr^{+++} + 7 H_2O$

(d) $3 H_2S + 2 H^+ + 2 NO_3^- \rightarrow 3 S + 2 NO + 4 H_2O(l)$

(e) $Cu + 4 H^+ + 2 NO_3^- \rightarrow Cu^{++} + 2 NO_2 + 2 H_2O(l)$

20-19. Using data in Table 19-1, determine ΔG at 25°C for each of the following reactions:

(a) $Zn(s) + 2 Cr^{+++}(1.0\ M) \rightarrow Zn^{++}(0.01\ M) + 2 Cr^{++}(0.01\ M)$

(b) $Sn^{++}(1.0\ M) + 2 Fe^{+++}(1.0\ M) \rightarrow Sn^{++++}(0.01\ M) + 2 Fe^{++}(0.01\ M)$

(c) $Sn^{++}(0.01\ M) + 2 Fe^{+++}(0.01\ M) \rightarrow Sn^{++++}(1.0\ M) + 2 Fe^{++}(1.0\ M)$

(d) $Cr_2O_7^{--}(1.0\ M) + 14 H^+(0.1\ M) + 6 Fe^{++}(1.0\ M)$
$$\rightarrow 6 Fe^{+++}(1.0\ M) + 2 Cr^{+++}(1.0\ M) + 7 H_2O(l)$$

(e) $Cu(s) + 4 H^+(0.1\ M) + 2 NO_3^-(0.1\ M)$
$$\rightarrow Cu^{++}(0.1\ M) + 2 NO_2(g, 0.1\ atm) + 2 H_2O(l)$$

20-20. ΔG at 298°K for the reaction and conditions shown by the equation:

$$Co(s) + 2 Fe^{+++}(0.1\ M) \rightarrow Co^{++}(0.2\ M) + 2 Fe^{++}(0.01\ M)$$

was found experimentally to be -51.74 kcal. Determine with the help of Table 19-1, $\mathscr{E}^\circ$ for the half-reaction, $Co \rightarrow Co^{++} + 2\ e^-$.

Appendix I

Mathematical Review

Without exception, students of chemistry will encounter mathematical problems which must be solved. Generally in a first-year course, these problems will involve simple arithmetic and very elementary algebra. In the discussion which follows it is assumed that students have a knowledge of simple mathematics and need only to be reminded of the operations involved in the solution of equations.

An algebraic statement is one representing an equality. Generally it contains several terms and usually at least one term whose numerical value is unknown. The equality

$$\frac{2X}{3} - 6 = 10$$

is such an algebraic statement. We are usually interested in solving for the unknown term X, that is, getting X on one side of the equality sign and the remainder of the terms on the other.

A. SOLUTION OF SIMPLE EQUATIONS

To solve the most commonly encountered equations, a few simple rules are employed, having to do with the transfer of quantities from one side of the equation to the other.

(a) First, the addition or subtraction of a number to both sides of the equation does not destroy the equality. In the example just given, the technique of transferring the term -6 to the right-hand side involves merely a change of sign, so that the equation now reads

$$\frac{2X}{3} = 10 + 6$$

This is obviously analogous to adding $+6$ to both sides of an equation and then, since $+6$ and -6 exactly cancel each other, the equation takes the form

$$\frac{2X}{3} - \cancel{6} + \cancel{6} = 10 + 6 \quad or \quad \frac{2X}{3} = 16$$

(b) Second, the multiplication or division of both sides of an equation by the same number does not alter the equality. In the above equality, let us next eliminate fractions by multiplying both sides of the equation by 3. In this event

$$\frac{2X \cdot \cancel{3}}{\cancel{3}} = 16 \cdot 3$$

Since the number 3 now appears in both the numerator and denominator on the left-hand side of the equation, they cancel; and the equation assumes the form

$$2X = 16 \cdot 3 \quad or \quad 2X = 48$$

The final step in solving for the unknown (X) is the division of both sides of the equation by 2, which yields the relation

$$\frac{\cancel{2}X}{\cancel{2}} = \frac{48}{2}$$

Again the number 2 appears in both the numerator and denominator of the left-hand term and therefore they cancel each other, yielding the relationship

$$X = \tfrac{48}{2} \quad or \quad X = 24$$

Thus $X = 24$ is the solution to the original equation. To check this, we merely substitute the number 24 for X in the original equation and determine whether the equality holds.

Check.

$$\frac{2X}{3} - 6 = 10$$

Substituting 24 for X

$$\frac{2 \cdot 24}{3} - 6 = 10$$

$$\frac{48}{3} - 6 = 10$$

Dividing 48 by 3

$$16 - 6 = 10$$

and

$$10 = 10$$

Therefore, we may conclude that 24 is a solution to the original equation.

B. SIGN CONVENTIONS

Sign conventions are of great importance in solving algebraic equations. In such operations, the use of negative numbers is commonly encountered. The addition of two numbers of like sign requires that the sum be of the same sign, for example, $-5 - 3 = -8$. The addition of numbers of opposite sign is carried out by subtracting the smaller from the larger and assigning the sign of the larger to the difference, for example, $-5X + 3X = -2X$. The multiplication of numbers of the same sign requires the product to have a positive sign, for example, $(-5)(-3) = +15$; whereas the product has a negative sign if the original numbers are of opposite sign, for example, $(-5)(3) = -15$. A similar rule applies to division; the quotient of either two positive or two negative numbers is positive, whereas the quotient of a positive and a negative number is negative.

C. SOLUTION OF FIRST-DEGREE EQUATIONS

First-degree equations are those in which the exponent of the unknown quantity is not greater than one. Such equations are easily solved using the principles outlined above; for example, solve the equation $(5X + 3 = X - 5)$ for X.

Solution.

$$5X + 3 = X - 5$$
$$5X - X = -5 - 3$$
$$4X = -8$$
$$X = -2$$

Similarly, solve the equation

$$\frac{6X - 3}{5} = \frac{6X}{4} \quad \text{for} \quad X$$

Solution.

$$\frac{6X - 3}{5} = \frac{6X}{4}$$

$$4(6X - 3) = (6X)5$$

$$24X - 12 = 30X$$

$$-12 = 30X - 24X$$

$$-12 = 6X$$

$$X = -2$$

D. SOLUTION OF QUADRATIC EQUATIONS

Occasionally a student will need to solve for an unknown in a simple quadratic such as $9X^2 - 25 = 0$. A quadratic equation is a second-degree equation, that is, one where the largest exponent of the unknown is 2. The quadratic equation above is a very simple one which is easily solved in the following manner.

Solution.

$$9X^2 - 25 = 0$$

$$9X^2 = 25$$

$$X^2 = \tfrac{25}{9}$$

$$X = \sqrt{\tfrac{25}{9}}$$

$$X = \pm\tfrac{5}{3}$$

An alternative method of solution of quadratic equations which are not so simple is shown by the solution to the general quadratic

$$aX^2 + bX + c = 0$$

where a and b are the coefficients of X^2 and X in the quadratic, and c is a constant. The solution to this equation is given by the relation

$$X = \frac{-b \pm \sqrt{b^2 - 4ac}}{2a}$$

In the example $9X^2 - 25 = 0$, $a = 9$, $b = 0$, and $c = 25$. The solution to the equation is

$$X = \frac{-0 \pm \sqrt{0^2 - (4)(9)(-25)}}{(2)(9)} = \frac{\pm\sqrt{(36)(25)}}{18} = \frac{\pm\sqrt{900}}{18}$$

$$X = \pm\tfrac{30}{18} = \pm\tfrac{5}{3}$$

PROBLEMS (Appendix I)

Solve the following algebraic equations for the numerical value of X.

I-1. $5X - 3 = 12$

I-2. $3X + 4 = 6X - 5$

I-3. $\dfrac{4X}{16} + 3 = 5$

I-4. $\dfrac{7X - 3}{2} = 9$

I-5. $\dfrac{3X}{4} = \dfrac{8X - 4}{10}$

I-6. $\dfrac{2X + 7}{18} = \dfrac{6X - 63}{45}$

I-7. $(7 - X)(4) = 16$

I-8. $\dfrac{(X + 2)(7)}{8} = 3X - 11$

I-9. $\dfrac{(5X - 7)(2)}{3} = \dfrac{(3X + 7)(8)}{4}$

I-10. $X^2 - 25 = 24$

I-11. $3X^2 - 5 = 7$

I-12. $X^2 - 5X = 6$

I-13. $3X^2 - 5X = 20$

I-14. $6X^2 - 7X - 10 = 0$

Appendix II

Exponential Numbers

A. EXPRESSING EXPONENTIAL NUMBERS AS DECIMAL NUMBERS

In science, very large numbers and very small numbers are expressed exponentially, usually as the product of a number and a power of 10. Two such numbers are the number of atoms in one gram atomic weight, 6.023×10^{23}, and the diameter of a hydrogen atom, 1.04×10^{-8} cm.

Consider first the meaning of a "power of 10." A power of 10 is the number of times a quantity must be multiplied by 10 to produce the equivalent decimal number. In the examples below, this principle is illustrated.

$$1 \times 10^3 = 1 \times 10 \times 10 \times 10 = 1000$$
$$5 \times 10^5 = 5 \times 10 \times 10 \times 10 \times 10 \times 10 = 500,000$$
$$5 \times 10^0 = 5$$

(10 raised to the 0 power means that 5 is multiplied by 10 zero times, hence $10^0 = 1$. Note that this operation is not the same as multiplication by zero.)

The power of 10 tells the number of places the decimal point must be moved to produce the desired number. If the exponent is positive, the decimal point is moved to the right exactly that number of places. If the exponent is negative, the decimal point is moved to the left that number of places in order to express the number as a decimal number.

$$2 \times 10^{-4} = 0.0002$$
$$7.6 \times 10^{+2} = 760$$

B. EXPRESSING DECIMAL NUMBERS AS EXPONENTIALS

To express a number as a power of 10, the number of places the decimal point must be moved indicates the power of 10 involved. Thus the number

21,000,000 is usually written as 2.1×10^7, but can also be written as 21×10^6 or 210×10^5, or 0.21×10^8 or 0.021×10^9, etc. Note that if the decimal point is moved to the right, the power of 10 is decreased by 1 for each decimal place moved and if the decimal point is moved to the left, the power of 10 is increased by 1 for each decimal place.

It is customary to express a given number in exponential form such that the first number to the left of the decimal point is between 1 and 10; thus 20,000,000 is usually expressed as 2×10^7 (where 2 is between 1 and 10) rather than 20×10^6, 0.2×10^8, etc.

To express numbers less than unity as powers of 10, the decimal point must be moved to the right and multiplied by negative powers of 10. These exponents must correspond to the number of decimal places moved. Consider the following series of numbers which are converted to powers of 10.

$$0.1 = 1 \times 10^{-1}$$
$$0.03 = 3 \times 10^{-2}$$
$$0.0000018 = 1.8 \times 10^{-6}$$

C. ADDITION OF NUMBERS WITH EXPONENTS

Numbers consisting of exponentials may be added or subtracted as ordinary numbers provided the exponents are the same for all. In the event that this is not the case, the numbers must be converted to a series having the same exponents; for example, consider the following series of numbers.

$$2.47 \times 10^4$$
$$1.580 \times 10^5$$
$$7.26 \times 10^4$$
$$8.0 \times 10^3$$

To add this column of figures, a common exponent must be selected for all, and in this instance it would be simplest if all were converted to the fourth power of 10. The numbers then become

$$2.47 \times 10^4$$
$$15.80 \times 10^4$$
$$7.26 \times 10^4$$
$$\underline{0.80 \times 10^4}$$

Answer 26.33×10^4

Thus the sum is 26.33×10^4 or 2.633×10^5.

D. SUBTRACTION OF NUMBERS WITH EXPONENTS

Subtraction is accomplished in a similar manner. For example, the difference between 4.88×10^{-2} and 0.081×10^{-1} is easily obtained by converting 0.081×10^{-1} to 0.81×10^{-2} and then subtracting 0.81 from 4.88

$$
\begin{array}{r}
4.88 \times 10^{-2} \\
-0.81 \times 10^{-2} \\
\hline
4.07 \times 10^{-2}
\end{array}
$$

Answer

Note that in addition and subtraction of numbers with exponents, the value of the exponent does not change unless the decimal point of the number is subsequently moved.

E. MULTIPLICATION OF NUMBERS WITH EXPONENTS

In multiplication of two numbers with exponents, the numbers are multiplied together as usual and the exponents of 10 are added algebraically to give a new exponent of 10 for the product.

(a) $3.3 \times 10^3 \times 2.0 \times 10^4 = (3.3 \times 2.0) \times (10^3 \times 10^4) = 6.6 \times 10^7$

(b) $8.0 \times 10^3 \times 5.0 \times 10^{-2} = (8.0 \times 5.0) \times (10^3 \times 10^{-2})$

$$= 40 \times 10^1 = 4.0 \times 10^2$$

(c) $1.6 \times 10^{-5} \times 4.0 \times 10^{-2}$

$$= (1.6 \times 4.0) \times (10^{-5} \times 10^{-2}) = 6.4 \times 10^{-7}$$

F. DIVISION OF NUMBERS WITH EXPONENTS

In division of numbers with exponents, the ordinary numbers are handled in the usual manner, but the exponent of 10 in the denominator is subtracted algebraically from the exponent of 10 in the numerator.

(a) $\dfrac{6 \times 10^4}{3 \times 10^1} = \left(\dfrac{6}{3}\right) \times \left(\dfrac{10^4}{10^1}\right) = 2 \times 10^3$

(b) $\dfrac{7.48 \times 10^4}{2.14 \times 10^{-3}} = 3.50 \times 10^7$

(c) $\dfrac{1.0 \times 10^{-5}}{4.0 \times 10^{-7}} = \dfrac{10 \times 10^{-6}}{4.0 \times 10^{-7}} = 2.5 \times 10^1$

It should be observed that the process of division by a power of 10 is analogous to multiplication by the negative of that power of 10. The ratio

(d) $\dfrac{5 \times 10^9}{10^4} = 5 \times 10^9 \times 10^{-4} = 5 \times 10^5$

and the ratio

(e) $\dfrac{6 \times 10^8}{3 \times 10^{-3}} = 2 \times 10^8 \times 10^3 = 2 \times 10^{11}$

G. EXTRACTING SQUARE ROOTS AND CUBE ROOTS

In extracting the square root of a number, the numbers without exponents are handled in the usual fashion and the exponent of 10 is divided by 2. In order not to have fractional powers of 10 for the answer, the number is adjusted so that the power of 10 is divisible by two an integral number of times prior to taking the square root.

(a) $\sqrt{4 \times 10^2} = \sqrt{4} \times \sqrt{10^2} = \pm 2 \times 10^1$

(b) $\sqrt{4 \times 10^{-2}} = \sqrt{4} \times \sqrt{10^{-2}} = \pm 2 \times 10^{-1}$

(c) $\sqrt{36 \times 10^8} = \sqrt{36} \times \sqrt{10^8} = \pm 6 \times 10^4$

(d) $\sqrt{6.4 \times 10^{-9}} = \sqrt{64 \times 10^{-10}} = \sqrt{64} \times \sqrt{10^{-10}} = \pm 8 \times 10^{-5}$

Similarly, to obtain the cube root of a number, the decimal number is handled in the usual way and the exponent of 10 is divided by 3. Again, in order to avoid fractional exponents, it is necessary to adjust the number such that the exponent of 10 is divisible by 3 an integral number of times.

(e) $\sqrt[3]{0.27 \times 10^{-4}} = \sqrt[3]{27 \times 10^{-6}} = \sqrt[3]{27} \times \sqrt[3]{10^{-6}} = 3 \times 10^{-2}$

(f) $\sqrt[3]{1.25 \times 10^{23}} = \sqrt[3]{125 \times 10^{21}} = \sqrt[3]{125} \times \sqrt[3]{10^{21}} = 5 \times 10^7$

PROBLEMS

II-1. Express the following numbers in exponential form:

(a) 22,400   (d) 0.00017

(b) 20,000,000   (e) 0.0821

(c) 31,500,000   (f) 0.00265

II-2. Express in decimal form:

(a) 1.5×10^3 (d) 365×10^2

(b) 8.4×10^{-5} (e) 0.006×10^5

(c) 0.34×10^{-4} (f) 65×10^0

III-3. Solve for X in the following:

(a) $X = (1.4 \times 10^4) + (4.0 \times 10^5) + (3.1 \times 10^3)$

(b) $X = (3.6 \times 10^{-6})(5.0 \times 10^{-8})$

(c) $X = (3 \times 10^6)(4 \times 10^7)$

(d) $X = (6 \times 10^{-9})(4 \times 10^7)$

(e) $X = (2.5 \times 10^{-3})(4.0 \times 10^6)(3.0 \times 10^{-4})$

(f) $X = \dfrac{6.3 \times 10^8}{2.1 \times 10^3}$

(g) $X = \dfrac{8.1 \times 10^{-9}}{3.0 \times 10^{-15}}$

(h) $X = \dfrac{(3.0 \times 10^6)(4 \times 10^{-8})}{(6 \times 10^{-4})(5 \times 10^3)}$

(i) $X^2 = 25 \times 10^6$

(j) $X^2 = 6.4 \times 10^{-7}$

(k) $X^3 = 6.4 \times 10^{-17}$

(l) $X^2 = \dfrac{(3 \times 10^6)(8 \times 10^{-4})}{(6 \times 10^{-10})(1.0 \times 10^8)}$

Appendix III

Logarithms

The logarithm of a number is the exponent which must be given 10 to produce the number; for example, the logarithm of 100 is 2 since this is the exponent which must be given 10 in order to produce 100.

$$10^2 = 100 \qquad \text{Therefore } \log 100 = \log 10^2 = 2$$

$$10^{-2} = 1/10^2 = 1/100 = 0.01 \qquad \text{Therefore } \log 0.01 = -2$$

$$10^6 = 1,000,000 \qquad \text{Therefore } \log 1,000,000 = \log 10^6 = 6$$

The logarithms of numbers which are not integral powers of 10 are not obvious. It is therefore necessary to resort to tables of computed values of logarithms to obtain the required information.

Logarithms of Numbers*

No.	0	1	2	3	4	5	6	7	8	9
1	0.000	0.041	0.079	0.114	0.146	0.176	0.204	0.230	0.255	0.279
2	0.301	0.322	0.342	0.362	0.380	0.398	0.415	0.431	0.447	0.462
3	0.477	0.491	0.505	0.519	0.532	0.544	0.556	0.568	0.580	0.591
4	0.602	0.613	0.623	0.634	0.644	0.653	0.663	0.672	0.681	0.690
5	0.699	0.708	0.716	0.724	0.732	0.740	0.748	0.756	0.763	0.771
6	0.778	0.785	0.792	0.799	0.806	0.813	0.820	0.826	0.833	0.839
7	0.845	0.851	0.857	0.863	0.869	0.875	0.881	0.887	0.892	0.898
8	0.903	0.909	0.914	0.919	0.924	0.929	0.935	0.940	0.945	0.949
9	0.954	0.959	0.964	0.969	0.973	0.978	0.982	0.989	0.991	0.996

* This table is for illustrative purposes only. A four-place table of logarithms is to be found in Appendix X.

To find the logarithm of a number, for example the number 8.5, we look down the first column to the number 8 and then across to the column heading 5. The number recorded at this point is 0.929 and this is the logarithm of 8.5. As a matter of interest, from the definition of a logarithm we see that $10^{0.929} = 8.5$.

To find the logarithm of a number not lying between 1 and 10, such as the number 4900, the number should be converted to a decimal number with only a single digit to the left of the decimal multiplied by the appropriate power of 10, that is, $4900 = 4.9 \times 10^3$. Since the logarithm of a product of two numbers is equal to the sum of the logarithms of the individual numbers, the logarithm of 4.9×10^3 becomes equal to the log of 4.9 plus the log of 10^3. In summary,

$$\begin{aligned} \log (4900) &= \log (4.9 \times 10^3) \\ &= \log 4.9 + \log 10^3 \\ &= 0.690 + 3 \\ &= 3.690 \end{aligned}$$

Similarly, the log of 0.00068 may be found

$$\begin{aligned} \log (0.00068) &= \log (6.8 \times 10^{-4}) \\ &= \log 6.8 + \log 10^{-4} \\ &= 0.833 - 4 \\ &= -3.167 \end{aligned}$$

Sooner or later, every student will be faced with the problem of finding a number whose logarithm is known; for example, find the number whose logarithm is -4, or stated algebraically, $\log X = -4$. Since this is an integral logarithm, the number X equals 10^{-4} or 0.0001. For a number where the logarithm is not integral, such as $\log X = 5.800$, we must again resort to a table of logarithms and proceed as follows. The logarithm is first written as the sum of two numbers, one of which is integral (called the "characteristic") and the other a decimal fraction (called the "mantissa"). The mantissa always has a positive value, whereas the characteristic, which locates the decimal point, may be either positive or negative.

$$\begin{aligned} \log X &= 5.800 \\ \log X &= 5 + 0.800 \end{aligned}$$

The number whose logarithm equals 5 is 10^5, and from the table, the number whose logarithm equals 0.800 is 6.3 or

$$5 = \log 10^5$$

and

$$0.800 = \log 6.3$$

on substitution,

$$\begin{aligned} \log X &= \log 10^5 + \log 6.3 \\ \log X &= \log (6.3 \times 10^5) \\ X &= 6.3 \times 10^5 \end{aligned}$$

6.3×10^5 is called the antilog of 5.800.

An example of finding the number whose logarithm is negative is the following:

$$\log X = -7.510$$
$$\log X = 0.490 - 8$$

(Converted to this form so the mantissa has a positive value.)

$$\log X = \log 3.1 + \log 10^{-8}$$
$$\log X = \log (3.1 \times 10^{-8})$$
$$X = 3.1 \times 10^{-8}$$

3.1×10^{-8} is called the antilog of -7.510.

A. MULTIPLICATION

As indicated above, the logarithm of the product of two numbers is the *sum* of the logarithms of the numbers. To carry out multiplication, we must (1) obtain the logarithms of the numbers to be multiplied, (2) obtain the sum of the logarithms, and (3) obtain the antilog of the sum. This process is illustrated by the following examples.

(a) Determine X if $X = (3)(50)(100)$

$$\log X = \log (3)(50)(100)$$
$$\log 3 = 0.477$$
$$\log 50 = 1.699$$
$$\log 100 = 2.000$$
$$\log X = \overline{4.176}$$
$$X = \text{antilog of } 4.176 = 1.5 \times 10^4$$

(b) $Y = (1.5 \times 10^{-2})(6.2 \times 10^6)$. Determine the product.

$$\log Y = \log (1.5 \times 10^{-2}) + \log (6.2 \times 10^6)$$
$$\log 1.5 \times 10^{-2} = \log 1.5 + \log 10^{-2} = 0.176 + (-2.000) = -1.824$$
$$\log 6.2 \times 10^6 = \log 6.2 + \log 10^6 = 0.792 + 6.000 \qquad = \quad 6.792$$
$$\log Y = \quad \overline{4.968}$$
$$Y = \text{antilog of } 4.968 = 9.3 \times 10^4$$

B. DIVISION

To carry out the operation of division, the logarithm of the divisor is *subtracted* from the logarithm of the dividend. The antilog of this difference is determined to obtain the quotient.

Determine X if

$$X = \frac{(4.7 \times 10^{-6})}{(9.4 \times 10^4)}$$

$\log X = \log (4.7 \times 10^{-6}) - \log (9.4 \times 10^4)$
$\log (4.7 \times 10^{-6}) = \log 4.7 + \log 10^{-6} = 0.672 + (-6.000) \quad = \quad -5.328$
$- [\log (9.4 \times 10^4)]$
$\qquad\qquad = -[\log 9.4 + \log 10^4] = -[0.973 + 4.000] = \quad \underline{-4.973}$
$$\log X = \quad -\overline{10.301}$$

In order to obtain a positive mantissa, this number (-10.301) may be expressed as $(0.699 - 11)$.

$$X = \text{antilog of } (0.699 - 11) = 5.0 \times 10^{-11}$$

C. ROOTS

To extract the nth root of a number, simply divide the logarithm of the number by n, and determine the antilog. For extracting square roots, $n = 2$, and for extracting cube roots, $n = 3$.

$$\sqrt{8100} = X$$

$$\log \sqrt{8100} = \frac{\log 8100}{2} = \frac{3.909}{2} = 1.955$$

$$X = \text{antilog of } 1.955 = 90$$

Similarly

$$\sqrt[3]{5800} = X$$

$$\log \sqrt[3]{5800} = \frac{\log 5800}{3} = \frac{3.763}{3} = 1.254$$

$$X = \text{antilog of } 1.254 = 18$$

D. POWERS

To raise a number to a power, obtain the logarithm, multiply the logarithm of the number by the power, and determine the antilog.

$$(210)^3 = X$$
$$\log (210)^3 = 3 \log 210 = (3)(2.322) = 6.966$$
$$X = \text{antilog of } 6.966 = 9.2 \times 10^6$$

PROBLEMS

III-1. Obtain the logarithm of the following numbers. (Use log tables in Appendix X.)
(a) 40,000   (c) 6.02×10^{23} (e) 0.0045
(b) 3.8×10^2 (d) 1.7×10^{-24} (f) 965×10^2

III-2. Find the number (antilog) corresponding to the following logarithm values
(a) 3.00   (c) 1.699   (e) 6.477
(b) -6.30 (d) -0.0223 (f) -3.699

III-3. Perform the following operations by logarithms.

(a) $(6.2 \times 10^3)(3.8 \times 10^2)$ (d) $\dfrac{6.6 \times 10^{-4}}{22 \times 10^3}$

(b) $(2.4 \times 10^{-4})(3.2 \times 10^6)$ (e) $\sqrt[3]{10.6 \times 10^2}$

(c) $\dfrac{1.8 \times 10^7}{8.1 \times 10^{-4}}$ (f) $(3.62 \times 10^{-4})^3$

Appendix IV

Significant Figures

A quantitative measurement of some property requires the placing of numerical values on that property and also a statement of the units in which the measurement is made. The number of digits used to designate the amount is referred to as the number of significant figures. Some very simple rules should be followed in their use.

1. Significant Figures

The significant figures in a number are those digits that give meaningful but not misleading information. Only the last digit contains an uncertainty. For example, the volume of a liquid may be recorded as 675 ml. This number has three significant figures, the last of which is uncertain. The uncertainty in the last digit depends upon the accuracy of measurement. If the measurement of volume is accurate to 1 ml, the volume may be specified as 675 $\pm$ 1 ml; that is, the volume is not less than 674 ml or more than 676 ml.

Consider another example in which the number of significant figures in a measurement depends upon the degree of refinement of the devices used in making the measurement. In determining atmospheric pressure by measuring the height of a mercury column supported by the atmosphere, the number of significant figures depends primarily upon the rulings on the scale; see Fig. IV-1.

The height of the mercury column is fixed and if measured with scale A, it is noted that the upper part of the meniscus is approximately $\frac{7}{10}$ of the distance between 69 and 70. Hence the height of the column is reported as 69.7 cm. On scale B, with more rulings, it is estimated that the meniscus falls $\frac{3}{10}$ of the distance between 69.7 and 69.8. Therefore this is recorded as 69.73 cm. If an instrument called a cathetometer (a device employing a telescope for measuring differences in heights accurately) were used, an

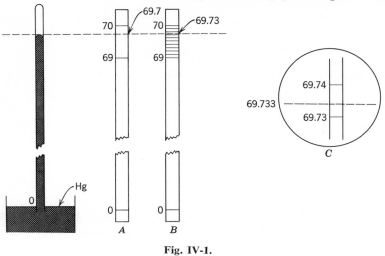

Fig. IV-1.

additional significant figure could be obtained and the height recorded as 69.733 cm (scale *C*).

The number 69.7 has three significant figures.
The number 69.73 has four significant figures.
The number 69.733 has five significant figures.

Similarly, the number 0.00697 has three significant figures, and the number 697 has three significant figures. However, the number 3970 may have three or four significant figures. The number is decided by the rules outlined in the following section.

2. Zeros

Zeros must be considered as a special case in relation to their position with respect to the decimal point.

(a) A zero between two significant figures is also a significant figure. Thus the number 107.8 consists of four significant figures.

(b) A zero to the right of a digit beyond the decimal point is included in the count of significant figures. The number 10.780 consists of five significant figures whereas the number 10.78 consists of only four.

(c) A zero is not significant if it occurs merely to fix the decimal point; for example, the zeros in 0.054 are not considered significant figures; in 5400 they may or may not be significant.

(d) To indicate the proper number of significant figures for the numbers 0.054 and 5400, the number is best represented as a power of 10 with one digit to the left of the decimal.

0.054 becomes 5.4×10^{-2} indicating only two significant figures. 5400 becomes 5.4×10^3 if there are two significant figures; 5.40×10^3 if there are three significant figures; and 5.400×10^3 if there are four significant figures.

3. Addition and Subtraction

In addition, no more digits are meaningful than the lowest digit common to all of the numbers to be summed. In the column of numbers below, the "units" digit is the lowest digit common to all numbers, and therefore their sum is "rounded" to the three significant figure number 535.

```
                                          Lowest common digit
  28.2
 505
   1.732
 534.932 = 535
```

In subtraction a similar rule applies. For example, the difference between 505 and 1.7 is 503.3, but since the "units" digit is again the lowest common digit, the result is 503.

In obtaining the difference $0.505 - 0.455 = 0.050$, the remainder contains only two significant figures. Notice that it is possible in subtraction to have fewer significant figures in the answer than in either of the original numbers. In this case, the 1/1000 digit is the smallest digit common to both three significant figure numbers, but the difference is a number containing only two significant figures. It is usually true that the difference of two numbers of nearly equal magnitude has fewer significant figures than either of the original numbers.

4. Multiplication and Division

In multiplication and division the same number of significant figures is retained as in the least accurate number. Occasionally the retention of one additional figure, but never more, is allowed if retaining only the prescribed number results in a loss of accuracy. A case where the number of digits retained is the same as in the least accurate number is

$$2.00 \times 102.14 = 204.28 = 204$$

The number 2.00 has only three significant figures, and hence the answer is rounded to the three significant figures 204.

An example where keeping an extra significant figure is permitted is

$$2.00 \times 62 = 124$$

The number 62 has only two significant figures but is known to 1 part in 62 or roughly 2%. Rounding the answer to 120 to retain only two significant figures results in a number uncertain to 10 parts in 120 or 8%. Under these conditions, retention of three significant figures is allowed.

Division is treated similarly to multiplication.

5. Rounding Off Unnecessary Figures

In rounding off unneeded figures and nonsignificant figures we consider the magnitude of the highest digit to be dropped.

(a) If this digit is 4 or less, the digit is dropped and the last digit retained is unchanged.

Thus, if 2.74 is to be rounded to two significant figures, it becomes 2.7. The number 2.748 is rounded to 2.7 and not to 2.8, and on rounding 0.56649 to three significant figures, it becomes 0.566, not 0.567.

(b) If the highest digit to be dropped is greater than 5, or is 5 followed by numbers other than zeros, the last digit retained is increased by one.

Thus in rounding 2.76 to two significant figures it becomes 2.8. In rounding 2.8502 to two significant figures it becomes 2.9.

(c) If the largest digit to be dropped is 5 followed only by zeros, the last digit retained is increased by one if it is odd and is left unchanged if it is even.

When rounded to two significant figures, 4.850 becomes 4.8 and 4.750 becomes 4.8. The number 4.8509 becomes 4.9, since the 5 is followed by digits other than zero.

6. Integral Numbers

Integral numbers are those that represent an entity, a body, or a class which is not subdivided into parts. In such instances, the concept of significant figures is without meaning; for example, the formula O_2 represents a molecule of oxygen and consists of exactly 2 atoms per molecule. In the equation for the volume of a sphere, $V = \frac{4}{3}\pi r^3$, the ratio $\frac{4}{3}$ is the ratio of integral numbers and, here also, the number of significant figures is meaningless. In a chemical equation such as $N_2 + 3\,H_2 \rightarrow 2\,NH_3$, the coefficients before the formulas tell that 1 molecule of N_2 reacts with 3 molecules of H_2 to form 2 molecules of NH_3. Similarly, the equation indicates that 1 mole of N_2 reacts with 3 moles of H_2 to form 2 moles of NH_3. These coefficients are integral numbers.

PROBLEMS (Significant Figures)

IV-1. Round each of the following numbers to three significant figures.
    (a) 1556   (f) 1.7451
    (b) 1.4532   (g) 1.7350
    (c) 1.7498   (h) 0.00372
    (d) 1.7448   (i) 4.630
    (e) 1.7450   (j) 451.4999

IV-2. Obtain the sum of each of the following sets of numbers to the proper number of significant figures.
    (a) 125 + 241 + 11.9 + 17.4
    (b) 42.8 + 29.5 + 0.99 + 1.75
    (c) 42.975 + 2.0025 + 1.6593 + 20.4701
    (d) 200 + 150 + 125
    (e) 0.001 + 0.550 + 0.05

IV-3. For each of the following pairs of numbers, obtain the difference, and express the answer to the correct number of significant figures.
    (a) 47.50 − 25.49   (c) 0.8001 − 0.8   (e) 28.7 − 21
    (b) 0.847 − 0.84   (d) 89.95 − 0.025

IV-4. Obtain the quotient for each of the following ratios and round to the correct number of significant figures.
    (a) 28.28/4.04   (c) 150.2/9.547   (e) 99/92
    (b) 0.9412/0.141   (d) 160.9/270   (f) 99/30

IV-5. Determine the products of the numbers listed below to the proper number of significant figures.
    (a) 44 × 2.01 =   (e) 0.0430 × 104.0 =
    (b) 98 × 75 =   (f) 105 × 1.25 =
    (c) 100 × 1.25 =   (g) 0.0057 × 125.2 =
    (d) 13.50 × 0.0751 =

Appendix V

Experimental Error

The experimental determination of any quantity, such as the measurement of the pressure of the atmosphere, is subject to error because it is impossible to carry out such measurements with absolute certainty. The extent of the error in determination of the barometric pressure is a function of the quality of the instrument, the temperature, and the experience and capability of the observer.

The *accuracy* of a determination is a measure of the approach of the experimental value to the true or generally accepted value. This is usually described by means of the per cent error, defined as

$$\text{Per cent error} = \frac{\text{Experimental value} - \text{True value}}{\text{True value}} \times 100$$

The *precision* of a determination refers to the reproducibility of the measurement and is usually described in terms of the average or mean deviation. This is computed by determining first the average of the individual measurements in the usual manner. Then the absolute value of the difference between each individual determination and the average value is calculated. These differences are summed and divided by the number of values in order to get the average or mean deviation.

Measurements	Deviation from Mean
49.9	0.2
49.7	0.0
49.8	0.1
49.5	0.2
4) 198.9	4) 0.5
49.7	0.1

The average value is, therefore, 49.7 and the average deviation is ±0.1 units. The number is generally reported as 49.7 ± 0.1 units.

Occasionally the deviation is expressed as the percentage deviation. This is simply obtained by dividing the average deviation by the average value of the quantity and multiplying by 100. In the above example, the percentage deviation is calculated as follows:

$$\text{Percentage deviation} = \frac{0.1}{49.7} \times 100 = 0.2\%$$

Appendix VI

Vapor Pressure of Water

Temperature (°C)	Pressure (mm of Hg)	Temperature (°C)	Pressure (mm of Hg)
0	4.6	24	22.4
2	5.3	25	23.8
4	6.1	26	25.2
6	7.0	27	26.7
8	8.0	28	28.3
10	9.2	29	30.0
12	10.5	30	31.8
15	12.8	35	42.1
16	13.6	40	55.3
17	14.5	50	92.5
18	15.5	60	149
19	16.5	70	234
20	17.5	80	355
21	18.6	90	526
22	19.8	100	760
23	21.1	110	1074

Appendix VII

Ionization Constants for Acids and Bases

Acid	Formula	K_a	Acid	Formula	K_a
Acetic	$HC_2H_3O_2$	1.7×10^{-5}	Hypobromous	HBrO	2.1×10^{-9}
Arsenic	H_3AsO_4	5.6×10^{-3}	Hypochlorous	HClO	3.0×10^{-8}
		1.7×10^{-7}	Hypoiodous	HIO	2.3×10^{-11}
		4×10^{-12}	Hydrogen sulfide		
Ascorbic	$H_2C_6H_6O_6$	8×10^{-5}	fide	H_2S	1.1×10^{-7}
		1.6×10^{-12}			1×10^{-14}
Benzoic	$HC_7H_5O_2$	6.4×10^{-5}	Iodic	HIO_3	1.9×10^{-1}
Bromoacetic	$HC_2H_2O_2Br$	2.0×10^{-3}	Lactic	$HC_3H_5O_3$	8.4×10^{-4}
Butyric	$HC_4H_7O_2$	1.5×10^{-5}	Nitrous	HNO_2	4.6×10^{-4}
Carbonic	H_2CO_3	4.3×10^{-7}	Oxalic	$H_2C_2O_4$	5.6×10^{-2}
		5.6×10^{-11}			5.2×10^{-5}
Chloroacetic	$HC_2H_2O_2Cl$	1.4×10^{-3}	Periodic	H_5IO_6	2.3×10^{-2}
Chlorobenzoic	$HC_7H_4O_2Cl$	1.5×10^{-4}	Phosphoric	H_3PO_4	7.5×10^{-3}
Formic	$HCHO_2$	1.6×10^{-4}			6.2×10^{-8}
Hydrocyanic	HCN	4.9×10^{-10}			2.2×10^{-12}
Hydrofluoric	HF	1×10^{-3}	Sulfurous	H_2SO_3	5×10^{-6}
					1.0×10^{-7}

Base	Formula	K_b	Base	Formula	K_b
Ammonia	NH_3	1.8×10^{-5}	Pyridine	C_5H_5N	1.6×10^{-9}
Hydrazine	N_2H_4	1.4×10^{-6}	Quinoline	C_9H_7N	6.3×10^{-10}
Hydroxylamine	NH_2OH	1.1×10^{-8}	Trimethylamine	C_3H_9N	5.4×10^{-5}

(Successive ionization constants of polyprotic acids, shown in order)

266

Appendix VIII

Solubility Product Constants

Compound	K_{sp}	Compound	K_{sp}
$Al(OH)_3$	1×10^{-32}	HgS	1×10^{-54}
AgBr	5×10^{-13}	Li_2CO_3	1.7×10^{-3}
AgCl	1.1×10^{-10}	$MgCO_3$	2.6×10^{-5}
AgI	1.2×10^{-17}	$Mg(OH)_2$	7×10^{-12}
Ag_2S	1.2×10^{-51}	$MgNH_4PO_4$	2.5×10^{-13}
$BaCO_3$	8×10^{-9}	MnS	1×10^{-16}
$BaCrO_4$	2.4×10^{-10}	NiS	1.4×10^{-24}
$BaSO_4$	1.0×10^{-10}	$PbCO_3$	3.3×10^{-14}
$CaCO_3$	1×10^{-8}	$PbCrO_4$	1.8×10^{-14}
CaC_2O_4	3.0×10^{-9}	$PbCl_2$	2×10^{-5}
$CaSO_4$	2.0×10^{-4}	$PbSO_4$	2×10^{-8}
CdS	1.6×10^{-28}	PbS	4×10^{-28}
CuS	1×10^{-40}	$SrCO_3$	1.6×10^{-9}
$Fe(OH)_3$	1×10^{-36}	$SrSO_4$	2.8×10^{-7}
FeS	3.7×10^{-19}	$Zn(OH)_2$	1.8×10^{-14}
Hg_2Cl_2	2×10^{-18}	ZnS	1×10^{-24}

Appendix IX

Dissociation Constants for a Few Complex Ions

Complex ion	K
$Ag(CN)_2^-$	1.4×10^{-20}
$Ag(NH_3)_2^+$	6.2×10^{-8}
$Al(OH)_4^-$	1×10^{-34}
$Cd(NH_3)_4^{++}$	2.8×10^{-7}
$Co(NH_3)_6^{+++}$	2.2×10^{-34}
$Cu(CN)_2^-$	1×10^{-27}
$Cu(NH_3)_4^{++}$	4.9×10^{-14}
$Fe(CN)_6^{----}$	1×10^{-35}
$Ni(NH_3)_6^{++}$	2×10^{-9}
$Zn(NH_3)_4^{++}$	3×10^{-10}
$Zn(OH)_4^{--}$	3×10^{-16}

Appendix X

Table of Logarithms

Natural Numbers	0	1	2	3	4	5	6	7	8	9	Proportional parts						
											1	2	3	4	5	6	7
10	0000	0043	0086	0128	0170	0212	0253	0294	0334	0374	4	8	12	17	21	25	29
11	0414	0453	0492	0531	0569	0607	0645	0682	0719	0755	4	8	11	15	19	23	26
12	0792	0828	0864	0899	0934	0969	1004	1038	1072	1106	3	7	10	14	17	21	24
13	1139	1173	1206	1239	1271	1303	1335	1367	1399	1430	3	6	10	13	16	19	23
14	1461	1492	1523	1553	1584	1614	1644	1673	1703	1732	3	6	9	12	15	18	21
15	1761	1790	1818	1847	1875	1903	1931	1959	1987	2014	3	6	8	11	14	17	20
16	2041	2068	2095	2122	2148	2175	2201	2227	2253	2279	3	5	8	11	13	16	18
17	2304	2330	2355	2380	2405	2430	2455	2480	2504	2529	2	5	7	10	12	15	17
18	2553	2577	2601	2625	2648	2672	2695	2718	2742	2765	2	5	7	9	12	14	16
19	2788	2810	2833	2856	2878	2900	2923	2945	2967	2989	2	4	7	9	11	13	16
20	3010	3032	3054	3075	3096	3118	3139	3160	3181	3201	2	4	6	8	11	13	15
21	3222	3243	3263	3284	3304	3324	3345	3365	3385	3404	2	4	6	8	10	12	14
22	3424	3444	3464	3483	3502	3522	3541	3560	3579	3598	2	4	6	8	10	12	14
23	3617	3636	3655	3674	3692	3711	3729	3747	3766	3784	2	4	6	7	9	11	13
24	3802	3820	3838	3856	3874	3892	3909	3927	3945	3962	2	4	5	7	9	11	12
25	3979	3997	4014	4031	4048	4065	4082	4099	4116	4133	2	3	5	7	9	10	12
26	4150	4166	4183	4200	4216	4232	4249	4265	4281	4298	2	3	5	7	8	10	11
27	4314	4330	4346	4362	4378	4393	4409	4425	4440	4456	2	3	5	6	8	9	11
28	4472	4487	4502	4518	4533	4548	4564	4579	4594	4609	2	3	5	6	8	9	11
29	4624	4639	4654	4669	4683	4698	4713	4728	4742	4757	1	3	4	6	7	9	10
30	4771	4786	4800	4814	4829	4843	4857	4871	4886	4900	1	3	4	6	7	9	10
31	4914	4928	4942	4955	4969	4983	4997	5011	5024	5038	1	3	4	6	7	8	10
32	5051	5065	5079	5092	5105	5119	5132	5145	5159	5172	1	3	4	5	7	8	9
33	5185	5198	5211	5224	5237	5250	5263	5276	5289	5302	1	3	4	5	6	8	9
34	5315	5328	5340	5353	5366	5378	5391	5403	5416	5428	1	3	4	5	6	8	9
35	5441	5453	5465	5478	5490	5502	5514	5527	5539	5551	1	2	4	5	6	7	9
36	5563	5575	5587	5599	5611	5623	5635	5647	5658	5670	1	2	4	5	6	7	8
37	5682	5694	5705	5717	5729	5740	5752	5763	5775	5786	1	2	3	5	6	7	8
38	5798	5809	5821	5832	5843	5855	5866	5877	5888	5899	1	2	3	5	6	7	8
39	5911	5922	5933	5944	5955	5966	5977	5988	5999	6010	1	2	3	4	5	7	8
40	6021	6031	6042	6053	6064	6075	6085	6096	6107	6117	1	2	3	4	5	6	8
41	6128	6138	6149	6160	6170	6180	6191	6201	6212	6222	1	2	3	4	5	6	
42	6232	6243	6253	6263	6274	6284	6294	6304	6314	6325	1	2	3	4	5	6	
43	6335	6345	6355	6365	6375	6385	6395	6405	6415	6425	1	2	3	4	5	6	
44	6435	6444	6454	6464	6474	6484	6493	6503	6513	6522	1	2	3	4	5	6	
45	6532	6542	6551	6561	6571	6580	6590	6599	6609	6618	1	2	3	4	5	6	
46	6628	6637	6646	6656	6665	6675	6684	6693	6702	6712	1	2	3	4	5	6	
47	6721	6730	6739	6749	6758	6767	6776	6785	6794	6803	1	2	3	4	5	5	
48	6812	6821	6830	6839	6848	6857	6866	6875	6884	6893	1	2	3	4	4	5	
49	6902	6911	6920	6928	6937	6946	6955	6964	6972	6981	1	2	3	4	4	5	
50	6990	6998	7007	7016	7024	7033	7042	7050	7059	7067	1	2	3	3	4	5	
51	7076	7084	7093	7101	7110	7118	7126	7135	7143	7152	1	2	3	3	4	5	
52	7160	7168	7177	7185	7193	7202	7210	7218	7226	7235	1	2	2	3	4	5	
53	7243	7251	7259	7267	7275	7284	7292	7300	7308	7316	1	2	2	3	4	5	
54	7324	7332	7340	7348	7356	7364	7372	7380	7388	7396	1	2	2	3	4	5	

0	1	2	3	4	5	6	7	8	9	Proportional parts								
										1	2	3	4	5	6	7	8	9
7404	7412	7419	7427	7435	7443	7451	7459	7466	7474	1	2	2	3	4	5	5	6	7
7482	7490	7497	7505	7513	7520	7528	7536	7543	7551	1	2	2	3	4	5	5	6	7
7559	7566	7574	7582	7589	7597	7604	7612	7619	7627	1	2	2	3	4	5	5	6	7
7634	7642	7649	7657	7664	7672	7679	7686	7694	7701	1	1	2	3	4	4	5	6	7
7709	7716	7723	7731	7738	7745	7752	7760	7767	7774	1	1	2	3	4	4	5	6	7
7782	7789	7796	7803	7810	7818	7825	7832	7839	7846	1	1	2	3	4	4	5	6	6
7853	7860	7868	7875	7882	7889	7896	7903	7910	7917	1	1	2	3	4	4	5	6	6
7924	7931	7938	7945	7952	7959	7966	7973	7980	7987	1	1	2	3	3	4	5	6	6
7993	8000	8007	8014	8021	8028	8035	8041	8048	8055	1	1	2	3	3	4	5	5	6
8062	8069	8075	8082	8089	8096	8102	8109	8116	8122	1	1	2	3	3	4	5	5	6
8129	8136	8142	8149	8156	8162	8169	8176	8182	8189	1	1	2	3	3	4	5	5	6
8195	8202	8209	8215	8222	8228	8235	8241	8248	8254	1	1	2	3	3	4	5	5	6
8261	8267	8274	8280	8287	8293	8299	8306	8312	8319	1	1	2	3	3	4	5	5	6
8325	8331	8338	8344	8351	8357	8363	8370	8376	8382	1	1	2	3	3	4	4	5	6
8388	8395	8401	8407	8414	8420	8426	8432	8439	8445	1	1	2	2	3	4	4	5	6
8451	8457	8463	8470	8476	8482	8488	8494	8500	8506	1	1	2	2	3	4	4	5	6
8513	8519	8525	8531	8537	8543	8549	8555	8561	8567	1	1	2	2	3	4	4	5	5
8573	8579	8585	8591	8597	8603	8609	8615	8621	8627	1	1	2	2	3	4	4	5	5
8633	8639	8645	8651	8657	8663	8669	8675	8681	8686	1	1	2	2	3	4	4	5	5
8692	8698	8704	8710	8716	8722	8727	8733	8739	8745	1	1	2	2	3	4	4	5	5
8751	8756	8762	8768	8774	8779	8785	8791	8797	8802	1	1	2	2	3	3	4	5	5
8808	8814	8820	8825	8831	8837	8842	8848	8854	8859	1	1	2	2	3	3	4	5	5
8865	8871	8876	8882	8887	8893	8899	8904	8910	8915	1	1	2	2	3	3	4	4	5
8921	8927	8932	8938	8943	8949	8954	8960	8965	8971	1	1	2	2	3	3	4	4	5
8976	8982	8987	8993	8998	9004	9009	9015	9020	9026	1	1	2	2	3	3	4	4	5
9031	9036	9042	9047	9053	9058	9063	9069	9074	9079	1	1	2	2	3	3	4	4	5
9085	9090	9096	9101	9106	9112	9117	9122	9128	9133	1	1	2	2	3	3	4	4	5
9138	9143	9149	9154	9159	9165	9170	9175	9180	9186	1	1	2	2	3	3	4	4	5
9191	9196	9201	9206	9212	9217	9222	9227	9232	9238	1	1	2	2	3	3	4	4	5
9243	9248	9253	9258	9263	9269	9274	9279	9284	9289	1	1	2	2	3	3	4	4	5
9294	9299	9304	9309	9315	9320	9325	9330	9335	9340	1	1	2	2	3	3	4	4	5
9345	9350	9355	9360	9365	9370	9375	9380	9385	9390	1	1	2	2	3	3	4	4	5
9395	9400	9405	9410	9415	9420	9425	9430	9435	9440	0	1	1	2	2	3	3	4	4
9445	9450	9455	9460	9465	9469	9474	9479	9484	9489	0	1	1	2	2	3	3	4	4
9494	9499	9504	9509	9513	9518	9523	9528	9533	9538	0	1	1	2	2	3	3	4	4
9542	9547	9552	9557	9562	9566	9571	9576	9581	9586	0	1	1	2	2	3	3	4	4
9590	9595	9600	9605	9609	9614	9619	9624	9628	9633	0	1	1	2	2	3	3	4	4
9638	9643	9647	9652	9657	9661	9666	9671	9675	9680	0	1	1	2	2	3	3	4	4
9685	9689	9694	9699	9703	9708	9713	9717	9722	9727	0	1	1	2	2	3	3	4	4
9731	9736	9741	9745	9750	9754	9759	9763	9768	9773	0	1	1	2	2	3	3	4	4
9777	9782	9786	9791	9795	9800	9805	9809	9814	9818	0	1	1	2	2	3	3	4	4
9823	9827	9832	9836	9841	9845	9850	9854	9859	9863	0	1	1	2	2	3	3	4	4
9868	9872	9877	9881	9886	9890	9894	9899	9903	9908	0	1	1	2	2	3	3	4	4
9912	9917	9921	9926	9930	9934	9939	9943	9948	9952	0	1	1	2	2	3	3	4	4
9956	9961	9965	9969	9974	9978	9983	9987	9991	9996	0	1	1	2	2	3	3	3	4

Appendix XI

Answers to Problems

Chapter 2

2-1. 1437.1 g
2-2. 1.24 g
2-3. 3034.9 m
2-4. 10^4 m^2
2-5. 10 Å
2-6. 1000 liters
2-7. (a) 10^{-5} m
    (b) 10^{-9} kg
    (c) 10^{-4} liters
    (d) 10^{-9} m^3
    (e) 50 liters
    (f) 1820 g
    (g) 3.5 in
    (h) 5.6 qt
    (i) 5.00×10^{-4} lb
    (j) 1.6×10^3 m
2-8. 0.45 μ; 450 mμ; 4.5×10^3 Å
2-9. (a) 113°
    (b) 37.0
    (c) 0°K
    (d) −148°F
    (e) −80°C
    (f) −18°C
    (g) −40°F
    (h) 0°K
    (i) 373°K
    (j) −459°F
2-11. −40°
2-12. 7.50 g/ml
2-13. 60.0 ml
2-14. 11 kg
2-15. 45%
2-16. 24 liters
2-17. 4.8×10^2 kg
2-18. 272 g
2-19. 63 liters
2-20. 2.7×10^4 g
2-21. 30 kg
2-22. 126°F
2-23. 4.50 g/ml

2-24. 270 ml
2-25. 4×10^9 liters or 8%
2-26. (a) −11.4°C
    (b) 410°F
    (c) 574°F
2-27. $\dfrac{9K - 2297}{5}$
2-28. 63 lb
2-29. 1.4×10^8 t/cm^3
2-30. 14 m
2-31. 8×10^9/cm^3
2-32. 7.2×10^5 t
2-33. 40 μ
2-34. 62 ft^2
2-35. 36%
2-36. 6.8×10^3 liters
2-37. (a) 6.5×10^{21} t
    (b) 5.5 g/ml
2-38. 3.2×10^6 yr

Chapter 3

3-1. (a) 30
    (b) 98
    (c) 99
    (d) 77
    (e) 80
    (f) 55
    (g) 104
    (h) 122
    (i) 328
3-2. 84
3-3. (a) 120
    (b) 178
    (c) 343
    (d) 70
    (e) 294
    (f) 136
    (g) 74
    (h) 78
    (i) 80

3-4. 59
3-5. (a) 9
 (b) 3
3-6. (a) 3
 (b) 4
3-7. 69.7
3-8. (a) 15.0
 (b) 3
3-9. 17.0
3-10. 0.005
3-11. 4.0
3-12. (a) 0.428
 (b) 0.675
 (c) 2.32
 (d) 4.18
 (e) 1.05
 (f) 0.45
3-13. (a) 5.87
 (b) 2.13
 (c) 1.85
 (d) 2.50
 (e) 2.50
 (f) 1.11
3-14. (a) 25.6 g
 (b) 32.8 g
 (c) 745 g
 (d) 102 g
3-15. (a) 1.7×10^{-24} g
 (b) 1.18×10^{-22} g
 (c) 3.85×10^{-22} g
3-16. H $= 4.8 \times 10^{24}$
 P $= 2.4 \times 10^{24}$
 O $= 8.4 \times 10^{24}$
3-17. H $= 1.2 \times 10^{23}$
 N $= 1.2 \times 10^{23}$
 O $= 3.6 \times 10^{23}$
3-18. 6.2 g
3-19. 26.0 g
3-20. 190
3-21. 18.9984
3-22. 10.81
3-23. 93.0% K^{39}
3-24. 15.9994
3-25. (a) 80
 (b) 171
 (c) 96
 (d) 102.5
 (e) 284
 (f) 181
3-26. 0.40 mole
3-27. 6.0 moles

3-28. 0.6 mole
3-29. (a) (i) 0.51 mole
 (ii) 0.313 mole
 (iii) 0.625 mole
 (iv) 0.171 mole
 (b) (i) H $= 1.53$ g atoms
 P $= 0.51$ g atom
 O $= 2.04$ g atoms
 (ii) Cu $= 0.313$ g atom
 S $= 0.313$ g atom
 O $= 1.25$ g atoms
 (iii) N $= 1.25$ g atoms
 O $= 1.88$ g atoms
 H $= 2.50$ g atoms
 (iv) Fe $= 0.342$ g atom
 C $= 0.513$ g atom
 O $= 1.54$ g atoms
3-30. 2.19 g
3-31. (a) 392 g
 (b) 39 g
 (c) 368 g
 (d) 0.855 g
3-32. (a) 4.18 moles
 (b) 0.507 mole
 (c) 0.517 mole
 (d) 0.705 mole
3-33. (a) 12.8 g
 (b) 115 g
 (c) 129 g
 (d) 1.01×10^{3} g
3-34. 0.4 g atom
3-35. (a) 1.00 g atom
 (b) 6.02×10^{23} atoms
3-36. Na $= 0.2$
 Cl $= 0.2$
3-37. 10 g atoms
3-38. 0.05 mole
3-39. 490 g
3-40. 0.4 mole
3-41. 78.4 g
3-42. (a) 0.05 mole
 (b) 0.10 g atom
 (c) 0.20 g atom
 (d) 3.0×10^{22}
 (e) 1.6 g
3-43. (a) 0.60 mole
 (b) 0.60 g atom
 (c) 8.4 g
 (d) 3.6×10^{23}
 (e) 3.0×10^{24}
3-44. 65.37

3-45. 118.8
3-46. 27.5
3-47. 59.3 g
3-48. 43.8
3-49. 3
3-50. 20.7
3-51. 24.35
3-52. 23.2
3-53. 60.5 Ga^{69}
3-54. 6.94
3-55. 72.63
3-56. 57.7% Sb^{121}
3-57. 236.85
3-58. 2.39
3-59. 69.5% Cu^{63}
3-60. 95.89

Chapter 4

4-1. (a) 22.33; 77.67
          (b) 75.74; 24.26
          (c) 92.24; 7.76
4-2. (a) 24.74; 34.76; 40.50
          (b) 49.10; 16.97; 33.93
          (c) 41.87; 18.97; 39.16
4-3. (a) 7.0 g
          (b) 13.2 g
          (c) 17.5 g
          (d) 10.6 g
4-4. 1.4 g
4-5. 15.0%
4-6. 80.0%
4-7. $MoCl_3$
4-8. HgI_2
4-9. (a) (b) BrCl
4-10. (a) $AlCl_3$
          (b) Al_2Cl_6
4-11. (a) C_2H_6O
          (b) HNO_2
          (c) $C_3H_6O_2$
          (d) $C_6H_5O_2N$
          (e) C_3H_7ON
4-12. 22.7%
4-13. 36.0%
4-14. $GdCl_3 \cdot 6 H_2O$
4-15. $HClO_4 \cdot H_2O$
4-16. (a) (b) XeF_6
4-17. (a) 19.7
          (b) 42.7
          (c) 50.7
          (d) 48.0

          (e) 17.1
          (f) 40.0
4-18. 480 g
4-19. 0.64 kg
4-20. 51.8 g
4-21. $Mg_2As_2O_7$
4-22. BF_3
4-23. C_6H_{12}
4-24. (a) CH_3
          (b) Cu_2O
          (c) $K_2S_2O_7$
          (d) B_2O_3
          (e) $CrCl_3O_9$
4-25. $[Al(CH_3)_3]_2$
4-26. (a) 55.9%
          (b) 36.1%
          (c) 11.7%
          (d) 15.2%
          (e) 45.2%
          (f) 23.7%
4-27. $NiSO_4 \cdot 7 H_2O$
4-28. $MnCl_2 \cdot 4 H_2O$
4-29. $Na_2CO_3 \cdot 10 H_2O$
4-30. $ZnCl_2 \cdot 4 NH_3$
4-31. $Li_2B_4O_7 \cdot 5 H_2O$
4-32. Ag_2O
4-33. $KClO_3$
4-34. $XeOF_2$
4-35. Al_2O_3
4-36. 22.7%
4-37. $KClO_3$
4-38. (a) $279
          (b) $176
          (c) $187
          (d) $220
4-39. Cu_2S
4-40. $XePtF_6$
4-41. 61.7%

Chapter 5

Answers are listed as numerical coefficients of substances in order of appearance in the balanced equation. When the order is not clear or when H^+, OH^-, or H_2O must be added, the answer is given as the sum of all coefficients.

5-1. 2, 2, 1
5-2. 2, 6, 2, 3
5-3. 3, 1, 2, 1

5-4. 1, 2, 1, 1
5-5. 2, 2, 2, 1
5-6. 2, 2, 4, 1
5-7. 1, 2, 1, 2
5-8. 1, 2, 1, 1, 1
5-9. 3, 2, 1, 6
5-10. 2, 2, 6, 2, 3
5-11. 2, 1, 1, 2, 2
5-12. 2, 5, 4, 2
5-13. 2, 2, 4, 1
5-14. 2, 3, 2, 2
5-15. 2, 1, 1, 1, 1
5-16. 1, 5, 6, 4
5-17. 2, 13, 8, 10
5-18. 3, 4, 1, 12
5-19. 1, 3, 1, 2
5-20. 1, 3, 1, 3
5-21. 1, 2, 1, 1, 1
5-22. 1, 1, 1, 1
5-23. 1, 2, 1, 1
5-24. 1, 3, 2
5-25. 1, 2, 2, 1, 1
5-26. 1, 6, 2, 3
5-27. 1, 2, 1, 1
5-28. 1, 6, 2, 3
5-29. 3, 2, 1, 3
5-30. 3, 4, 1, 8
5-31. 2, 3, 2, 3
5-32. 10, 8, 2, 5, 1, 2, 8
5-33. 5, 2, 6, 5, 2, 2, 8
5-34. 6, 10, 3, 10, 5
5-35. 2, 3, 6, 3, 2, 2
5-36. 8, 5, 8, 5, 8, 12
5-37. 1, 4, 1, 1, 2
5-38. 3, 4, 1, 3, 1, 2
5-39. 3, 6, 1, 3, 1, 3
5-40. 2, 3, 16, 4, 3, 11
5-41. 3, 8, 20, 12, 20, 16
5-42. 1, 2, 2, 1, 1
5-43. 2, 11, 22, 2, 6, 11, 8
5-44. 3, 10, 28, 3, 9, 28, 5
5-45. 1, 15, 20, 2, 15, 3, 20, 30
5-46. (a) 43
    (b) 22
    (c) 36
    (d) 37
    (e) 30
    (f) 16
    (g) 26
5-47. (a) 13
    (b) 15

    (c) 13
    (d) 23
    (e) 28
    (f) 6
    (g) 24
    (h) 14
    (i) 12
    (j) 15
    (k) 9
    (l) 28
    (m) 8
    (n) 9
5-48. 4, 1, 11, 4, 1, 4
5-49. 1, 8, 11, 1, 7, 8
5-50. 1, 6, 14, 2, 6, 7
5-51. 1, 6, 14, 2, 3, 7
5-52. 1, 3, 8, 2, 6, 7
5-53. 2, 3, 10, 2, 6, 8
5-54. 2, 3, 10, 1, 2, 8, 5
5-55. 8, 12, 5, 4, 8, 12, 5
5-56. 3, 6, 5, 1, 3
5-57. 3, 8, 6, 1
5-58. 1, 3, 3, 1, 6, 6
5-59. 2, 1, 3, 1, 6, 2, 8
5-60. 1, 3, 3, 3, 1
5-61. 4, 10, 1, 4, 1, 3
5-62. 1, 4, 1, 1, 2
5-63. 2, 5, 3, 1, 2, 10, 8
5-64. 2, 3, 4, 3, 5, 2
5-65. 1, 1, 1, 2, 1
5-66. 2, 1, 2, 1
5-67. 2, 5, 14, 5, 7, 2
5-68. 3, 7, 7, 3, 7, 2
5-69. 2, 4, 2, 1, 4, 2
5-70. 1, 1, 1, 1, 2, 2
5-71. 5, 1, 11, 5, 1, 22
5-72. 2, 10, 1, 10, 5
5-73. 2, 3, 12, 2, 3, 5
5-74. 1, 4, 1, 1, 1, 1, 1, 2
5-75. 3, 3, 6, 1
5-76. 6, 1, 14, 6, 2, 7
5-77. 5, 2, 6, 5, 2, 3
5-78. 1, 10, 1, 10, 3, 5
5-79. 3, 4, 28, 6, 9, 28, 4
5-80. 10, 6, 2, 5, 8, 2
5-81. 1, 4, 1, 1, 1, 1, 2
5-82. 1, 1, 2, 1, 2, 4
5-83. 3, 2, 1, 3, 2, 1
5-84. 2, 2, 1, 2, 2, 3
5-85. 1, 20, 28, 2, 5, 40
5-86. 1, 1, 4, 1, 1, 2

Chapter 6

6-1. $\frac{1}{4}$
6-2. 330 ml
6-3. 1200 ft^3
6-4. 8.52 g/liter
6-5. 477°C
6-6. 420 ml
6-7. 560 ml
6-8. 250 ml
6-9. 71 ml
6-10. 915 ml
6-11. 125 ml
6-12. 360 ml
6-13. 273 ml
6-14. 1 atm
6-15. 3.0 liters
6-16. 200 ml
6-17. 815 mm
6-18. 150 ml
6-19. 9.4 liters
6-20. 100 atm
6-21. 71.4 atm
6-22. 1 m^3
6-23. 1500 mm
6-24. 1527°C
6-25. 3 atm
6-26. 101 ml
6-27. $p_{O_2} = \frac{2}{5}$ atm
  $p_{N_2} = \frac{3}{5}$ atm
  $p_{total} = 1$ atm
6-28. 159 mm; 0.3 mm
6-29. 48 ft^3
6-30. 1.75 atm
6-31. 0.078 liter
6-32. $p_A = 15$ atm
  $p_B = 25$ atm
  $p_C = 60$ atm
6-33. $p_{CO_2} = 100$ mm
  $p_{H_2} = 250$ mm
  $p_{O_2} = 200$ mm
6-34. $p_{N_2} = 568$ mm
  $p_{O_2} = 132$ mm
6-35. (a) H$_2$
  (b) 5
6-36. $\frac{9}{8}$
6-37. 16
6-38. 0.7 micromole/hr
6-39. 36
6-40. 180 liters
6-41. 40 cm from A end

6-42. 1200 ml
6-43. 10 liters
6-44. (a) 125 ft^3
  (b) 150 ft^3
6-45. 160
6-46. 113
6-47. 58
6-48. 6.02×10^{21}
6-49. 140
6-50. 4.5 liters
6-51. 1.43 g
6-52. 120
6-53. 13.4
6-54. 6.02×10^{21}
6-55. 67.2 liters
6-56. 4.5×10^{23}
6-57. 1.3 liters
6-58. (a) 0.125 moles
  (b) 192
6-59. 184°C
6-60. 97.8 ml
6-61. 5.0×10^{23}
6-62. (a) 150 atm
  (b) 223 moles
6-63. 98 g
6-64. 490 atm
6-65. 1.88 kg
6-66. 74 atm
6-67. 0.5 liter
6-68. 100 atm
6-69. 546 ml
6-70. 380 ml
6-71. (a) 277 ml
  (b) 12.6
6-72. 7.5 liters
6-73. 990
6-74. 727°C
6-75. 0.98 g
6-76. 0.36; 0.25; 0.23; 0.16; 0.18; 0.7; 0.2
6-77. 240
6-78. $2 \times 6 \times 10^{23}$
6-79. 327°C
6-80. 340 ml
6-81. 52
6-82. 89.6 liters
6-83. 1.5×10^{22}
6-84. (a) 5.0 liters
  (b) 3.6 liters
6-85. 1227°C

6-86. 0.17 ml
6-87. 73.5 atm
6-88. 39
6-89. 546 ml
6-90. $\frac{1}{12}$ atm
6-91. 660 mm
6-92. 1.00 g/liter
6-93. 128
6-94. (a) 3.04 liters
    (b) 4.34 g
6-95. 20 atm
6-96. 12.3 moles
6-97. 4.9 atm
6-98. 227°C
6-99. 86
6-100. 1.105 g/liter
6-101. 760 mm
6-102. 11.2 atm
6-103. 15 cm
6-104. 22.4 liters
6-105. 1.2×10^{17}
6-106. 20 min
6-107. (a) C_3H_4
    (b) C_6H_8
6-108. 9.00 g
6-109. 90.0 liters
6-110. (a) 23.2
    (b) 69.6
6-111. 30.4 liters
6-112. 4.1×10^{-3} atm
6-113. He/O_2 about 3
6-114. 49 atm
6-115. (a) 4.3 atm
    (b) $p_{H_2} = 1.23$ atm
      $p_{H_2O} = 4.92$ atm
6-116. 22.5 liters
6-117. (a) 3.3×10^{-7} cm
    (b) 5.4×10^{-4}
6-118. (a) 27.9
    (b) 2
6-119. (a) 20.17
    (b) 20.17
    (c) 20.18
6-120. (a) 1.293 g/liter
    (b) 1.085 g/liter
6-121. (a) 32.0
    (b) N_2H_4
6-122. (a) 76.0
    (b) CS_2
6-123. 6.02×10^{23}

Chapter 7

7-1. (a) 0.60 mole
    (b) 0.40 mole
    (c) 19.2 g
    (d) 512 g
7-2. (a) 45 moles
    (b) 1440 g
    (c) 54 moles
    (d) 0.78 g
7-3. (a) 125 moles
    (b) 100 moles
    (c) 5 moles
    (d) 4 moles
    (e) 160 g
    (f) 176 g
    (g) 2 moles
    (h) 52 g
7-4. (a) 25 moles
    (b) 425 g
    (c) 1000 g
    (d) 68 g
7-5. (a) 2 moles
    (b) 64 g
    (c) 1 mole
    (d) 98 g
7-6. (a) 34.0 g
    (b) 11.1 g
    (c) 16.4 g
7-7. (a) 83.0 g
    (b) 21.4 g
    (c) 37.8 g
    (d) 60.6 g
7-8. (a) 78.5 g
    (b) 13.4 g
    (c) 70.3 g
7-9. (a) 44.9 kg
    (b) 69.8 kg
7-10. (a) 45 kg
    (b) 140 kg
    (c) 165 kg
    (d) 60 kg
7-11. 1.12 liters
7-12. 8.96 liters
7-13. (a) 80 liters
    (b) 143 liters
7-14. (a) 11.2 liters
    (b) 7.5 liters
    (c) 28.2 liters
7-15. 1010 liters

7-16. (a) 2, 5, 4, 6
    (b) 12.5 liters
    (c) 10 liters
7-17. (a) 1, 6, 4
    (b) 120 liters
    (c) 160 liters
7-18. (a) 80 liters
    (b) 160 liters
    (c) 38 liters
7-19. (a) 490 g
    (b) 600 g
    (c) 2.0 moles
    (d) 126 g
7-20. (a) 2 moles
    (b) 22 g
    (c) 0.5 mole
    (d) 1000 g
    (e) 504 g
    (f) 3 moles
7-21. 72 lb
7-22. 0.35 liter
7-23. (a) 0.87 kg
    (b) 0.56 kg
7-24. (a) 192 g
    (b) 298 g
7-25. (a) 607 g
    (b) 192 liters
    (c) 192 liters
    (d) 17.1 g
    (e) 1.13 kg
7-26. (a) 1963 liters
    (b) 1257 liters
    (c) 9820 liters
7-27. (a) 768 g
    (b) 538 liters
    (c) 750 ft^3
7-28. (a) 168 liters
    (b) 1 liter
    (c) 240 g
    (d) 16 liters
7-29. 395 g
7-30. (a) 44.8 liters
    (b) 495 liters
7-31. 19.4 %
7-32. 92.5 %
7-33. (a) 20.8 g
    (b) C_2H_5
7-34. C_2H_6O
7-35. 0.1 mole
7-36. 10 %
7-37. 11,200 liters

7-38. 4.77 g; 2.27 g
7-39. 30.7 liters
7-40. 600 g
7-41. (a) $\frac{5}{6}$ mole
    (b) 15.8 g; 28.0 g
    (c) 7.00 liters
    (d) 61 ml
7-42. 5.00 kg; 2.91 kg; 0.48 kg
7-43. 6.72×10^4 liters
7-44. 196 kg
7-45. (a) Pb_3O_4
    (b) 1, 4, 3, 4
7-46. (a) Fe_2O_3
    (b) 69.9 %
    (c) 3.78 g
7-47. 71.9 %
7-48. MgO = 75.6 %
7-49. Cd = 48.7 %
7-50. 55.5 %; 58.0 %
7-51. 35.453
7-52. 6.940
7-53. XeF_4
7-54. (a) 3.9 atm
    (b) 0.30 mole
    (c) 2.9 atm
7-55. (a) PbS 20 %
       ZnS 40 %
    (b) SO_2 1.974 g

Chapter 8

8-1. (a) 1
    (b) 4
    (c) 1
    (d) 4
    (e) 2
    (f) 1
8-2. 6, 8, 12
8-3. (a) 1.86 Å
    (b) 4.29 Å
    (c) 23.7 cm^3
8-4. (a) 4.09 Å
    (b) 10.5 g/cm^3
8-5. (a) 4.95 Å
    (b) 1.87×10^{-22} cm^3
    (c) 56.3 cm^3
    (d) 1.52 g/cm^3
8-6. (a) 1
    (b) 6.78 g/cm^3
    (c) 17.5 cm^3

8-7.   (a) 4
    (b) 1.041×10^{-22} cm³
    (c) 15.66 cm³
    (d) 7.32 g/cm³
8-8.   (a) 4 NaCl
    (b) 2.814 Å
    (c) 3.980 Å
    (d) 1.781×10^{-22} cm³
8-9.   (a) 2.67 Å
    (b) 5.34 Å
    (c) 3.78 Å
    (d) 1.52×10^{-22} cm³
8-10. (a) 3.66 Å
    (b) 4.23 Å
    (c) 6
    (d) 7.57×10^{-23} cm³
8-11. (a) 6.54 Å
    (b) 2.87 g/cm³
8-12. (a) 4.56 Å
    (b) 57.1 cm³
8-13. (a) 4.30 Å
    (b) 3.72 Å
8-14. (a) 3.022 Å; 3.181 Å
    (b) 4, 2
8-15. 2.63 Å
8-16. (a) 1.25 Å
    (b) 12
8-17. 3.25 Å; 3.38 Å
8-18. (a) 1.68×10^{-22} cm³
    (b) 3.90 Å
    (c) 12
    (d) 6.05 g/cm³
8-19. (a) 5.924 Å
    (b) 2.079×10^{-22} cm³
    (c) 4.188 Å
    (d) 31.30 cm³
8-20. 6.020×10^{23}
8-21. 5.15 Å
8-22. 4, 4
8-23. 6.023×10^{23}
8-24. (a) 0.523
    (b) 0.680
    (c) 0.741
8-25. Answers in order without units
    (a) 3.28, 14.83
    (b) 6.000, 3.00, 4.12
    (c) 4.85, 3.43, 48.51
    (d) 4.810, 3.35
    (e) 4.67, 3.30, 43.27, 2.75
    (f) 4.02, 2.84
    (g) 3.91, 2.77, 25.69

    (k) 4.286, 3.71, 4.49
    (l) 4.567, 3.95
    (m) 3.834, 3.31, 33.9
    (n) 3.33, 2.88, 22.2, 9.9
    (o) 2.74, 6.0
    (p) 3.90, 35.7
8-26. 1.76%; 0.18%; 0.44%

Chapter 9

9-1.   20 g
9-2.   200 g; 800 g
9-3.   50 g
9-4.   1.0
9-5.   (a) 0.625
    (b) 0.255
    (c) 0.247
    (d) 0.073
9-6.   88.2 g
9-7.   0.025
9-8.   47 g
9-9.   0.08
9-10. (a) 9.8 g
    (b) 34.2 g
    (c) 5.6 g
    (d) 9.6 g
9-11. 7.20 liters
9-12. 180 ml
9-13. 2.0
9-14. 0.25 liter
9-15. (a) 0.1 mole
    (b) 0.1 mole
    (c) 0.4; 0.1
9-16. (a) 0.160
    (b) 0.080
9-17. 0.184 g
9-18. 9.8 g
9-19. 0.08 equiv
9-20. (a) 0.16
    (b) 0.04
9-21. 29 g
9-22. 0.0234; 0.966
9-23. 0.075; 0.925
9-24. 0.017
9-25. 0.0222
9-26. 0.538
9-27. 2.5
9-28. 16 kg
9-29. 8.7 g
9-30. 0.25
9-31. 0.80

9-32. 39.5 g
9-33. 3.8 g
9-34. 0.050
9-35. (a) 0.125
    (b) 0.25
9-36. 9.45 g
9-37. 0.5; 1.0
9-38. 1.25; 1.25
9-39. (a) 19.6 g
    (b) 0.200 mole
    (c) 0.400 equiv
9-40. (a) 1.00; 1.00
    (b) 0.408; 0.816
    (c) 0.234; 0.468
    (d) 0.117; 0.702
    (e) 0.408; 1.22
9-41. (a) 1.58 g
    (b) 0.18 g
    (c) 4.1 g
    (d) 24.6 g
    (e) 4.9 g
9-42. (a) 0.5 liter
    (b) 1.5 liters
    (c) 180 ml
    (d) 30 ml
9-43. (a) 0.005; 0.01
    (b) 0.1; 0.1
    (c) 0.6; 0.6
    (d) $\frac{2}{3}$; 2
    (e) 0.25; 0.5
9-44. Urea = 0.118
9-45. (a) 16.86%
    (b) 0.0221
9-46. (a) 0.04388
    (b) 2.32
9-47. (a) 40.0%
    (b) 5.32
    (c) 10.64
    (d) 6.80
    (e) 0.109
9-48. 1.16 g/ml
9-49. (a) 19.0 ml
    (b) 17.4 ml
    (c) 5.21 ml
9-50. (a) 4.3; 0.07
    (b) 4.0; 0.065
    (c) 18.5; 0.25
9-51. 200 g
9-52. (a) 0.72
    (b) 11.4%
9-53. 46.4 g

9-54. (a) 2.92
    (b) 2.81
    (c) 0.05
    (d) 1.42
    (e) 32.0
9-55. (a) 25.3
    (b) 4.0
    (c) 4.54
    (d) 0.077
9-56. $O_2 = 33.4\%$
    $N_2 = 66.6\%$

Chapter 10

10-1. 23.24 mm
10-2. 22.57 mm
10-3. 134
10-4. 11 g
10-5. $-7.44°C$
10-6. 102.1°C
10-7. 102.1°C
10-8. $-22.3°C$
10-9. 90
10-10. $-2.95°C$
10-11. 90
10-12. 0.50
10-13. 0.48 atm
10-14. 9.5×10^4
10-15. 7.9 atm
10-16. 1.4×10^4
10-17. (a) $-0.186°C$
    (b) 100.05°C
10-18. (a) $-1.12°C$
    (b) 100.31°C
10-19. 0.75
10-20 2%
10-21. $-0.93°C$
10-22. 4%
10-23. $-3.44°C$
10-24. 2
10-25. 150
10-26. 101.30°C
10-27. 880 g
10-28. 500 g
10-29. (a) 2.2
    (b) 1.47
10-30. 0.27
10-31. $-19.7°C$
10-32. 200
10-33. 25
10-34. 63.7°C
10-35. 200

10-36. 263
10-37. $C_4H_6O_4$
10-38. 4.9°C
10-39. (a) 31.42 mm
 (b) 17.3 atm
10-40. 240
10-41. (a) 1.32°C
 (b) −3.87°C
10-42. $C_6H_5NO_2$
10-43. 29.6 atm
10-44. 7.1 atm
10-45. 26 atm

Chapter 11

11-1. 24.9 kcal
11-2. 0.054
11-3. 55.86
11-4. 118.7
11-5. 139.6
11-6. 36 kg
11-7. 5.44 kg
11-8. 25°C
11-9. 7.71 kcal
11-10. 204.4; 3
11-11. 75 g
11-12. −202.2 kcal
11-13. 99.0 kcal
11-14. 47.2 kcal
11-15. −108.8 kcal
11-16. −59.7 kcal
11-17. 35.8°C
11-18. 20 g
11-19. 670 g
11-20. −30.8 kcal
11-21. −16.4 kcal
11-22. 111 kcal
11-23. (a) −542 kcal
 (b) −56 kcal
 (c) −3.0 kcal
 (d) −77 kcal
 (e) −10 kcal
 (f) −13 kcal
 (g) −27 kcal

Chapter 12

12-1. 0.16
12-2. 120 ml
12-3. 160 ml
12-4. (a) 0.32
 (b) 0.16

 (c) 0.392
12-5. 0.05; 0.10
12-6. $A = 2.00\%$
 $B = 4.0\%$
 $C = 6.74\%$
 $D = 3.36\%$
12-7. 122
12-8. 59.0
12-9. (a) 0.04
 (b) 90%
12-10. (a) 2.96 g
 (b) 145 g
 (c) 0.666
 (d) 0.25
12-11. (a) 12.7
 (b) 50 ml
12-12. (a) 5 equiv/mole, 1 equiv/mole
 (b) 3 equiv/mole, 2 equiv/mole
 (c) 1 equiv/mole, 1 equiv/mole
 (d) 3 equiv/mole, 2 equiv/mole
 (e) 6 equiv/mole, 1 equiv/mole
12-13. (a) 3.16 g; 15.2 g
 (b) 5.27 g; 21.5 g
 (c) 33.2 g; 28.4 g
 (d) 2.1 g; 1.7 g
 (e) 3.57 g; 15.4 g
12-14. (a) 0.127; 0.0026
 (b) 0.0076; 0.0019
 (c) 0.0013; 0.0014
 (d) 0.019; 0.024
 (e) 0.011; 0.0026
12-15. 11.2%
12-16. (a) 0.02
 (b) 2.0
 (c) 1.33
12-17. (a) 0.20
 (b) 23.3 g
12-18. 80 ml
12-19. 28 ml
12-20. 0.216
12-21. 1.08 g
12-22. 0.1125 N
12-23. 500 ml
12-24. 45 ml
12-25. 75 ml
12-26. 0.75
12-27. 0.40
12-28. 0.988
12-29. (a) 0.004
 (b) 0.163
 (c) 0.122

12-30. (a) 0.670
(b) 4.02%
12-31. (a) 8.08
(b) 14.2%
12-32. 75
12-33. (a) 31.6 g
(b) 2
12-34. (a) 0.00254
(b) 0.00254
(c) 0.0635
12-35. (a) 0.005
(b) 0.005
12-36. 0.20
12-37. (a) 31.6
(b) 63 g
(c) 0.125
12-38. 0.01
12-39. (a) 63.5 g
(b) 158
(c) 10.0%
12-40. 18%
12-41. 4.00%

Chapter 13

13-1. (a) 12.8 hr
(b) 0.273
13-2. (a) 0.67
(b) 0.312 hr
13-3. (a) 0.0392/min
(b) 17.7 min
(c) 0.0314 (mole/liter)/min
(d) 0.0157 (mole/liter)/min
13-4. (a) 0.0183/sec
(b) 75.8 sec
13-5. (a) 0.648/hr
(b) 0.103
(c) 0.097 (mole/liter)/hr
(d) 0.032 (mole/liter)/hr
13-6. (a) 0.156/min
(b) 9.33×10^{-3}
(c) 0.125
(d) 0.0220 (mole/liter)/min
13-7. (a) $R = k$ [complex]
(b) 1.3×10^{-4}/min
13-8. (a) $R = k$[NO_2][CO]
(b) Second
(c) 1.9 (mole/liter)/hr
(d) 2.3×10^{-8} (mole/liter)/hr
13-9. (a) $R = k$[N_2O_5]
(b) First
(c) 9×10^{-2} (mole/liter)/sec

13-10. (a) $R = k$
(b) Zero
13-11. (a) $R = k$[HI]^2
(b) Second
(c) 1.2×10^{-6} (mole/liter)/min
13-12. $t_{1/2} = \dfrac{0.693}{k}$
13-14. 1.5×10^{-2}/hr
13-15. 7.62×10^{-6} (mole/liter)/min
13-16. (a) 248 yr
(b) 0.00489 g
13-17. 13,800 yr
13-19. (a) $R = k$[NO]^2[Cl_2]
(b) 8×10^{-6} (mole/liter)/min
(c) 6.4×10^{-5} (mole/liter)/min
13-20. (a) 40.6 min
(b) 21.0 min
(c) 5.8×10^{-4} (mole/liter)/min

Chapter 14

14-1. 4
14-2. 0.5
14-3. $\frac{4}{3}$
14-4. 11.3
14-5. 6.7
14-6. 8.0 moles/liter
14-7. 0.0016
14-8. 1.0 moles/liter
14-9. 0.10 mole/liter
14-10. 0.047
14-11. 3.7 moles/liter; 0.35
14-12. 0.75 mole/liter
14-13. 0.25
14-14. (a) 0.6 mole/liter
(b) 0.23
14-15. (a) 0.5
(b) [H_2] = [Br_2] = 0.011 mole/liter
[HBr] = 3.98 moles/liter
(c) $p_{H_2} = p_{Br_2} = 1.15$ atm
$p_{HBr} = 424$ atm
14-16. 5; 0.5
14-17. 18 cm
14-18. 3.4 moles
14-19. $p_{CO} = p_{H_2O} = 2.4$ atm
14-20. (a) 0.570
(b) [Br_2] = 0.94 mole/liter
[CO] = 0.19 mole/liter
[$COBr_2$] = 0.31 mole/liter

14-21. $[SO_3] = 0.49$
$[NO] = 0.34$
$[SO_2] = 0.31$
$[NO_2] = 0.11$
14-22. (a) $[H_2] = [I_2] = 0.240$
mole/liter
$[HI] = 0.02$ mole/liter
(b) 96
14-23. (a) 4.0
(b) 0.5 mole
14-24. (a) $[N_2] = 1.00$ mole/liter
$[H_2] = 2.99$ moles/liter
$[NH_3] = 5.67$ moles/liter
(b) 1.20
(c) 2.43 moles
14-25. 1.35×10^{-3}
14-26. (a) 9.04
(b) 9.04 atm
14-27. (a) 0.3421
(b) 0.5637
14-28. 4.8×10^{-2}
14-29. (a) $K_p = 122$
(b) $7\% \, CO_2$
14-30. 7.41 g
14-31. 5 atm
14-33. (a) 0.0625
(b) 114.4 atm
(c) $p_{NH_3} = 9.8$ atm
$p_{H_2} = 98.4$
$p_{N_2} = 6.2$ atm
(d) 1.6×10^{-5}
14-34. $[PCl_3] = [Cl_2] = 0.029$
$[PCl_5] = 0.021$
14-35. $[PCl_5] = 0.022$
$[PCl_3] = 0.028$
$[Cl_2] = 0.033$
14-36. $[H_2] = 0.0419$
$[I_2] = 0.0116$
$[HI] = 0.0162$
14-37. 2.2 g
14-38. (a) 1.99 mg
(b) 5×10^{-5} mg
14-39. 66
14-40. 0.57 g

Chapter 15

15-1. 4.0×10^{-4} mole/liter
15-2. 2.7×10^{-7}
15-3. 6.0×10^{-4} mole/liter
15-4. 1.3×10^{-3} mole/liter
1.3% dissociated

15-5. 4.8×10^{-3} mole/liter
9.6% reacted
15-6. (a) 2.8×10^{-5} mole/liter
(b) 5.7×10^{-6} mole/liter
15-7. 4×10^{-4}
15-8. 5.0×10^{-10}
15-9. (a) 0.082 mole/liter
(b) 5.2×10^{-5} mole/liter
15-10. (a) 3.4×10^{-2} mole/liter
(b) 5×10^{-6} mole/liter
15-11. (a) 0.001 mole/liter;
1×10^{-11} mole/liter
(b) 1×10^{-4} mole/liter;
1×10^{-10} mole/liter
(c) 1×10^{-5} mole/liter;
1×10^{-9} mole/liter
(d) 2.5×10^{-2} mole/liter;
4×10^{-13} mole/liter
(e) 3×10^{-5} mole/liter;
3.3×10^{-10} mole/liter
(f) 2×10^{-8} mole/liter;
5×10^{-7} mole/liter
(g) 2.5×10^{-13} mole/liter;
4×10^{-2} mole/liter
(h) 1×10^{-12} mole/liter;
1×10^{-2} mole/liter
(i) 1×10^{-13} mole/liter;
0.1 mole/liter
15-12. (a) 10.78; 3.22
(b) 10.98; 3.02
(c) 2.54; 11.46
15-13. 0.20
15-14. 5.5×10^{-5}
15-15. 1.0×10^{-7}
15-16. 3.2×10^{-6}
15-17. 13.5
15-18. 0.53
15-19. 6.57
15-20. 3.17
15-21. 8.1
15-22. (a) 5.6×10^{-3} mole/liter
(b) 3.2×10^{-4} mole/liter
15-23. (a) 2.1×10^{-5} mole/liter
(b) 1.05×10^{-5} mole/liter
15-24. (a) 4.1×10^{-3} mole/liter
(b) 3.4×10^{-5} mole/liter
15-25. (a) 1.4×10^{-3} mole/liter
(b) 1.8×10^{-5} mole/liter
15-26. 4.87
15-27. NH_4^+; H^+
15-28. 8.35

15-29. 5.44
15-30. (a) 9.27
     (b) 0.82 mole
15-31. 0.32 mole
15-32. 1.3×10^{-4}
15-33. 2.6×10^{-3}
15-34. (a) 4.14×10^{-3}
     (b) 0.013
     (c) 0.04
15-35. 3.3
15-36. (a) 2.2×10^{-2}
     (b) 2.0×10^{-1}
     (c) 5.1×10^{-1}
15-37. 0.27
15-38. 4×10^{-6}
15-39. (d)
15-40. (a) 5×10^{-12} mole/liter; 11.3
     (b) 1×10^{-12} mole/liter; 12.0
     (c) 6.7×10^{-13} mole/liter; 12.2
     (d) 1.25×10^{-12} mole/liter; 11.9
     (e) 5×10^{-10} mole/liter; 9.3
15-41. (a) 2×10^{-13} mole/liter; 12.7
     (b) 1×10^{-11} mole/liter; 11.0
     (c) 1.25×10^{-8} mole/liter; 7.9
     (d) 1.0×10^{-14} mole/liter; 14.0
     (e) 2.5×10^{-11} mole/liter; 10.6
15-42. 3.4
15-43. 4.35 g
15-44. 8.0
15-45. 1.6×10^{-5}
15-46. 5×10^{-10}
15-47. 5×10^{-10}
15-48. (a) 3.2
     (b) 2×10^{-11} mole/liter
15-49. (a) 10.5
     (b) 2.8×10^{-4} mole/liter
     (c) 2×10^{-12} mole/liter
15-50. 3.6×10^{-7}
15-51. 3.82
15-52. 9.9
15-53. 1.7×10^{-5}
15-54. 1.0×10^{-3}
15-55. 0.50
15-56. (a) 10.36
     (b) 2.3×10^{-4} mole/liter
     (c) 0.164 mole/liter
15-57. 9.09

Chapter 16

16-1. $PbCO_3$
16-2. 1.72×10^{-4}

16-3. 1.4×10^{-18}
16-4. 4.00×10^{-28}
16-5. (a) 2×10^{-4} mole/liter
     (b) 0.07 g/liter
16-6. (a) 0.00136 mole/liter
     (b) 0.626 g/liter
16-7. 8×10^{-7} mole/liter
16-8. No, trial product $= 1.5 \times 10^{-8}$
16-9. Yes, trial product $= 9 \times 10^{-18}$
16-10. Yes, trial product $= 3 \times 10^{-8}$
16-11. (a) 8.1×10^{-8} mole/liter
     (b) 2.0×10^{-12} mole/liter
16-12. (a) 4.2×10^{-3} mole/liter
     (b) 3.6×10^{-3} mole/liter
16-13. 6×10^{-3} mole/liter
16-14. (a) 1×10^{-4} mole/liter
     (b) 0.0367 g/liter
16-15. 2×10^{-9} mole/liter
16-16. (a) 4×10^{-16} mole/liter
     (b) 9.4×10^{-14} g/liter
     (c) 4.5×10^{-9} mole/liter
16-17. 2.8×10^{-3} mole/liter
16-18. 1.6×10^{-2} mole/liter
16-19. (a) 9.3×10^{-5} mole/liter
     (b) 4.0×10^{-3} mole/liter
16-20. (a) 1×10^{-4} mole/liter
     (b) 1×10^{-2} mole/liter
16-21. 8×10^{-3} mole/liter
16-22. 0.056; Ag^+
16-23. (a) 5×10^{-5} mole/liter
     (b) 2.12
16-24. 5×10^{-21}; 0.8
16-25. 1.3×10^{-2} mole/liter
16-26. 6.8×10^{-4} mole/liter
16-27. 7×10^{-7}; 2.5×10^{-8}
16-28. 5×10^{-12} mole/liter
16-29. 2×10^{-5}
16-30. 4×10^{-12}
16-31. 9.8×10^{-9}
16-32. 6.4×10^{-15} mole/liter
16-33. (a) 2×10^{-7} mole/liter
     (b) 8×10^{-4} mole/liter
16-34. (a) 1.2×10^{-11} mole/liter
     (b) 3×10^{-6} mole/liter
16-35. Yes, trial product $= 7 \times 10^{-13}$
16-36. Yes, trial product $= 3 \times 10^{-5}$
16-37. 1.0×10^{-4} mole/liter
16-38. 3.4×10^{-2} mole/liter
16-39. 3.3×10^{-7} mole/liter
16-40. No, trial product $= 5 \times 10^{-17}$
16-41. 4.0×10^{-7}

16-42. 4×10^{-4} mole/liter
16-43. 1.2×10^{-17} mole/liter
16-44. Hg, Ag, Sb, Bi, Cd, Zn, Tl
16-45. 2.4×10^{-8} mole/liter
16-46. (a) $[Pb^{++}] = 4 \times 10^{-6}$
mole/liter
$[Sr^{++}] = 8 \times 10^{-5}$
mole/liter
(b) 1.511 g; 0.903 g
16-47. (a) 1.2×10^{-13} mole/liter
(b) 1×10^{-4} mole/liter
16-48. 6.9×10^{-12}
16-49. 9.5×10^{-33}
16-50. 3.3
16-51. $[Ag^+] = 1.1 \times 10^{-4}$ mole/liter
at $[Cl^-]$ of 0.1×10^{-5} mole/liter

Chapter 17

17-1. (a) 2.2×10^{-11}
(b) 8.5
17-2. 4.8
17-3. (a) 5.5×10^{-10}
(b) 2.2×10^{-5} mole/liter
(c) 4.5×10^{-10} mole/liter
(d) 4.6
(e) 2.2×10^{-5} mole/liter
17-4. 10.9
17-5. (a) 4.2
(b) 6.3×10^{-5} mole/liter
17-6. (a) 5.3×10^{-5}
(b) 3×10^{-3}
17-7. 8.2
17-8. 5.0
17-9. 2.8×10^{-7}
17-10. 1.25×10^{-2}
17-11. 4.9×10^{-14}
17-12. 1.1×10^{-47}
17-13. About 1.00 mole/liter
3.2×10^{-4} mole/liter
1.3×10^{-3} mole/liter
17-14. (a) $2.1 \times 10^{-3}\%$
(b) about 1.00 mole/liter
2.14×10^{-5} mole/liter
8.6×10^{-5} mole/liter
17-15. (a) 1.0×10^{-16} mole/liter
7.9×10^{-3} mole/liter
about 1.0 mole/liter
(b) 7.9×10^{-3} mole/liter
17-16. (a) 2.2×10^{30}
(b) 1.14 moles/liter
17-17. 8.6

17-18. 4.8
17-19. 5×10^{-4}
17-20. 0.80
17-21. 9.8×10^{-15}
17-22. 7.5×10^{-4} mole/liter
17-23. (a) 2×10^{-4} mole/liter
(b) 4×10^{-6}
17-24. 4.7×10^{-4} mole/liter
17-25. (a) 0.06 mole/liter
(b) 0.08 mole/liter
17-26. 0.40; 5.0×10^{-30}
17-27. (a) $K_w = K_a K_h$
(b) 10^{-8}
17-28. (a) 4.4×10^{12}
(b) $\cong 1.25$

Chapter 18

18-1. 2 mg
18-2. 1.35×10^{10} yr
18-3. 1.3×10^4 yr
18-4. 1.6×10^4 yr
18-5. 20 min
18-6. (a) $-4, -2$
(b) $0, +1$
(c) $0, 0$
(d) $0, -1$
18-7. 10^{20} atoms
18-8. 8 mo
18-9. 7×10^5 atoms
18-10. 5.0 hr
18-11. 3.5×10^{18} atoms
18-12. 12 hr
18-13. (a) 8
(b) 6
18-14. (a) 22.4 liters
(b) 22.4 atm
(c) infinite time
18-15. 1.1×10^{19} atoms
18-16. 5 hr
18-17. 13 min
18-18. 49.9 min
18-19. 1.57×10^4 yr
18-20. 3.15×10^{18}
18-21. 1.0×10^{-3}
18-22. (a) 5.67×10^{14}
(b) 1.53×10^3 c
18-23. 2.03×10^{13} sec
18-24. 1.8 hr
18-25. 2.72×10^9 yr
18-26. 10,700 yr
18-27. 19.2 yr

18-28. 1.84 liters
18-29. (a) 1.64×10^{-9} c
  (b) 1640 pc
  (c) 92.9 yr
18-30. (a) $_2He^4$
  (b) $_0n^1$
  (c) $_He^0$
  (d) $_2He^4$
  (e) 3_0n^1
18-33. 2.24×10^{-11} g
18-34. 15.2 g
18-35. (a) 16 g
  (b) 2.3×10^6 dis/sec
18-36. 4.5×10^{21} ergs
18-37. 5.6×10^{27} Mev
18-38. (a) 0.01882 amu
  (b) 17.5 Mev; 6.68×10^{-13} cal
  (c) 4×10^{11}
18-39. 1. (a) 1.8×10^{-4} amu
  (b) 0.17 Mev
  (c) 3.9×10^6 kcal/g atom
  2. (a) 6.17×10^{-3} amu
  (b) 5.74 Mev
  (c) 1.33×10^8 kcal/g atom
  3. (a) 5.13×10^{-3} amu
  (b) 4.78 Mev
  (c) 1.10×10^8 kcal/g atom
18-40. (a) 6.92 Mev/nucleon
  (b) 8.51 Mev/nucleon
  (c) 8.39 Mev/nucleon
  (d) 8.60 Mev/nucleon
  (e) 8.73 Mev/nucleon
  (f) 7.57 Mev/nucleon

Chapter 19

19-1. 4825 coulombs
19-2. 33.0 g
19-3. (a) 0.5585 g Fe
    0.5871 g Ni
    0.3466 g Cr
  (b) 0.1787 amp
19-4. 32.2 min
19-5. (a) 2.45 liters
  (b) 64.3 min
19-6. 1.60×10^{-19} coulomb
19-7. 24.4 g
19-8. (a) 0.59 volt
  (c) Ni to Cu
  (d) 1.0×10^{20}
19-9. (a) No
  (b) about 10^{-91}

19-10. (a) 1.20 volts
  (c) Sn to Pt
  (d) about 10^{41}
19-11. (a) -0.740 volt
  (b) -0.307 volt
  (c) 0.362 volt
  (d) 0.356 volt
19-12. 1.15 volts
19-13. (a) 3×10^{-9}
  (b) 3×10^{28}
  (c) 2×10^{16}
  (d) 10^{21}
  (e) 2×10^{39}
19-14. 4.27 g
19-15. 0.67 amp
19-16. 5.4 g Ag; 1.59 g Cu;
  1.47 g Ni; 0.75 g Sc;
  9.85 g Au; 4825 coulombs
19-17. 40.2
19-18. 8.4 liters H_2; 4.2 liters O_2
19-19. (a) 1.56 volts
  (b) 0.51 volt
  (c) 0.43 volt
  (d) 1.52 volts
  (e) 0.46 volt
19-21. (a) Ni, Al
  (b) Ag^+, Br_2, Cu^{++}
19-22. (a), (c), (d)
19-23. -0.240 volt
19-24. (a) 10^{20}
  (b) 10^{16}
  (c) 10^{-36}
  (d) 10^{-18}
  (e) 10^{18}
  (f) 10^{61}
19-25. (a) 3×10^8
  (b) 10^{28}
  (c) 10^{53}
  (d) 10^{24}
  (e) 10^{66}
19-26. (a) 0.76 volt
  (b) -0.91 volt

Chapter 20

20-1. 212.8 kcal
20-2. -312.0 kcal/mole
20-3. 413 cal
20-4. (a) 2.4 cal/mole-deg
  (b) 0.7 cal/mole-deg
  (c) -32.7 cal/mole-deg
  (d) -21.5 cal/mole-deg

20-5. $\Delta H = -372.9$
$\Delta E = -371.4$

20-6. (a) 56.25 kcal; 53.88 kcal
(b) 29.6 kcal; 28.4 kcal

20-7. 357°C

20-8. $p_{CO} = p_{Cl_2} = 7.1$ atm

20-9. $K = 1.8 \times 10^{-7}$
$\Delta G = 9.2$ kcal

20-10. $K_p = 29$

20-11. $\Delta H = 14.16$ kcal/mole
$\Delta E = 12.9$ kcal/mole
$\Delta S = 22.5$ cal/mole-deg
$\Delta G = 0$

20-12. $\Delta G = -22.8; -94.3;$
-91.8 kcal

20-13. (a) -195.6 kcal
(b) -82.4 kcal
(c) -94.3 kcal
(d) -350.8 kcal

20-14. -7.2 cal/mole-deg

20-15. $\Delta H = -67.7$ kcal
$\Delta S = -20.7$ cal/deg
$\Delta G = -61.5$ kcal

20-16. (a) -15.5 kcal
(b) -23.0 kcal
(c) 9.52 kcal
(d) -1.14 kcal
(e) -1.01 kcal
(f) -5.2 kcal
(g) -9.9 kcal

20-17. (a) 2.0
(b) 10^{90}
(c) 1.3×10^{37}
(d) 5×10^{17}
(e) 1.2×10^{18}
(f) 10^{135}
(g) 5×10^{5}
(h) 8.7×10^{2}
(i) 10^{102}

20-18. (a) -16.2 kcal
(b) -28.6 kcal
(c) -77.5 kcal
(d) -113.5 kcal
(e) -21.7 kcal

20-19. (a) -24.3 kcal
(b) -36.8 kcal
(c) -20.4 kcal
(d) -94.4 kcal
(e) -17.6 kcal

20-20. $+0.27$

Appendix I

I-1. 3
I-2. 3
I-3. 8
I-4. 3
I-5. 8
I-6. 80.5
I-7. 3
I-8. 6
I-9. -7
I-10. ± 7
I-11. ± 2
I-12. $-1, 6$
I-13. $3.55, -1.88$
I-14. $2, -\frac{5}{6}$

Appendix II

II-1. (a) 2.24×10^{4}
(b) 2×10^{7}
(c) 3.15×10^{7}
(d) 1.7×10^{-4}
(e) 8.21×10^{-2}
(f) 2.65×10^{-3}

II-2. (a) 1500
(b) 0.000084
(c) 0.000034
(d) 36500
(e) 600
(f) 65

II-3. (a) 4.17×10^{5}
(b) 1.8×10^{-13}
(c) 1.2×10^{14}
(d) 0.24
(e) 3.0
(f) 3.0×10^{5}
(g) 2.7×10^{6}
(h) 0.04
(i) $\pm 5 \times 10^{3}$
(j) $\pm 8 \times 10^{-4}$
(k) 4×10^{-6}
(l) $\pm 2 \times 10^{2}$

Appendix III

III-1. (a) 4.6021
(b) 2.5798
(c) 23.7796
(d) -23.77
(e) -2.347
(f) 4.985

III-2. (a) 1000
        (b) 5.01×10^{-7}
        (c) 50
        (d) 0.95
        (e) 3.0×10^6
        (f) 2.0×10^{-4}

III-3. (a) 2.36×10^6
        (b) 7.68×10^2
        (c) 2.2×10^{10}
        (d) 3.0×10^{-8}
        (e) 10.2
        (f) 4.74×10^{-11}

Appendix IV

IV-1.   (a) 1560
        (b) 1.45
        (c) 1.75
        (d) 1.74
        (e) 1.74
        (f) 1.75
        (g) 1.74
        (h) 0.00372
        (i) 4.63
        (j) 451

IV-2.   (a) 395
        (b) 75.0
        (c) 67.107
        (d) 500
        (e) 0.60

IV-3.   (a) 22.01
        (b) 0.01
        (c) 0
        (d) 89.92
        (e) 8

IV-4.   (a) 7.00
        (b) 6.68
        (c) 15.73
        (d) 0.60
        (e) 1.08
        (f) 3

IV-5.   (a) 88
        (b) 7.4×10^2
        (c) 1×10^2
        (d) 1.014
        (e) 4.47
        (f) 131
        (g) 0.71

Index